**Enabling the Internet of Things:**
**Fundamentals, Design, and Applications**

# Enabling the Internet of Things: Fundamentals, Design, and Applications

*Muhammad Azhar Iqbal*
*Southwest Jiaotong University, China*

*Sajjad Hussain*
*University of Glasgow, UK*

*Huanlai Xing*
*Southwest Jiaotong University, China*

*Muhammad Ali Imran*
*University of Glasgow, UK*

WILEY

This edition first published 2021


*Registered Offices*
John Wiley & Sons, Inc., 111 River Street, Hoboken, NJ 07030, USA
John Wiley & Sons Ltd, The Atrium, Southern Gate, Chichester, West Sussex, PO19 8SQ, UK

*Editorial Office*
The Atrium, Southern Gate, Chichester, West Sussex, PO19 8SQ, UK

For details of our global editorial offices, customer services, and more information about Wiley products visit us at www.wiley.com.

Wiley also publishes its books in a variety of electronic formats and by print-on-demand. Some content that appears in standard print versions of this book may not be available in other formats.



*Library of Congress Cataloging-in-Publication Data*
Names: Iqbal, Muhammad Azhar, author. | Hussain, Sajjad, author. | Huanlai, Xing, author. | Imran, Muhammad Ali, author.
Title: Enabling the internet of things : fundamentals, design, and applications / Muhammad Azhar Iqbal, Sajjad Hussain, Huanlai Xing, Muhammad Ali Imran.
Description: First edition. | Hoboken, NJ : Wiley, 2021. | Series: Wiley - IEEE | Includes bibliographical references and index.
Identifiers: LCCN 2020040726 (print) | LCCN 2020040727 (ebook) | ISBN 9781119701255 (cloth) | ISBN 9781119701477 (adobe pdf) | ISBN 9781119701484 (epub)
Subjects: LCSH: Internet of things.
Classification: LCC TK5105.8857 .I77 2021 (print) | LCC TK5105.8857 (ebook) | DDC 004.67/8--dc23
LC record available at https://lccn.loc.gov/2020040726
LC ebook record available at https://lccn.loc.gov/2020040727

Cover Design: Wiley
Cover Image: © ivanastar/Getty Images

Set in 9.5/12.5pt STIXTwoText by SPi Global, Chennai, India
Printed and bound by CPI Group (UK) Ltd, Croydon, CR0 4YY

10 9 8 7 6 5 4 3 2 1

# Contents

## About the Authors

**Muhammad Azhar Iqbal** received his MS degree in Computer Software Engineering from the National University of Sciences and Technology, Pakistan, in 2007 and completed his PhD in Communication and Information Systems from Huazhong University of Science and Technology, China, in 2012. Currently, he is working as Lecturer in School of Information Science and Technology at Southwest Jiaotong University, China. Previously, he has served as Associate Professor in Computer Science department at Capital University of Science and Technology, Pakistan. He has experience of teaching various basic and advanced courses related to the domain of computing and mobile/wireless communication and networks. His research interests include wireless ad hoc networks, Internet of Things (IoT), and large-scale simulation modeling and analysis of computer networks in Cloud.

**Sajjad Hussain** is a Senior Lecturer in Electronics and Electrical Engineering at the University of Glasgow, UK. He has served previously at Electrical Engineering Department, Capital University of Science and Technology (CUST), Islamabad, Pakistan as Associate Professor. Sajjad Hussain did his masters in Wireless Communications in 2006 from Supelec, Gif-sur-Yvette and PhD in Signal Processing and Communications in 2009 from University of Rennes 1, Rennes, France. His research interests include 5G self-organizing networks, industrial wireless sensor networks and machine learning for wireless communications. Sajjad Hussain is a senior member IEEE and fellow Higher Education Academy.

**Huanlai Xing,** received his B.Eng. degree in communications engineering from Southwest Jiaotong University, Chengdu, China, in 2006; his M.Eng. degree in electromagnetic fields and wavelength technology from Beijing University of Posts and Telecommunications, Beijing, China, in 2009; and his PhD degree in computer science from University of Nottingham, Nottingham, UK, in 2013. He is an Associate Professor with School of Information Science and Technology, Southwest Jiaotong University. His research interests include edge and cloud computing, network function virtualization, software defined networking, and evolutionary computation.

**Muhammad Ali Imran** Fellow IET, Senior Member IEEE, Senior Fellow HEA is Dean University of Glasgow UESTC and a Professor of Wireless Communication Systems with research interests in self organised networks, wireless networked control systems and the wireless sensor systems. He heads the Communications, Sensing and Imaging CSI research group at University of Glasgow and is the Director of Glasgow-UESTC Centre for Educational Development and Innovation. He is an Affiliate Professor at the University of Oklahoma, USA and a visiting Professor at 5G Innovation Centre, University of Surrey, UK. He has over 20 years of combined academic and industry experience with several leading roles in multi-million pounds funded projects. He has filed 15 patents; has authored/co-authored over 400 journal and conference publications; has edited 7 books and authored more than 30 book chapters; has successfully supervised over 40 postgraduate students at Doctoral level. He has been a consultant to international projects and local companies in the area of self-organised networks.

# Preface

## Objectives

The emerging paradigm of Internet-of-Things (IoT) plays a consequential role to improve almost all aspects of human life, i.e. domestic automation, transportation, education, health, agriculture, industry, etc. The simple conception of IoT as a network of identifiable connected smart things is fundamentally based on the integration of various diversified technologies including pervasive computing, sensor technology, embedded system, communication technologies, sensor networking, Internet protocols, etc. for the provisioning of intelligent computing services. In our experience we have noticed that although the simple idea of IoT is easy to comprehend, at the undergraduate level, students are unable to describe the importance and placement of IoT components in an IoT system. This book tries to provide the basic, precise, and accurate demonstration of IoT building blocks as well as their role in various IoT systems. The objective of this book is to provide a good starting point for undergraduate students who have basic prior knowledge of Internet architecture. At an abstract level, this book is an effort to partially fill the gap associated with the understanding of IoT concepts through the designing of the IoT system prototypes in Packet Tracer. We believe that after implementing IoT system prototypes in Packet Tracer, students will find it easier to grasp complete details of IoT systems.

## Key Feature

Concerning the building of IoT foundations, this book can be used as a textbook at the undergraduate level. The key feature of this book is that it targets core aspects of IoT and provides its readership a better perspective both in terms of basic understanding of IoT technologies as well as the designing of IoT systems in Packet Tracer. To the best of our knowledge, this book can be considered as the first attempt to design simple IoT systems using Blockly programming language.

## Audience

This book is suitable for undergraduate students enrolled in the IoT course. This book assumes that the reader has a good understanding of Computer Networks and basic programming concepts. Students are comprehensively facilitated in this book to explain IoT essentials besides the guidance of designing IoT systems in Packet Tracer.

## Approach

At the end of each chapter, review questions in the form of case studies have been asked to explore students' clarity about IoT concepts discussed in that particular chapter. In this book, the design and implementation of IoT systems at an abstract level are presented in Blockly language.

## Organization of the Book

To address the issues related to the understanding of IoT fundamentals at the undergraduate level, this book is structured as follows:

Chapter 1 is exclusively written to introduce the evolution, vision, definition, characteristics, enablers, architectures of the IoT paradigm, and its distinction from other related technologies. This chapter builds the foundation for the understanding of IoT systems and is considered a prerequisite for the following chapters.

The primary focus of Chapter 2 is to establish an understanding of the IoT building blocks along with the necessary details related to various IoT hardware and software technologies. Besides, this chapter also provides a concise design and implementation perspective of IoT systems in Packet Tracer.

The contents of Chapter 3 are oriented along the lines of sensing principles and understanding of various aspects related to the design and implementation of wireless sensors and sensor networks. The layer-level functionality of wireless sensor networks in this chapter explains the effective communication requirements of sensors in IoT systems.

Chapter 4 describes the basics of IoT gateways in terms of its architecture and functionalities. In addition, this chapter also elaborates how IoT gateways having advanced features of data filtering and analytics support Edge computing and how Edge computing-based solutions provide benefits to specific IoT-based real-life applications.

Chapter 5 discusses the mapping of IoT protocols to layered IoT architecture and provides in-depth details of various infrastructure, service discovery, and application layer protocols of IoT protocol stack in terms of their functionality and usage in a real-life scenario.

Chapter 6 focuses on the description and explanation of components and employment of Cloud and Fog architectures in different IoT systems.

Chapter 7 introduces real-life application domains (i.e. domestic automation, smart transportation, smart agriculture and farming, smart manufacturing and industry automation, energy conservation, etc.) where the IoT technologies play a vital role to improve the standard of human life through the automation of these systems.

In Chapter 8, the classification of IoT attacks, as well as constraints and requirements of IoT systems, are discussed. Moreover, the discussion about security threats at each layer of IoT architecture is also the part of this chapter.

Chapter 9 illustrates the nature of social relationships between IoT devices, explains the functionality of the components of social IoT architecture, and provides an understanding of the applicability of social aspects of smart devices in IoT applications.

Chapters 10 and 11 are devoted to the design and implementation details of IoT projects in Packet Tracer exploiting constructs of Blockly programming language.

# Acknowledgments

We want to appreciate the efforts of the reviewing team at Wiley publishers for providing us the feedback and opportunity to publish this book. We would like to acknowledge the cooperation extended by our colleagues at Southwest Jiaotong University, China and the University of Glasgow, United Kingdom. We would also like to thank our students Sana Aurengzeb and Muhammad Talha (at Capital University of Science and Technology, Pakistan) for providing their support for the implementation of IoT system prototypes in Packet Tracer, which are part of Chapters 10 and 11 of this book. Finally, we want to acknowledge the most important contribution of our families for showing patience and understanding for the time we spent away from them while writing this book.

# 1

# Internet of Things (IoT) Fundamentals

**LEARNING OBJECTIVES**

After studying this chapter, students will be able to:

- describe the evolution of the IoT concept.
- state the vision and definition of IoT.
- explain the basic characteristics of IoT.
- distinguish the IoT from other related technologies.
- elaborate the IoT enablers.
- explain the IoT architectures.
- articulate the pros and cons of IoT.
- apply the IoT architecture concepts for specific IoT applications.
- understand the implementation aspect of IoT architecture.

## 1.1 Introduction

In our daily lives, the augmented practice of Information and Communication Technologies (ICT) plays a paramount role in the development of emerging information societies. In developed countries, ICT is being employed to develop various innovative applications and services to address the challenges of sustainable societies, thus improving the quality of human lives. In the modern era, a plethora of things are being connected to each other using underlying network technologies with an aim to promote the paradigm of the Internet of Things (IoT). IoT is a network of uniquely identifiable connected *things* (also known as devices, objects, and items) offering intelligent computing services [1]. Things in IoT are also known as Smart Things that provide feasibility in performing the execution of daily life operations in a rational way. Moreover, IoT also positively assists the communication process among human beings. IoT comprises diversified technologies including pervasive computing, sensor technology, embedded system, communication technologies, sensor networking, Internet protocols, etc. which eventually underpin the economic growth of modern societies. The fundamental notion behind IoT is the provision of seamless ubiquitous connectivity among things and human beings. The basic idea of IoT

*Enabling the Internet of Things: Fundamentals, Design, and Applications*, First Edition.
Muhammad Azhar Iqbal, Sajjad Hussain, Huanlai Xing, and Muhammad Ali Imran.

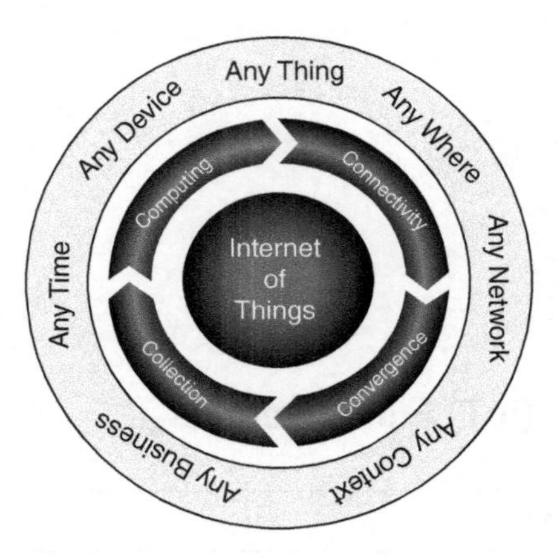


**Figure 1.1** The concept of As and Cs in the IoT.

can be conceived as a representation of various As and Cs, as shown in Figure 1.1 [2]. In Figure 1.1, the As reflect the concept of ubiquity or globalization (i.e. any device, anywhere, anytime, any network etc.) and the Cs mirror the main characteristics of IoT (i.e. connectivity, computing, convergence, etc.). IoT, in essence, can be seen as an addition of the third dimension named "Thing" to the plane of ICT world, which is fundamentally based on two dimensions of Place and Time as shown in Figure 1.2. This "anything" dimension ultimately boosts the ubiquity by enabling new forms of communication of humans and things and between things themselves [3].

## 1.2 Evolution of IoT Concept

The concept of ubiquitous computing through smart devices dates back to the early 1980s when a Coke machine at Carnegie Mellon University was connected to the Internet and able to report its inventory of cold drinks [4, 5]. Similarly, Mark Weiser in 1991 [6] provided the contemporary vision of IoT through the terminologies of ubiquitous computing and pervasive computing. Raji in 1994 elaborated the concept of home appliance automation to entire factories [7]. In 1999, Bill Joy presented *six web* frameworks wherein device-to-device communication could be formed [8]. Neil Gershenfeld in 1999 used a similar notion in his popular book *When Things Start to Think* [9]. In the same year, the term "Internet of Things" was promoted by Kevin Ashton during his work on Radio Frequency Identification (RFID) infrastructure at the Auto-ID Center of Massachusetts Institute of Technology (MIT) [10]. In 2002, Kevin was quoted in Forbes Magazine with his saying "We need an *Internet for things*, a standardized way for computers to understand the real world" [11]. The article was entitled as *The Internet of Things*, which was the first-ever official document with the use of this term in a literal sense.

The evolution of IoT with reference to the technological progress in Internet conception is shown in Figure 1.3. The typical Internet introduced in the early 1990s was only concerned with the generation of static and dynamic contents on the World Wide Web (WWW). Later on, large-scale production and enterprise-level business

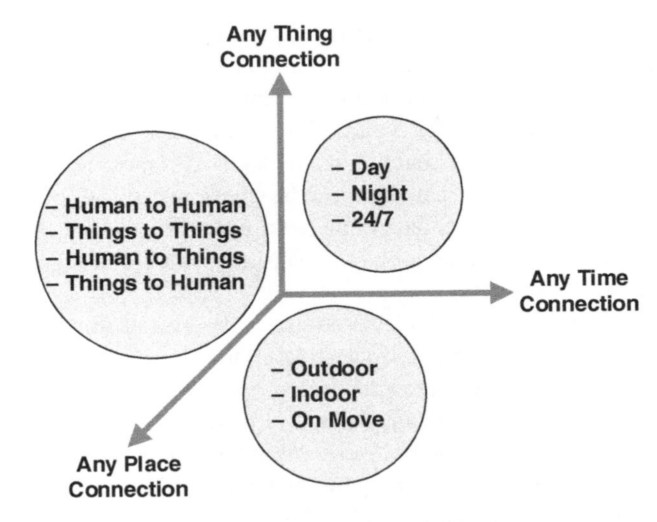


**Figure 1.2** Thing as a new dimension to endorse IoT. Source: Peña-López [3].

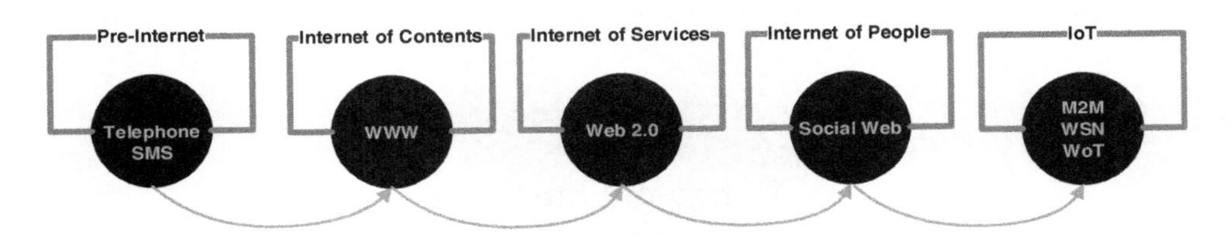


**Figure 1.3** Technological progression in IoT.

collaborations initiated the creation of web services which laid the foundation of Web 2.0. Nevertheless, with the proliferation of affordable smartphones and tablets, social network apps become dominant on the Internet. In current situation, advancements in embedded system, Machine-to-Machine (M2M) communication, Cyber Physical Systems (CPS), Wireless Sensor Network (WSN), and Web of Things (WoT) technology enabled the communication of things over the Internet. The overall technological progression related to IoT is shown in Figure 1.3.

## 1.3 IoT Vision

The conventional WWW offers the convenience of information searching, e-mail conversation, and social networking. The emerging trend of IoT comes up with a vision of expanding these abilities through interactions with a wide spectrum of electronic appliances. In general, the IoT vision can be seen in terms of things centric and Internet centric. The things-centric vision encompasses the advancements of all technologies related to the notion of "Smart Things." On the other hand, the Internet-centric vision involves the advancement of network technologies to establish the connection of interactive smart things with the storage, integration, and management of generated data. Based on these views, the IoT system can be seen as a dynamic distributed network of smart things to produce, store, and consume the required information [12]. The IoT vision demands significant advances in different fields of ICT (i.e. digital identification technology, communication technology, networking technology, computing technology, and distribution system technology), which are in fact the enabling technologies or fundamental elements of IoT [13, 14]. More specifically, the IoT paradigm can be envisioned as the convergence of three elementary visions, i.e. Things-oriented vision, Network-oriented vision, and Semantic-oriented vision [15, 16]. This convergence of three visions with abilities and technologies is shown in Figure 1.4.

Things-oriented vision at the initial level promotes the idea of things network through unique identifiable Electronic Product Code (EPC). Things-oriented vision in the present form is evolved into smart sensor networks. In Internet-oriented vision, Internet Protocol for Smart Object (IPSO) communities is formed to realize the challenging task of smart sensor communication. Considering unique identification through Internet Protocol (IP) addressing, IPSO communities are working for the interoperability of smart things (having sensors) to IP protocol technologies. Finally, the Semantic-oriented vision provides the solution to deal with the huge amount of data generated by the IoT devices. IoT architectural layers and associated protocols have been structured in these three envisions [17].

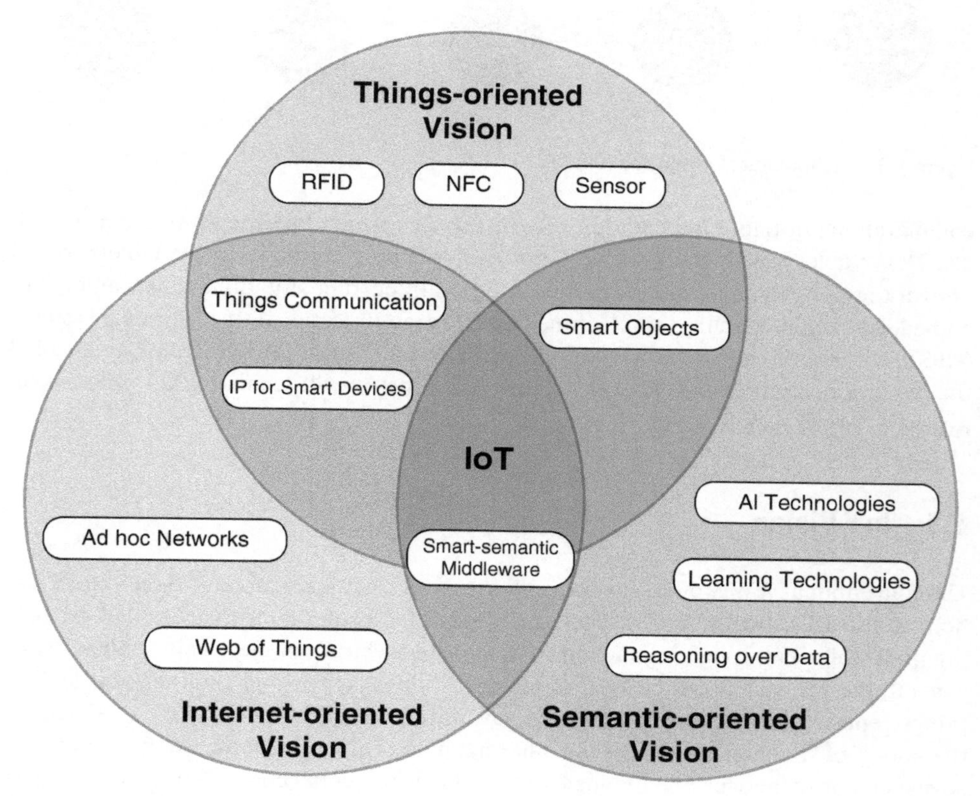


**Figure 1.4** IoT as convergence of three visions. Source: Adapted from Atzori et al. [15].

## 1.4 IoT Definition

Considering the facts of similarity with peer technologies and envision the convergence of three different visions, it is not an easy job to provide a precise definition of IoT. In simple words, IoT could be deemed as a system wherein things are connected in such a manner that they can intelligently interact with each other as well as to humans. However, to better comprehend IoT, a number of standard organization and development bodies have provided their own definitions [13, 15, 18, 19]. A few IoT definitions presented by different standard organizations are illustrated in Table 1.1 [20].

**Table 1.1** IoT definitions by standard organizations.

| Standard organization | IoT definition |
|---|---|
| Institute of Electronic and Electric Engineering (IEEE) | "The Internet of Things (IoT) is a framework in which all things have a representation and a presence in the Internet. More specifically, the IoT aims at offering new applications and services bridging the physical and virtual worlds, in which Machine-to-Machine (M2M) communications represents the baseline communication that enables the interactions between Things and applications in the Cloud." |
| Organization for the Advancement of Structured Information Standards (OASIS) | "System where the Internet is connected to the physical world via ubiquitous sensors." |
| National Institute of Standards and Technology (NIST) | "Cyber Physical systems (CPS) – sometimes referred to as the Internet of Things (IoT) – involves connecting smart devices and systems in diverse sectors like transportation, energy, manufacturing, and healthcare in fundamentally new ways. Smart Cities/Communities are increasingly adopting CPS/IoT technologies to enhance the efficiency and sustainability of their operation and improve the quality of life." |
| International Standard Organization (ISO) | "It is an infrastructure of interconnected objects, people, systems, and information resources together with intelligent services to allow them to process information of the physical and the virtual world and react." |
| Internet Engineering Task Force (IETF) | "In the vision of IoT, "things" are very various such as computers, sensors, people, actuators, refrigerators, TVs, vehicles, mobile phones, clothes, food, medicines, books, etc. These things are classified as three scopes: people, machines (for example, sensor, actuator, etc.) and information (for example, clothes, food, medicine, books, etc.). These 'things' should be identified at least by one unique way of identification for the capability of addressing and communicating with each other and verifying their identities. In here, if the 'thing' is identified, we call it the 'object'." |
| International Telecommunication Unit (ITU) | "IoT is type of network that is available anywhere, anytime, by anything and anyone." |

## 1.5 IoT Basic Characteristics

Considering all perspectives of modern-day IoT systems, a few generic and vital characteristics are shown in Figure 1.5 and explained in Table 1.2 [21, 22].

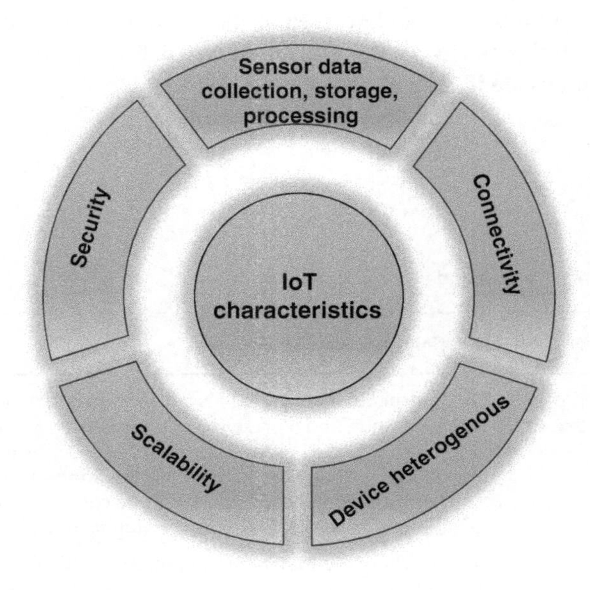


**Figure 1.5** Fundamental IoT characteristics.

**Table 1.2** Description of fundamental characteristics of IoT.

| IoT characteristic | Description |
|---|---|
| Sensor Data Acquisition, Storage, Filtering and Analysis | The plethora of distributed Sensors (or smart things) gather observation of physical environment/entity and direct to Cloud for storage and analytics with an ultimate objective to improve business workflow |
| Connectivity | IoT has made possible the interconnectivity of Physical and Virtual things with the help of the Internet and global communication infrastructure (that is built using wired and wireless technologies) |
| Device Heterogeneity and Intelligence | The interoperability of devices (based on different hardware and network platforms) with the provisioning of ambient intelligence at the hardware/software level supports intelligent interactions |
| Scalability | The plethora of IoT devices connectivity shifts human interactions to device interactions |
| Security | The security paradigm is required to be implemented at the network level as well as the end-devices level to ensure the security of data |

## 1.6 IoT Distinction

From the evolutionary perspective of IoT, it seems that IoT in different eras has been regarded as another name of a particular technology. Therefore, the term IoT is associated with other technologies in literature, i.e. embedded system, M2M communication, CPS, WSN, and WoT. However, the IoT concept is not attributable to any single technology.

### 1.6.1 IoT Versus Embedded Systems

Table 1.3 shows the differences between embedded systems and IoT.

### 1.6.2 IoT Versus M2M

Table 1.4 shows the differences between M2M and IoT.

### 1.6.3 IoT Versus CPS

CPS and IoT are highly overlapped; therefore, it is very difficult to demarcate the boundary between their differences. Both IoT and CPS encompass embedded devices that are able to transmit physically sensed data over the network. However, the use of these terms has been exploited by different communities on the basis of perceived criteria. Table 1.5 shows the differences between CPS and IoT.

**Table 1.3** Difference between embedded systems and IoT.

| Embedded system | IoT |
|---|---|
| Embedded systems include electronic devices that are usually standalone in nature and independently run on the Internet | IoT is a system that includes Internet connectivity-reliant devices for communication |
| Embedded systems are a combination of hardware and software (firmware) | IoT systems are a combination of computer hardware, software, and networking capabilities that are embedded into things of our daily lives |
| Embedded systems firmware mostly needs no modifications once the device is delivered to the clients | IoT requires continuous update |
| Example: ECG machine in a healthcare service that analyzes health parameters associated with humans is an example of embedded systems | Example: ECG machine connected to the Internet and able to transfer human health parameters on a remote server is an example of IoT devices |
| Embedded systems are a subset of IoT | IoT is a broader term including different technologies, i.e. embedded systems, networking, and information technology |

**Table 1.4** Difference between M2M and IoT.

| M2M | IoT |
|---|---|
| In M2M, mostly communication type is point to point | In IoT, communication takes place at IP networks |
| Middleware not necessarily required for data delivery | Middleware is responsible for data delivery |
| Mostly, M2M devices do not rely on Internet Connection | In IoT, most of the devices require Internet connectivity |
| M2M devices have limited options to integrate with other devices due to corresponding communication standard requirements | In IoT, multiple communications demand unlimited integration options |
| M2M is a subset of IoT | IoT is a broader term which includes M2M as well as various other technologies |

**Table 1.5** Difference between CPS and IoT.

| CPS | IoT |
|---|---|
| The term CPS is usually preferred over IoT by the engineering communities. The computer scientists working with an embedded system also used this term | The term IoT is frequently preferred over CPS by the network and telecommunications communities and the computer scientists doing research in the areas of next-generation networks and future Internet advancements |
| In the United States, the CPS term is preferred over IoT | In the European Union, the term IoT is preferred over CPS |
| CPS is considered as a system | IoT is considered as devices on the Internet |
| Development of effective, reliable, accurate, and real-time control system is the primary goal of CPS | BigData collection, storage, management, analysis, and sharing over Quality of Service (QoS) networks are primary goals of IoT |

### 1.6.4 IoT Versus WSN

Table 1.6 shows the differences between WSN and IoT.

### 1.6.5 IoT Versus WoT

Table 1.7 shows the differences between WoT and IoT.

According to literature, the terms *embedded systems, M2M, CPS, WSNs,* and *WoT* are occasionally interchangeable with IoT; however, these are not synonyms of the term IoT. IoT is likely to be the prevailing term over all these terms. The conceptual relationship of IoT with other related technologies [23] is shown in Figure 1.6. Figure 1.6 illustrates that IoT is essentially the outcome of various existing technologies used for the collection, processing, inferring, and transmission of data.

**Table 1.6** Difference between WSNs and IoT.

| WSN | IoT |
|---|---|
| WSN refers to a set of dedicated sensors to monitor, record, and transmit physical parameters of an entity or environment to a central location | IoT system includes all uniquely identifiable physical things/devices (i.e. home appliances, vehicles, etc.) embedded with electronics, software, sensors, and actuators, with ubiquitous connectivity to each other over the Internet. Moreover, sensor data processing and analysis is also part of IoT |
| WSN is a subset of IoT | IoT is a broader term and includes various technologies other than WSNs |
| Example: A large collection of sensors (optionally connected) used to monitor the moisture in a field likely to be considered as WSNs | Example: A fridge having the capability of sensing and transmitting the temperature reading to the Internet is an example of a smart device in the IoT system |

**Table 1.7** Difference between WoT and IoT.

| WoT | IoT |
|---|---|
| WoT system involves the incorporation of IoT entities over the web | IoT is a network of smart things/objects/devices, people, systems, and applications |
| WoT includes web-based applications over the network layer of IoT architecture | IoT applications include all sorts of applications such as web-based, android-based applications |
| Example: Embedded systems to connect objects over the web for communication with other objects | Example: Network of wireless devices and objects |

## 1.7 IoT General Enablers

Mark Weiser in 1999 [13, 24] considered the general trends in technology and provided the imagination that future IT developments would not be dependent on a particular technology but would be based on the confluence of computing technologies, which ultimately results in Ubiquitous Computing. In his depiction, the world of ubiquitous computing consists of real-life objects which are capable to sense, communicate, analyze, and act according to the situation. In general, miniaturization, portability, ubiquitous connectivity, integration of a diverse range of emerging devices, and pervasive availability of digital ecosystems (i.e. Cloud) are the general enablers that play a significant role for enabling IoT systems [1, 25, 26]. Precisely, the five stages of IoT functional view or information value loop are related to data creation, data communication, data aggregation, and data analysis and are necessary actions to achieve set goals [14]. Each stage of the IoT information value loop is empowered by a specific technology. For example, an observed action in the environment creates data that is passed to different networks for communication. Communicated data is of heterogeneous nature and is required to follow standards before aggregated for the purpose of comprehensive analysis [27]. Considering this functional view, IoT systems

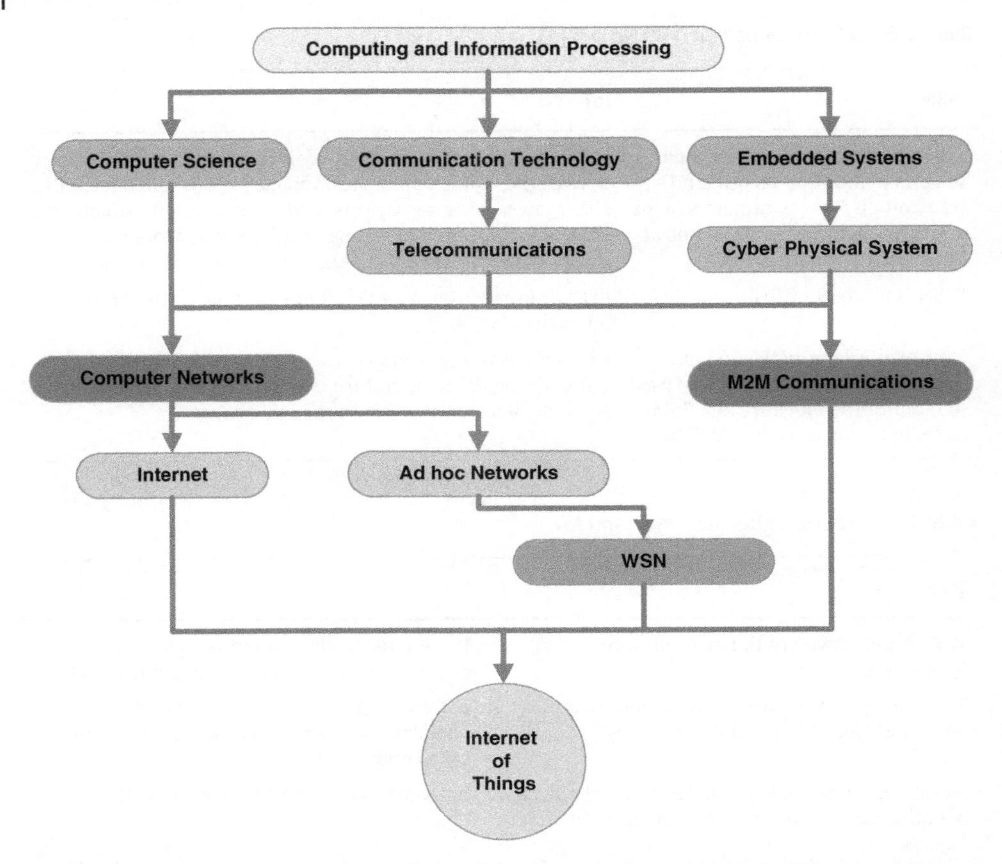


**Figure 1.6** IoT relationship with peer technologies. Source: Adapted from Manrique et al. [23].

are technologically dependent upon the sensors, networks, standard aggregations, artificial intelligence, and augmented behavior as discussed in Table 1.8. Therefore, in terms of technological advancements, the following five technologies are the main reasons for the enabling of IoT [1, 13, 14, 25, 28].

The working of IoT systems revolves around the paradigm of identification, communication, interaction of anything along with the analysis on data originated from anything. Following this paradigm, details about these technologies are part of the next subsections.

### 1.7.1 Identification and Sensing Technologies

Nowadays, our living/working environment demands the use of electronics in various ways including computers, projectors, cameras, tags, and sensors/actuators. Smart identification of sensor/actuators and detection of physical sensation are two of the basic system-level characteristics of IoT systems [12, 26].

Identification in terms of naming and matching of smart things and services in IoT is essential, and a number of identification methods are in use (i.e. Ubiquitous Code [UCode], EPCs, Universal Product Code [UPC], Quick Response Code [QR Code], European Article

Table 1.8 IoT enabling technologies.

| Technology | Description |
|---|---|
| Identification and Sensing Technologies | Include the development of devices (sensors) that converts any physical stimulus into an electronic signal |
| Wireless Communication and Networking | Include (network) devices that are able to communicate electronic signals |
| Aggregation Standardizations | Include technical standards that enable efficient data processing and allow interoperability of aggregated data sets |
| Augmented Intelligence | Include analytical tools that improve the ability to describe and predict relationships among sensed data |
| Augmented Behavior | Include technologies and techniques that improve compliance with prescribed action |

Number [EAN], , etc.). Nevertheless, addressing (including IPv4 or IPv6) of IoT objects is also important to refer to its address in communication networks. Recognizing the difference between the object's identification and object's address is essential because identification methods are not globally unique. However, the addressing, in this case, is required to globally identify an object. Identification methods offer unique identity of objects within the network, and public IP addressing provides the unique identity to the smart things over the Internet [1].

Sensing in IoT involves the originating of data from interrelated smart things through the use of sensors and actuators. A sensor is basically an electronic device responsible to produce electrical, optical, and digital data deduced from the physical environment that further electronically transformed into useful information for intelligent devices or people [14]. Actuators are the technological complement of sensors that are responsible to convert an electric signal to nonelectric energy.

### 1.7.2 Wireless Communication and Networking

Wireless communication and wireless networking are the core of Wireless Identification and Sensing Technologies (WIST), which play a vital role in the IoT. WIST refers to RFID-based sensors and WSNs. RFID systems are the vital components of IoT. RFID stands for Radio Frequency IDentification, and it is a wireless communication technology, which uses electromagnetic fields to automatically identify tags that are attached to physical objects [29]. In general, it is stated that RFID technology has its roots in Identification of Friend or Foe (IFF) systems, used in the Second World War [30]. Basic digital identification codes have been used in IFF systems [31], which are transmitted between an interrogator and a responder to identify planes belonging to the enemy or allies. Similar to IFF technology, RFID systems utilize radio waves to identify physical objects in real-time through digital tag reading. Basically, a typical RFID consists of the following three components [29] (shown in Figure 1.7):

- An RFID Tag (also known as Transponder or Smart Label) composed of an antenna, (optional) battery, and semiconductor chip

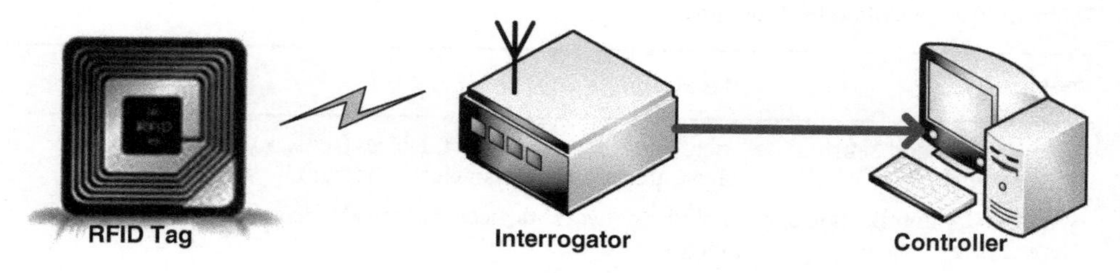


**Figure 1.7** Building blocks of RFID system.

- An Interrogator (also known as Reader or read/write device) having RF module, control module, and an antenna
- A Controller (also known as Host or Workstation) to store required information in a database

The RFID tag and interrogator in the RFID system do not require line of sight and communicate with each other through radio waves. Within the transmission range, the Interrogator reads the required information (i.e. serial number, manufacturer, location, usage history, maintenance schedule, etc.) stored on the RFID tag and directs this information toward the Controller that ultimately uses this information for various purposes.

RFID tags can be of two types, i.e. Active Tags (Tags having on-board power source) and Passive Tags (Tags without an on-board power source). Active RFID Tags have greater capabilities (i.e. large memory, long read range, high data transmission rate, lower infrastructure cost, etc.) than Passive Tags but are more complex and expensive. Active Tags use battery power and are able to transmit information over a longer range. On the other hand, to transmit information, Passive RFID Tags are able to derive power from the signal received by the Interrogator. RFID usage covers a wide spectrum of application areas, i.e. object identification, asset tracking, manufacturing, supply chain management, payment systems, and location identification. Two basic power harvesting approaches, i.e. Electromagnetic Wave Capture and Magnetic Induction have been implemented [32]. (Internal details of RFID tags have been discussed in the chapter on Sensing Principles of Sensors.)

Contrary to the RFID devices, sensors in the WSN have cooperative capabilities to sense and transfer data [33]. In fact, sensed data is required to be collected to store it in a database, data warehouse, or Cloud for analysis to make in-time right decisions. These databases, data warehouses, or Cloud are located far from the place of actual data creation. Therefore, the information created by the sensors is required to be transmitted to these locations for aggregation and analysis. Typically, this kind of transmission of data in IoT involves multiple types of wireless communication and network technologies. Wireless communication in IoT emphasizes the way how heterogeneous devices are able to communicate with each other in a sustainable way that they can understand. On the other hand, wireless networking involves the interconnectivity of devices for the efficient transmission of sensed data. Various types of networks are involved in the transmission of data from the place of origination to the destination. WSN comprises many tiny sensors, which are distributed in an ad hoc manner but work in collaboration to measure and transfer certain physical phenomena to the required destination (also known as a sink). Considering the requirements of different scenarios, Infrared (IR), Radio Frequency (RF), and optical (Laser) are three popular communication schemes used in WSNs. In addition, WSNs follow layered architecture and consists of protocols and algorithms with self-organizing capabilities. Generally, commercial WSNs technology is based on the IEEE 802.15.4 standard that ultimately provides the definition of Physical (PHY) and MAC layers for low-power communications. For the seamless connectivity to the Internet, upper layers of TCP/IP protocol stack have not been specified for WSNs. Therefore, a number of energy-efficient routing and transport layer solutions have been proposed in the literature for the efficient and reliable transmission of sensed data [34, 35].

Currently, the integration of RFID technology having sensing capabilities (Wireless Identification and Sensing Platform [WISP] project [36]) enables new types of IoT applications. WISP devices are able to sense different physical quantities (i.e. temperature, light, liquid level, acceleration, etc.) and are able to harvest energy through received reader's signal. This WISP technology permits the creation of RFID sensor networks that ultimately requires no batteries [37].

RFID, sensors, and RFID sensor are connected to the Internet through heterogeneous network devices, i.e. Bluetooth, Access Points (APs), Wi-Fi routers, Gateways, etc. Therefore, a unique IP address is required for all smart things on the Internet. IP is responsible for the provisioning of unique IP addressing over the Internet. Due to the greater scalability, IPv6 has been considered as one of the main enablers of IoT.

Smart things require continuous connectivity and need to be connected to various heterogeneous networks through switches, routers, gateways, etc. Therefore, the right choice of network technology is essential. Depending on the range and/or rate of data transmission, a number of network technologies are available, i.e. USB, Ethernet, Bluetooth, ZigBee, Near Field Communication (NFC), Wi-Fi, WiMax, 2G/3G/4G (Long Term Evolution [LTE]) [14, 28], etc. These technologies can be classified as Wired (including USB and Ethernet) and Wireless (including Bluetooth, NFC, Wi-Fi, WiMax, and 2G/3G/4G [LTE]). Connectivity type and network type of different communication technologies are given in Table 1.9.

**Table 1.9** Wireless technologies.

| Technology | Connectivity type | Network type |
|---|---|---|
| USB | Wired | Personal Area Network |
| Ethernet | Wired | Local Area Network |
| Bluetooth/Bluetooth Low Energy | Wireless | Personal Area Network |
| ZigBee | Wireless | Personal Area Network |
| Near Field Communication (NFC) | Wireless | Personal Area Network |
| Wi-Fi | Wireless | Local Area Network |
| WiMax | Wireless | Metropolitan Area Network |
| 2G/3G/4G, LTE/LTE-Adv. | Wireless | Wide Area Network |

Factors (other than Data Rate and Transmission Range) that drive the adoption of network technology for a particular IoT system includes Internet transit prices, IPv6 adoption, power efficiency, security/privacy [14, 15], etc.

### 1.7.3 Aggregation Standardization

Aggregation refers to the gathering of sensed data in a way that eases the process of handling, processing, and storage of data. Aggregation, besides providing the ease of handling, is also helpful to extract meaningful conclusions for future decision-making. Within the context of data aggregation in IoT, Standardization is one of the most important issues. So far, relational databases and SQL have been considered for storing and querying of structured data. However, no standard is available to handle unstructured data. IoT promises the scalability of billions of devices that ultimately demand common standards in order to communicate and aggregate the data of heterogeneous nature. The existing Internet standards had been developed without the consideration of IoT vision. Correspondingly, IoT systems have been developed using proprietary protocols that eventually make the communication problematic among IoT devices. Standardization is inevitable within the domain of IoT and is essential to guarantee interoperability, scalability, alike data semantics, security, and privacy [38, 39]. Several standards are required to be followed to realize data aggregation in IoT. However, Technology Standards and Regulatory Standards are two broad categories of standards, which are related to the process of aggregation [14].

Technology standards include network protocols (set of rules dealing with the identification and connectivity among devices), communication protocols (set of rules with the provision of a common language for devices' communication), and data-aggregation protocols (set of rules that assist the aggregation and processing of sensed data).

Hitherto, not a single or universal standardization body exists to make IoT technology standards. However, few standardization organizations are active at different level, i.e. international, regional, and national level (described in Table 1.10) [40].

On the other hand, regulatory standards are important in the evolution of IoT and deal with the ownership, use, and sale of the data. Envisioning the scale of emerging IoT application, the US Federal Trade Commission defined recommendations called Fair Information Practice Principles (FIPPs), which must be considered. For example, rules in FIPPs state that:

- Before data collection, concerned users must be notified and given options to choose about the usage of their personal information.
- After the usage of the required information, data must be deleted.
- Organizations must care about the security and privacy of collected data.

However, until now, it is undecided about the main organization which would be responsible for the implementation of regulatory standards for IoT applications [14].

### 1.7.4 Augmented Intelligence

Analysis of collected data demands the practice and advancements of different augmented cognitive technologies. Augmented intelligence enables the automation of systems to

**Table 1.10** Standardization organizations at different levels.

| Organization | Active at |
|---|---|
| NoSQL | International Level |
| MapReduce and Hadoop Distributed File System (HDFS) | International Level |
| Institute of Electric and Electronic Engineers (IEEE) | International Level |
| Internet Engineering Task Force (IETF) | International Level |
| International Telecommunication Unit (ITU-T) | International Level |
| One M2M | International Level |
| European Telecommunications Standards Institute (ETSI) | Regional Level |
| Korean Agency for Technology and Standards (KATS) | National Level |
| Telecommunication Standards Development Society, India | National Level |
| Global ICT Standardization Forum for India (GISFI) | National Level |
| Bureau of Indian Standards (BIS) | National Level |

Source: Based on Pal et al. [40].

perform descriptive (amenable representation of data to recognize insights), predictive (to foresee future consequences), and prescriptive (related to optimization) analysis [41]. SAS Visual [42] and Tableau [43] are examples of tools that are helpful in (big) data analysis through visualization, which is an unavoidable aspect of business analytics. Predictive analysis performs analysis on historical data to find future trends through the use of machine learning approaches by avoiding explicit programming instructions. Hadoop, Spark, Neo4j, etc. are different tools that have been proposed to support predictive analysis on (big) data. However, these technologies need to be more matured because in many practical applications, it is very difficult to forecast a future trend even if there exists a strong correlation between entities. Prescriptive analysis techniques improve prescribed accuracy in decision optimizations. Computer vision, natural-language processing, and speech recognition are a few examples of cognitive technologies that are playing an important role in predictive and prescriptive analytics. Computer vision techniques are mostly used to process images for different types of diagnoses and predictions of medical diseases. Natural language processing and speech recognition techniques are preferred to perform analysis related to the expressions and transcription of words in text and accent in speech. Applications include voice control computer systems, spam e-mail detection, medical dictation, etc. Availability of BigData generated through IoT devices, high demands of crowdsourcing, advancements in analysis tools and real-time data processing are the main driving factors for augmented intelligence [14].

### 1.7.5 Augmented Behavior

Augmented behavior involves the actions that are required to perform while considering all phases of the information value loop, i.e. from sensing to data analysis. Following the changes in people's behavior and organizational processes, augmented behavior supports the manifestation of suggestive actions with the use of advanced technologies (i.e. M2M and Machine to Human [M2H]). At this phase of the information loop, IoT concerns transferred from data science to behavioral sciences. Advancements in M2M and M2H are main driving forces that support the cognitive and actuation abilities of machines to understand the environment and act logically, respectively [14].

## 1.8 IoT Architectures

In the Internet, communication is based on a layering stack of TCP/IP protocols. Similarly, the IoT paradigm is a multilayer technology that supports meaningful communication of billions of smart things equipped with a processor, sensor/actuator, and communicator. Considering basic IoT elements (as shown in Figure 1.8), the IoT essentially connects a diverse range of hardware devices to a plethora of application domains. The heterogeneity of application and hardware domains imposes varied significant challenges that are essential to meet for the successful deployment of simple and complex IoT systems [44]. In addition to heterogeneity, considering all time ubiquitous connectivity, IoT needs to address a diverse range of issues including scalability, interoperability, security/privacy, and QoS for high traffic/storage needs that ultimately affect the architecture of IoT systems. A number

**Figure 1.8** IoT elements.

of IoT architectures have been proposed in the literature. These architectures are varied not only with each other's functionalities but also in technical terminologies. Interoperability between different IoT systems is limited as the proposed architectures have not yet converged to a single reference architecture [44, 45]. Therefore, there is a need for a layered architecture that is central to all IoT projects. In this chapter, from the pool of proposed IoT architectures, the functionality of each layer of the following IoT architectures has been explicated:

- Three-layer IoT architecture
- Five-layer IoT architecture
- Six-layer IoT architecture
- Seven-layer architecture

### 1.8.1 Three-layer IoT Architecture

The simplest IoT architecture consists of three layers, i.e. perception, network, and application layers [1, 28, 46] as shown in Figure 1.9.

#### 1.8.1.1 Perception Layer

The perception layer at the bottom of IoT architecture is responsible for the collection of various types of information through physical sensors or components of smart things (i.e. RFID, sensors, objects with RFID tags or sensors, etc.). Moreover, the perception layer transmits the processed information to the upper network layer via service interfaces. The main challenge at the perception layer is related to the recognition and perception of environmental factors through the use of low-power and nanoscale technology in smart things.

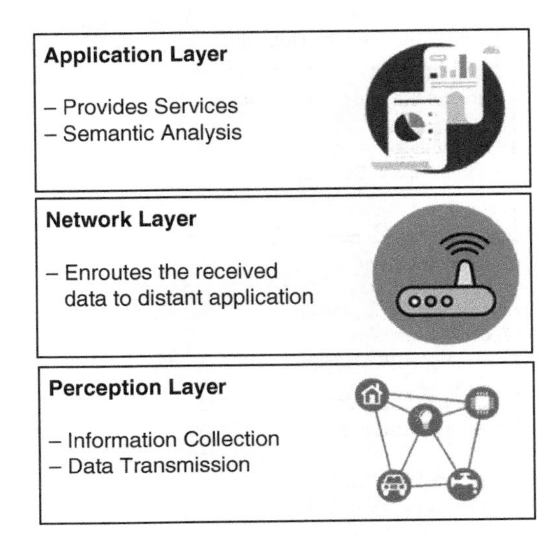


**Figure 1.9** Three-layer IoT architecture.

#### 1.8.1.2 Network Layer

The middle layer in three-layer IoT architecture is Network (also known as transmission) layer. The network layer accepts processed information from the perception layer and forward the received data to distant application interface(s) by using integrated networks, the Internet and other communication technologies. A number of communication technologies (i.e. Wireless Local Area Networks (WLAN), Wi-Fi, LTE, Bluetooth Low Energy [BLE], Bluetooth, 3G/4G/5G, etc.) are integrated with IoT gateways that handle heterogeneous types of data to or from different things to applications and vice versa. In addition to network operations, the Network layer in some cases enhances to perform information operations within the Cloud.

#### 1.8.1.3 Application Layer

The application layer at the top of the three-layer IoT architecture is responsible for the provisioning of services requested by the users, e.g. temperature, moisture, humidity, air pressure, light intensity measurements, etc. In addition to the user-requested services, the application layer provides data services (i.e. Data warehousing, BigData storage, data mining, etc.) to perform semantic data analysis. Smart health, intelligent transportation

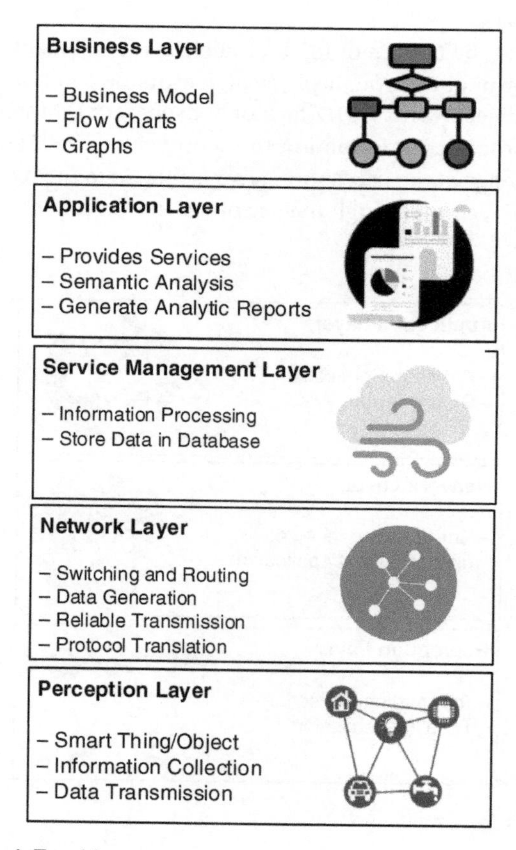


**Figure 1.10** Five-layer IoT architecture.

system, smart building, smart industry, and smart city are some of the applications with smart user interfaces at the application layer.

### 1.8.2 Five-Layer IoT Architecture

Object (Perception), Object Abstraction (Network), Service Management (middleware), Application, and Business are the names of the five layers in five-layer IoT architecture [1, 47] as shown in Figure 1.10. Each layer is briefly explained in the following sections.

#### 1.8.2.1 Object (Perception) Layer

The object layer primarily deals with the identification, collection, and processing of object-specific information (i.e. temperature, humidity, motion, chemical changes, etc.) through a diverse range of physical sensors. The object layer is also known as the perception layer or device layer. Physical sensors at this layer are based on different sensing principles (i.e. capacitance, induction, piezoelectric effect, etc.) and are responsible to digitize and transfer sensed data to Object Abstraction layer through secured channels. BigData is initialized at this layer.

#### 1.8.2.2 Object Abstraction (Network) Layer

Object Abstraction Layer or Network layer is responsible for secure data transmission from physical sensors to information processing systems by using various technologies, i.e. Wi-Fi, Infrared, ZigBee, BLE, WiMax, GSM, 3G/4G/5G, etc. In other words, the Network layer transfers the sensed information from the perception layer to the Service Management layer of the IoT layering stack.

#### 1.8.2.3 Service Management (Middleware) Layer

The smart things in IoT implement a diverse range of services, and each smart thing is connected and capable to communicate with smart objects that have implemented the same type of services. The service management layer provides pairing of services with its requesters' applications and enables IoT application programmers to deal with heterogeneous data created by smart things with different hardware specifications. This layer includes the processing of received data before transmitting to the application layer.

#### 1.8.2.4 Application Layer

The application layer in five-layer IoT architecture is responsible for the provisioning of services requested by the users, e.g. temperature, moisture, humidity, air pressure, light intensity measurements, etc. In addition to the user-requested services, the application layer provides data services (i.e. Data warehousing, BigData storage, data mining, etc.) to perform semantic data analysis. Smart health, intelligent transportation system, smart building, smart industry, and smart city are some of the applications with smart user interfaces at application layers.

#### 1.8.2.5 Business Layer

The business layer is responsible to manage overall activities/services of the IoT system through the creation of flowcharts, business models, and graphs on received processed data from the application layer. In addition, based on BigData analysis, this layer supports automatic decision-making as well as the making of smart business strategies.

A few authors have suggested another five-layer SoA-based architecture consisting of Objects, Object Abstraction, Service Management, Service Composition, and Application layers [15].

### 1.8.3 Six-layer Architecture

The six-layer architecture comprises of Focus Layer, Cognizance Layer, Transmission Layer, Application Layer, Infrastructure Layer, and Competence Business Layer as shown in Figure 1.11. This architecture model is proposed to design the integration of more than one IoT system (focusing on different subject areas) and analyzing their implications on business value [48, 49].

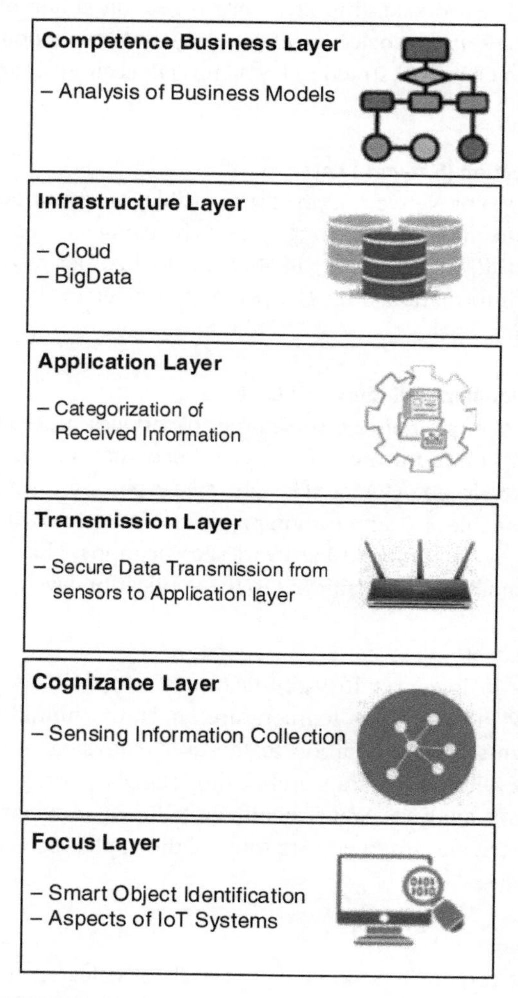


**Figure 1.11** Six-layer IoT architecture.

#### 1.8.3.1 Focus Layer

The modules at this layer are responsible for the identification of smart objects while focusing on the aspects of IoT systems under consideration.

#### 1.8.3.2 Cognizance Layer

This layer consisting of sensors, actuators, and data monitoring modules is responsible for the collection of sensing information from smart objects (identified in the Focus layer).

#### 1.8.3.3 Transmission Layer

This layer is responsible for the transmission of sensed data from the cognizance layer to the application layer.

#### 1.8.3.4 Application Layer

This layer is responsible for the categorization of received information on the basis of application modes.

#### 1.8.3.5 Infrastructure Layer

This deals with the availability of service-oriented technologies, i.e. Cloud, BigData, data mining, etc.

#### 1.8.3.6 Competence Business Layer

This layer includes the analysis of business models of IoT systems.

### 1.8.4 Seven-layer Architecture

Seven-layer IoT architecture comprises seven layers including Things, Edge Computing, Data Accumulation, Data Abstraction, Application, People collaboration and processes layer (as shown in Figure 1.12). The architecture provides the simplest way to understand the functionality of IoT systems [50]. The functionality of each layer is described in the following:

#### 1.8.4.1 Layer 1: Things Layer

The Things layer comprises endpoint devices of IoT systems including smart things (with sensors and controllers) and smart mobile devices (i.e. smartphones, tablets, Personal Digital Assistant [PDA], etc.) to send and receive information. The Things layer supports a diverse range of devices in terms of form, size, and sensing principles; the layer is capable to gather data and conversion of analog observations to digital signals.

#### 1.8.4.2 Layer 2: Connectivity

Considering a diverse range of communication and networking protocols, the connectivity layer is responsible for the in-time transmission of observed data within and between smart things of level 1 and across different networks. In other words, horizontal communication between smart things of level 1 and switching/routing and secure data transmission at different network levels are the basic functionalities of this layer. Although, communication and connectivity through existing IP-enabled network standards is the main focus of IoT

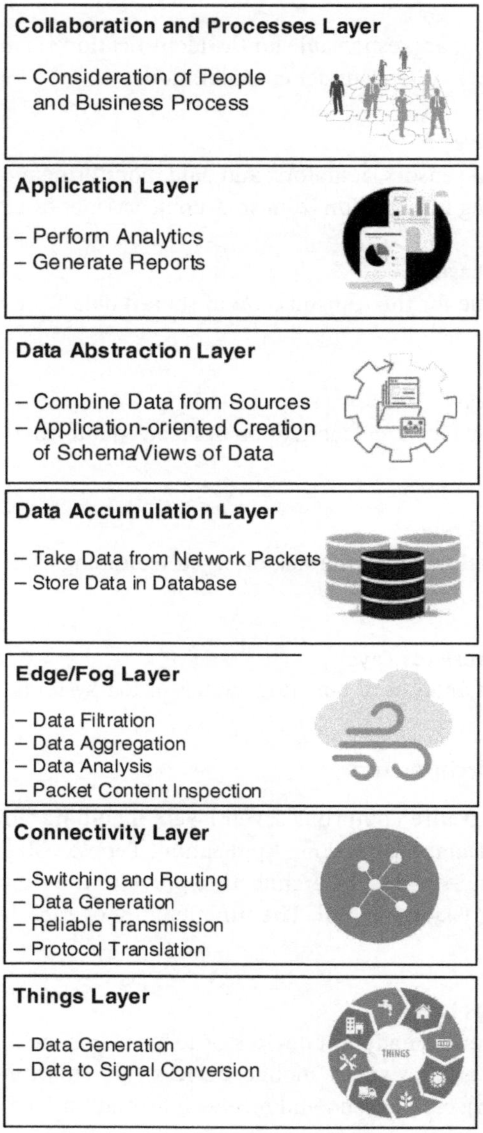

**Figure 1.12** Seven layer IoT architecture.

reference architecture, however, the involvement of non-IP-enabled devices demands gateway standardization.

#### 1.8.4.3 Layer 3: Edge/Fog Computing

Edge/Fog Computing layer is responsible for the conversion of heterogeneous network data flows into information that is appropriate in terms of storage and analysis. According to the notion of early information processing in intelligent IoT systems, this layer initiates limited processing on the received data at the edge of the network, which is mostly referred to as Fog computing. Data formatting, reduction, decoding, and evaluation are the basic functionalities of this layer. The focus of this layer is vertical communication between level 1 and level 4. IoT gateway is an example device at this level.

#### 1.8.4.4 Layer 4: Data Accumulation

Data accumulation or placement of moving data on disk is done at this layer. In other words, at this layer, event-based data is converted to query-based data for processing. Considering the interests of higher layers in available accumulated data, this layer performs filtering or selective storing to reduce data.

#### 1.8.4.5 Layer 5: Data Abstraction Layer

The main focus at the data abstraction layer is related to the rendering and storage of data in such a way that reconciles all the differences in data formats and semantics for the development of simple and performance-enhanced applications.

#### 1.8.4.6 Level 6: Application Layer

Considering the application requirements, the interpretation of level 5 data is done at this layer. Applications are diverse in nature (including system management and control applications, business applications, mission-critical applications, analytical applications, etc.); therefore, relevant data interpretation demands vary from application to application. If data is efficiently organized at layer 5, then information processing overhead gets reduced at this layer, which ultimately supports parallel activities at end devices.

#### 1.8.4.7 Layer 7: Collaboration and Processes

In IoT, different people with different aims use the same application. Therefore, in IoT, the ultimate objective is not the creation of applications but the empowerment of people to do work in a better way. In collaboration and communication for business, processes mostly transcend multiple IoT applications.

## 1.9 Advantages and Disadvantages of IoT

The pros and cons associated with developed and upcoming IoT systems are described in Table 1.11.

## Review Questions

**1.1** How is IoT distinct from other peer technologies, i.e. M2M, CPS, and WSNs?

**1.2** With the help of a diagram, explain the concept of the Information Value Loop. How is it related to IoT?

**Table 1.11** Pros and cons of IoT.

| Advantages | Disadvantages |
|---|---|
| Enhanced comfort and convenience through IoT-based ambient assisted living (AAL) applications improve the quality of life | Interoperability and compatibility of heterogeneous devices in IoT systems |
| In IoT-based systems, device-to-device interactions provide better efficiency in terms of fast reception of accurate results that ultimately save time | The complexity of IoT-based systems results in more failures |
| Automation of daily activities through IoT devices provides a better quality of services | There exist risks of increased unemployment in societies due to the adoption of IoT-based systems in the industrial sector |
| Optimum utilization of resources in IoT systems saves money | The ubiquitous and pervasive nature of IoT systems has increased the risks of losing security and privacy |

**1.3** Explain the difference between technology standards and regulatory standards.

**1.4** Is it possible to differentiate Augmented Behavior from Augmented Intelligence? If Yes, then how?

**1.5** Explain the driving factors that support the use of standardization and augmented intelligence.

**1.6** What is the importance of layered architecture? Identify the components and functionality of each layer of the seven-layer IoT architecture.

**1.7** A company launched an IoT-based application called *CureMe* for the monitoring and management of healthcare of chronic disease patients in a city. The *CureMe* application is designed for:

**A** Personal usage as the mobile platform is capable to generate alerts for caregivers when something goes wrong about the vital sign readings of chronic disease patients

**B** Government agencies with the provisioning of the web platform to analyze disease trends in the city

Considering the implementation perspective of an IoT system, provide a clear architecture diagram of *CureMe* application. All five IoT building blocks are required to be shown in this IoT-based healthcare architecture diagram. Moreover, working details of all involved components (for *CureMe* Application) are required to be explained.

**1.8** For good quality milk production, monitoring cows' activity in the farm is one of the main factors that demand nonstop (24×7) monitoring of every single cow on the farm. To address this challenge, a company launched an IoT-based application (called Ida – Intelligent Dairy Farmers Assistant) for dairy farmers. Ida combines sensor technology, machine learning, and Cloud computing to translate obtained cows' data from the farm into meaningful information that can be used to support decisions made by farmers every day. Therefore, Ida not only monitors the activities of all cows at the farm but also uses AI to learn the behavior of cows to generate information that is useful for the farmers. Thinking about the implementation perspective of an Ida-like IoT system, it is required from you to answer the following questions:

**A** Consider the following figure. Mark the boundary of any five parts of cow and label with respective five types of sensors that can be used to monitor cow activities that ultimately affect the quality of milk.

**Source: Financial Times www.ft.com.**

**B** Provide a clear IoT-based system architecture diagram of the Ida-like system.

## References


1 Al-Fuqaha, A., Guizani, M., Mohammadi, M. et al. (2015). Internet of things: a survey on enabling technologies, protocols, and applications. *IEEE Communication Surveys and Tutorials* 17 (4): 2347–2376.

2 Gupta, R. and Gupta, R. (2016). ABC of Internet of Things: advancements, benefits, challenges, enablers and facilities of IoT. In: *IEEE Symposium on Colossal Data Analysis and Networking (CDAN)*, 1–5. IEEE.

3 Peña-López, I. (2005). *ITU Internet report 2005: the Internet of Things.*

4 Ornes, S. (2016). Core concept: the internet of things and the explosion of interconnectivity. *Proceedings of the National Academy of Sciences* 113 (40): 11059–11060.

5 *The "Only" Coke Machine on the Internet* (2018). Carnegie Mellon University, www.cs.cmu.edu/~coke/history_long.txt.

6 Weiser, M. (1999). The computer for the 21st century. *Mobile Computing and Communications Review* 3 (3): 3–11.

7 Raji, R.S. (1994). Smart networks for control. *IEEE Spectrum* 31 (6): 49–55.

**8** Pontin, J. (2005). *ETC: Bill Joy's Six Webs*, MIT Technology Review.

**9** Gershenfeld, N.A. and Gershenfeld, N. (2000). *When Things Start to Think*. Macmillan.

**10** Mattern, F. and Floerkemeier, C. (2010). From the internet of computers to the internet of things. In: *From Active Data Management to Event-Based Systems and More*, 242–259. Springer.

**11** Schoenberger, C.R. (2002). The internet of things. *Forbes*: 155160–155160.

**12** Miorandi, D., Sicari, S., De Pellegrini, F. et al. (2012). Internet of things: vision, applications and research challenges. *Ad Hoc Networks* 10 (7): 1497–1516.

**13** Gubbi, J., Buyya, R., Marusic, S. et al. (2013). Internet of things (IoT): a vision, architectural elements, and future directions. *Future Generation Computer Systems* 29 (7): 1645–1660.

**14** Holdowsky, J., Mahto M., Raynor M.E. et al. (2015). *Inside the internet of things (IoT)*. Deloitte Insights. August, 2015.

**15** Atzori, L., Iera, A., and Morabito, G. (2010). The internet of things: a survey. *Computer Networks* 54 (15): 2787–2805.

**16** Parnian, A.R., Parsaei, M.R., Javidan, R. et al. (2017). Smart objects presence facilitation in the internet of things. *International Journal of Computer Applications* 168 (4): 25–31.

**17** Weyrich, M. and Ebert, C. (2016). Reference architectures for the internet of things. *IEEE Software* 1: 112–116.

**18** Atzori, L., Iera, A., and Morabito, G. (2017). Understanding the internet of things: definition, potentials, and societal role of a fast evolving paradigm. *Ad Hoc Networks* 56: 122–140.

**19** Borgia, E. (2014). The internet of things vision: key features, applications and open issues. *Computer Communications* 54: 1–31.

**20** Minerva, R., Biru, A., and Rotondi, D. (2015). Towards a definition of the internet of things (IoT). *IEEE Internet Initiative* 1: 1–86.

**21** Rose, K., Eldridge, S., and Chapin, L. (2015). *The Internet of Things: An Overview–Understanding the Issues and Challenges of a More Connected World*. The Internet Society (ISOC).

**22** Mukhopadhyay, S.C. and Suryadevara, N.K. (2014). Internet of things: challenges and opportunities. In: *Internet of Things*, 1–17. Springer.

**23** Manrique, J.A., Rueda-Rueda, J.S., Portocarrero, J.M. (2016). Contrasting internet of things and wireless sensor network from a conceptual overview. IEEE International Conference on Internet of Things (iThings) and IEEE Green Computing and Communications (GreenCom) and IEEE Cyber, Physical and Social Computing (CPSCom) and IEEE Smart Data (SmartData), 252–256.

**24** Weiser, M., Gold, R., and Brown, J.S. (1999). The origins of ubiquitous computing research at PARC in the late 1980s. *IBM Systems Journal* 38 (4): 693–696.

**25** Want, R., Schilit, B.N., and Jenson, S. (2015). Enabling the internet of things. *Computer* 1: 28–35.

**26** Raj, P. and Raman, A.C. (2017). *The Internet of Things: Enabling Technologies, Platforms, and Use Cases*. Auerbach Publications.

**27** Kejriwal, S. and Mahajan, S. (2016). *Smart Buildings: How IoT Technology Aims to Add Value for Real Estate Companies*. Deloitte Center for Financial Services.

**28** Lin, J., Yu, W., Zhang, N. et al. (2017). A survey on internet of things: architecture, enabling technologies, security and privacy, and applications. *IEEE Internet of Things Journal* 4 (5): 1125–1142.

**29** Hunt, V.D., Puglia, A., and Puglia, M. (2007). *RFID: A Guide to Radio Frequency Identification*. Wiley.

**30** Violino, B. (2005). The history of RFID technology. *RFID Journal* 16.

**31** Bowden, L. (1985). The story of IFF (identification friend or foe). *IEE Proceedings A (Physical Science, Measurement and Instrumentation, Management and Education, Reviews)* 132 (6): 435–437.

**32** Kaur, M., Sandhu, M., Mohan, N. et al. (2011). RFID technology principles, advantages, limitations & its applications. *International Journal of Computer and Electrical Engineering* 3 (1): 151.

**33** Ruiz-Garcia, L., Lunadei, L., Barreiro, P. et al. (2009). A review of wireless sensor technologies and applications in agriculture and food industry: state of the art and current trends. *Sensors* 9 (6): 4728–4750.

**34** Akkaya, K. and Younis, M. (2005). A survey on routing protocols for wireless sensor networks. *Ad Hoc Networks* 3 (3): 325–349.

**35** Jones, J. and Atiquzzaman, M. (2007). Transport protocols for wireless sensor networks: state-of-the-art and future directions. *International Journal of Distributed Sensor Networks* 3 (1): 119–133.

**36** Radoslav, L. (2012). *Wireless Identification and Sensing Platform*. PON Press.

**37** Philipose, M., Smith, J.R., Jiang, B. et al. (2005). Battery-free wireless identification and sensing. *IEEE Pervasive Computing* 4 (1): 37–45.

**38** Sen, J. (2018). *Internet of Things: Technology, Applications and Standardization*. IntechOpen.

**39** Pal, A. and Purushothaman, B. (2016). *IoT Technical Challenges and Solutions*. Artech House.

**40** Pal, A., Rath, H.K., Shailendra, S. et al. (2018). *IoT Standardization: the Road Ahead. Internet of Things-Technology, Applications and Standardization*, 53–74. IntechOpen.

**41** Davenport, T. and Harris, J. (2017). *Competing on Analytics: Updated, with a New Introduction: The New Science of Winning*. Harvard Business Press.

**42** Abousalh-Neto, N.A. and Kazgan, S. (2012). Big data exploration through visual analytics. In: *2012 IEEE Conference on Visual Analytics Science and Technology (VAST)*, 285–286. IEEE.

**43** Murray, D.G. (2013). *Tableau your Data!: Fast and Easy Visual Analysis with Tableau Software*. Wiley.

**44** Krco, S., Pokric, B., and Carrez, F. (2014). Designing IoT architecture (s): a European perspective. In: *IEEE World Forum on Internet of Things (WF-IoT)*. IEEE.

**45** Wang, W., De, S., Toenjes, R. et al. (2012). A comprehensive ontology for knowledge representation in the internet of things. In: *IEEE 11th International Conference on Trust, Security and Privacy in Computing and Communications*, 1793–1798. IEEE.

**46** Mahmoud, R., Yousuf, T., Aloul, F. et al. (2015). Internet of things (IoT) security: current status, challenges and prospective measures. In: *IEE 10th International Conference for Internet Technology and Secured Transactions (ICITST)*, 336–341. IEEE.

**47** Khan, R., Khan, S.U., Zaheer, R. et al. (2012). Future internet: the internet of things architecture, possible applications and key challenges. In: *IEEE 10th International Conference on Frontiers of Information Technology*, 257–260. IEEE.

**48** Kumar, N.M., Dash, A., and Singh, N.K. (2018). Internet of Things (IoT): an opportunity for energy-food-water nexus. In: *IEEE International Conference on Power Energy, Environment and Intelligent Control (PEEIC)*, 68–72. IEEE.

**49** Kumar, N.M. and Mallick, P.K. The Internet of Things: insights into the building blocks, component interactions, and architecture layers. *Procedia Computer Science 132*: 109–117.

**50** CISCO Draft (2014). *The Internet of Things Reference Model.* (White Paper), Available at: http://cdn.iotwf.com/resources/71/IoT_Reference_Model_White_Paper_June_4_2014.pdf.

# 2

# IoT Building Blocks – Hardware and Software

**LEARNING OBJECTIVES**

After studying this chapter, students will be able to:

- identify the basic IoT building blocks.
- discuss smart thing components and capabilities.
- understand the basics of Packet Tracer with reference to IoT.
- discuss the basics of IoT gateway, Cloud, and analytics.
- apply Packet Tracer to draw application-specific IoT architecture.

## 2.1 IoT Building Blocks

IoT is the integration of various technologies that have their own significance within a complete IoT system. The basic building blocks (i.e. Smart Thing [Object/Device], IoT Gateway, IoT Cloud, IoT Analytics, and IoT Applications) [1, 2] are shown in Figure 2.1 and are required to be apprehended.

## 2.2 The Smart Things

In IoT, a smart thing could be any physical object that is created either as a manufactured artifact or by placing embedded electronic tags or sensors to any non-smart physical thing. These smart things are also known as smart devices and hold the potential of exchanging data in the IoT system. The smart thing conceptual model (shown in Figure 2.2) represents the synergy between the physical things and the digital world [2]. A smart thing is essentially a physical object with some electronic device. With reference to a smart thing in IoT, the attached electronic device may have three subtypes, i.e. tag, sensor, and actuator. Tags provide digital identification (in terms of Barcode, Aztech code, QRcode, Radio Frequency Identification [RFID], etc.) to smart things and are read by reader sensors. Simple sensors are embedded or attached to smart things and are responsible for identity reading, state monitoring of physical things, and communication. Advanced sensors are onboard sensors

*Enabling the Internet of Things: Fundamentals, Design, and Applications*, First Edition.
Muhammad Azhar Iqbal, Sajjad Hussain, Huanlai Xing, and Muhammad Ali Imran.

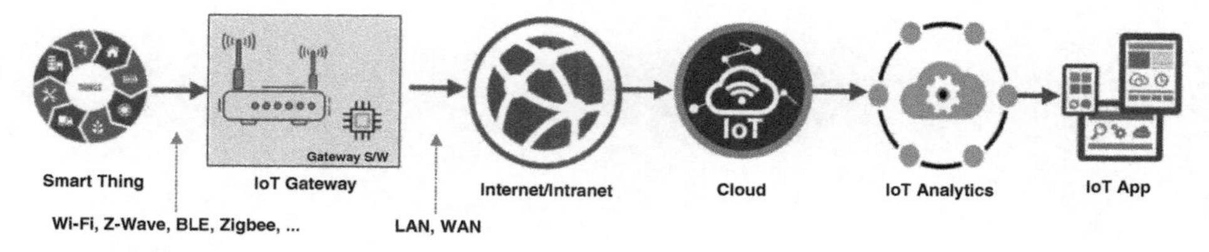


**Figure 2.1** Building blocks of an IoT system.

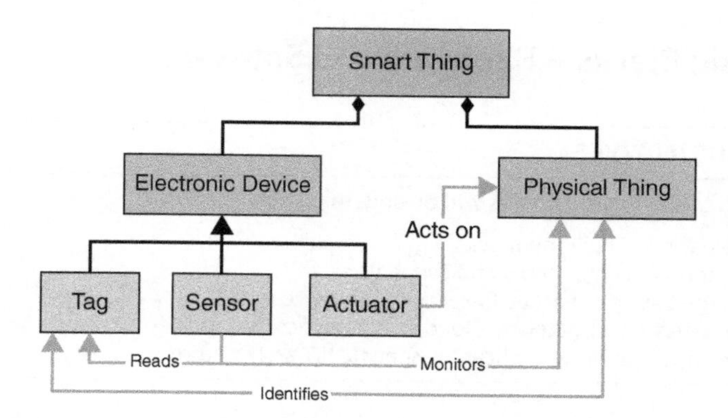


**Figure 2.2** Smart thing conceptual model. Source: Adapted from Serbanati et al. [2]

with intelligent algorithms and software techniques for the observations of a specific environment and monitoring of states of physical things. Actuators act on the physical thing to update its state.

In summary, smart things have unique digital identities with sensing (and/or actuator) capability, embedded processor, communicator, and/or actuator [3] as shown in Figure 2.3.

### 2.2.1 Smart Thing Sensor

The sensor is one of the most essential components of smart things that induce the sensing capability in smart things to perceive a change in the ambient conditions occurring around their environment. In general, a sensor is a device that is able to receive and respond to a stimulus (for example, variation in any natural phenomenon, i.e. temperature, pressure, humidity, motion, position, displacement, sound, force, flow, light, chemical presence, etc.). Technically, a sensor is a device that translates the received stimulus (a quantity, property, or condition/state of a physical object) into an electrical signal. For example, for temperature and pressure sensors, heat and atmospheric pressure are converted to electrical signals. Therefore, the input of a sensor is some kind of observation (non-electric) related to the change in the physical property of an object, and the output is an electrical signal in the terms of variation in charge, voltage, current, etc. that can be described by Frequency, Amplitude, and Phase. The sensor output is ultimately required to be compatible with electronic circuits. Commonly used sensors in IoT smart things include temperature sensor, humidity sensor, proximity sensor, motion sensor, gyroscope

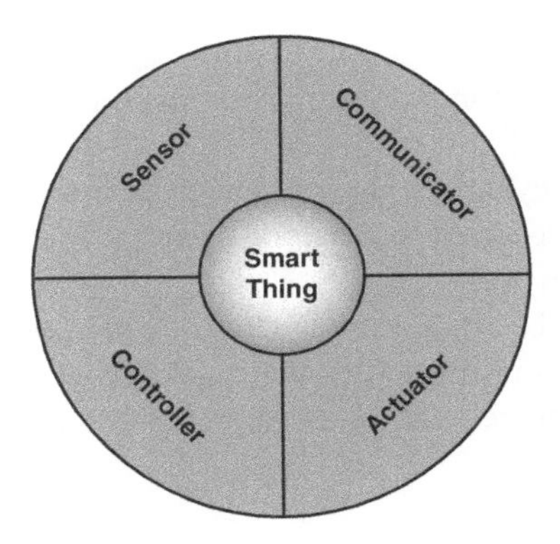


**Figure 2.3** Smart thing components.

sensor, rain sensor, gas and smoke sensor, pressure sensor, light sensor, iris motion sensor, alcohol sensor, etc.

### 2.2.2 Smart Thing Communicator

Besides the sensing capability, the smart things in IoT contain low power transmitting components that can transmit sensed data to respective network technology devices. Different types of wireless networks including Wi-Fi, Bluetooth, ZigBee, Long Range Wide Area Network (LoRAWAN), etc. enable the connectivity of smart things to the IP-based network infrastructure through IoT Gateways [4]. Details about the communication stack of these devices are available in Chapter 5.

### 2.2.3 Smart Thing Actuator

Actuators as a complement to sensors are required to take actions, which are based on some sensor readings. In simple words, actuators are electronic/mechanical components or devices of an IoT system that perform direct/indirect physical actions on the environment to control a certain mechanism [5]. Actuators fundamentally require two things, i.e. energy and control signal for their proper functioning. In typical IoT systems, after decision-making over received sensed information in the control center, commands are ultimately sent back to actuators. Upon the reception of control signals, the actuator converts the received energy and is able to respond through mechanical means. In general, actuators can be categorized into various types, i.e. electric-based actuators, magnetic-based actuators, mechanical-based actuators, hydraulic actuators, and pneumatic actuators.

Electric actuators use electric power motors to actuate the equipment in terms of mechanical torque. For example, a Solenoid valve is an electric actuator that is used to control water flow in pipes.

Thermal/Magnetic Actuators are actuated by the implication of thermal or magnetic energy. One of the uses of these actuators is the shaping of memory alloys (also known as SMAs).

Mechanical actuators are used in gears, pulleys, chains, etc. and are based on the principle of converting rotary motion into linear motion. Rack and Pinion actuator is an example of a mechanical actuator.

Hydraulic actuators use hydraulic power to facilitate mechanical motion by converting the mechanical motion into linear, oscillatory, or rotary motion. The radial engine is an example of a hydraulic actuator.

Pneumatic actuators are based on the principle of converting energy produced by compressed air into linear/rotary motion. Pneumatic rack and pinion actuators are examples of these types of actuators that are used to control valves of water pipes.

### 2.2.4 Smart Thing Controller

To make things smart, typically IoT applications demand more than just adding a sensor to a physical thing, i.e. a Microcontroller (MCU).

#### 2.2.4.1 Microcontroller (MCU)

An MCU contains one or more processors, memory, programmable input/output peripherals on a single integrated circuit. These MCUs are different from microprocessors (available in Personal Computers [PCs]) and basically designed for embedded applications. The cost-effectiveness of MCUs supports their usage for the addition and enhancement of computing capabilities of a physical thing. The following are the significant features of MCUs:

- MCUs have a specific amount of RAM.
- MCUs have flash memory to store offline data.
- MCUs have input/output pins (ranging from 1s to 100s) to connect sensors/actuators to MCU.
- MCUs have Ethernet/Wi-Fi ports for Internet connectivity.
- MCUs are generally available in a number of bits that ultimately affect the speed of computation.
- MCUs have power supply pins to supply power to attached sensors.

Concerning operating system (OS) on an MCU, three options (i.e. Bare metal, Real-Time Operating System [RTOS], and Linux) are available. Bare metal means the absence of OS that is efficient but provides less support for software developers. RTOS provides guarantees against time in which computation/processing is required to be completed. Linux provides no timing guarantees but easy to program and provides a real computer environment.

#### 2.2.4.2 Development Boards

MCUs are mostly available on printed circuit boards known as development boards, which also contain supporting components (i.e. power source, support for connecting sensors, and sometimes onboard sensor). These development boards are useful for the prototyping of an IoT system and enable users to quickly connect sensors. Moreover, the accompanying software of these development boards facilitates the deployment of code for these sensor devices. There are a number of development boards, and MCUs are available from different companies, i.e. Arduino, Raspberry Pi, Samsung, etc. Although selection of right development board and MCU ultimately depends on the nature of the application, however, different factors also play important role, i.e.:

- Easy availability and compatibility of development board to support sensors of your application.
- Sufficient memory to execute your IoT application.
- Energy-efficient architecture and cost of development board to implement IoT system.

#### 2.2.4.3 Packet Tracer and MCUs

The Packet Tracer is a simulation tool that supports students and network administrators to study computer network behavior before its actual deployment [6]. Packet Tracer 7.0, in addition to conventional computer network devices, brings support for the simulation of IoT systems [7] through the availability of:

- Smart Things (related to the smart home, smart city, smart industry, and smart grid) that can be connected to each other over computer networks.
- IoT sensors and actuators (that sense and manipulate the environment, respectively).
- IoT (home) gateway in terms of network interface.
- Development boards (MCU and Single Boarded Computers [SBC]) that can be programmed with reference to IoT systems. MCU and SBC are generally used in simulations to control functions of IoT devices.

Few IoT smart things and components available in Packet Tracer have been shown in Figure 2.4.

Therefore, now concerning the design understanding of an IoT system, Packet Tracer is helpful in two ways: first, provisioning of a virtual environment that mimics a real-life scenario, and secondly, it provides an integrated development environment for the programming of MCUs for the evaluation of proposed algorithms before actual implementation in real-world scenarios. Using Packet Tracer, computer science students program the MCU in three languages, i.e. Blockly, Java, or Python to develop an IoT system. In this book, we have selected Blockly because of its abstract nature. Chapter 10 of this book provides details related to the implementation of the IoT system in Packet Tracer using Blockly language. However, in this section, a brief introduction of Packet Tracer with reference to IoT system deployment is given. In Packet Tracer, depending on the scenario, typical IoT systems can be designed or configured in three ways:

- First, IoT smart things can be directly connected to the gateway as shown in Figure 2.5.
- Second, IoT smart things can be either directly connected to the gateway or indirectly through MCU as shown in Figure 2.6. However, basic IoT components, i.e. sensors and actuators (due to lack of ethernet port), can only be connected to Gateway through MCU. Gateway is dependent on the programmed MCU to receive the status of the attached sensors and actuators.
- Third, IoT components and smart things can be connected to MCU for automatic working without using gateway connectivity as shown in Figure 2.7.

**Figure 2.4** IoT components and smart things available in Cisco Packet Tracer.

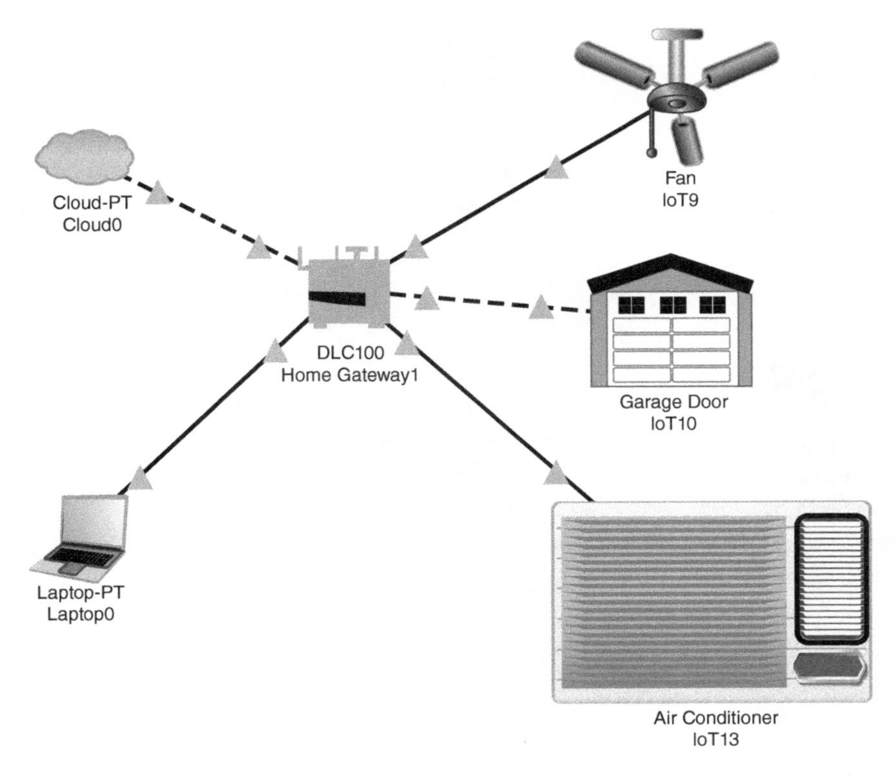


**Figure 2.5** IoT smart things directly connected to home gateway.

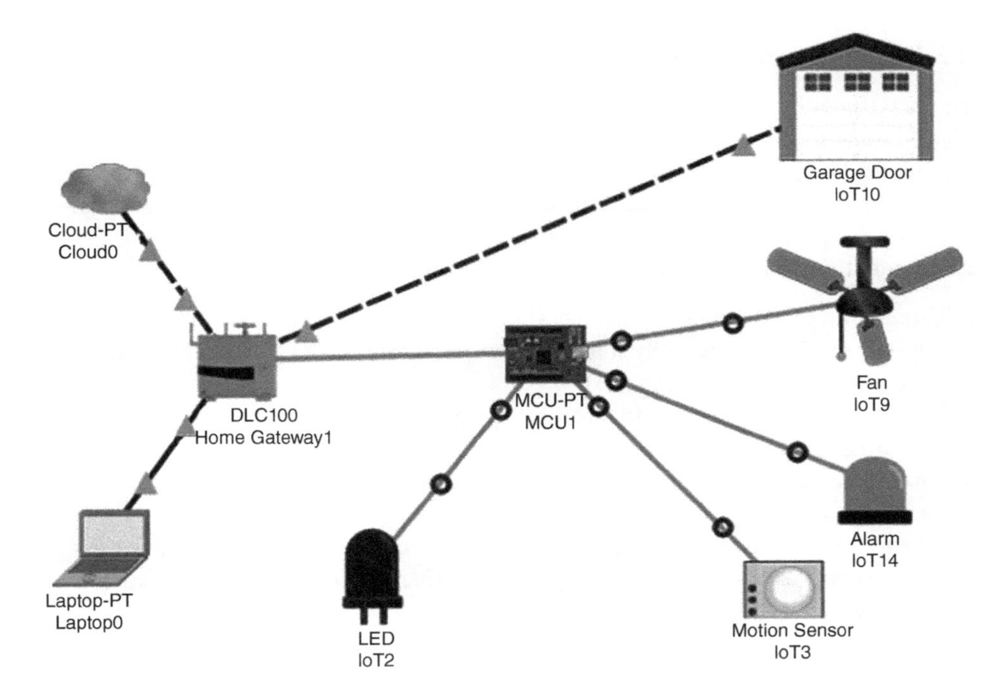


**Figure 2.6** IoT components and smart things connected to MCU.

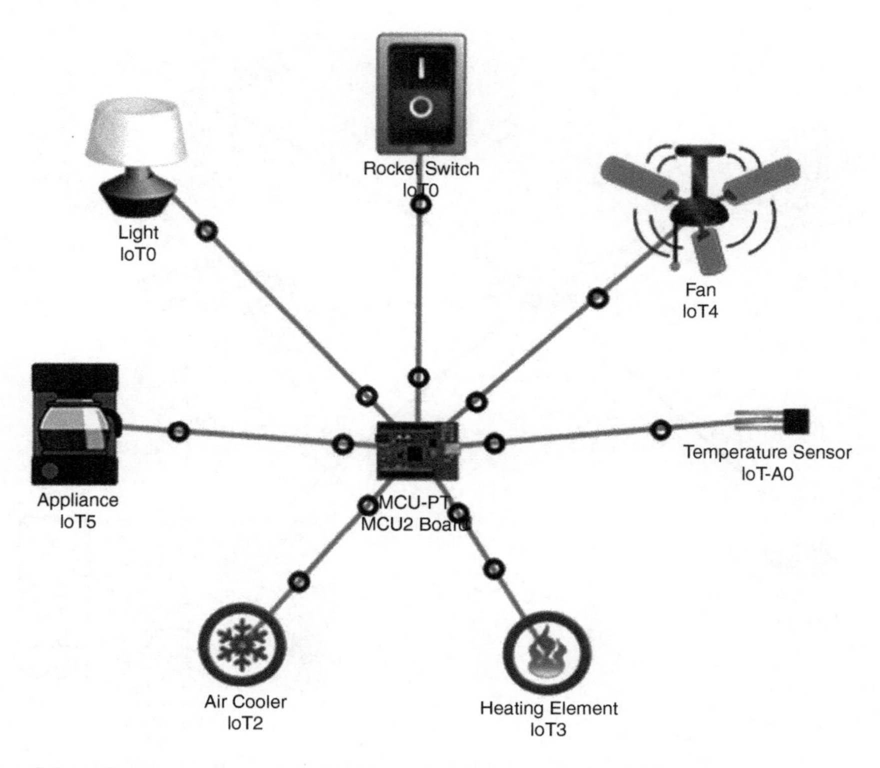


**Figure 2.7** IoT components and smart things directly connected to MCU.

### 2.2.5 Smart Thing Capabilities

The smartness of a(n) thing/object is dependent on the *capabilities* it possesses. Smart object capabilities can be divided into *core capabilities*, *enhanced capabilities*, and *advanced capabilities* [8, 9].

Core capabilities define the essential features of Smart Things; without these features an object or thing cannot be considered as a smart object or thing. Core capabilities include digital identification, retention, communication, and energy harvesting.

- Through unique and immutable digital identification of smart objects in IoT, the smart objects can identify themselves in IoT systems for the sake of information access and object presence in a digital context.
- Retention refers to the memory of smart things to store information about itself as well as of environment observation.
- Communication capability is necessary for information exchange within a network of smart objects in the IoT system.
- All smart objects require energy to perform basic processing and to carry out assigned tasks. Smart objects harvest energy from either external sources or autonomous generation.

Enhanced capabilities include sensing/actuating, processing, networking, shielding, and logging. However, self-awareness and self-management are considered advanced capabilities of smart objects:

- Sensing and actuating capabilities refer to the gathering of sensed information from the environment and provoking of an essential stimulus, respectively. In general, sensing is associated with actuating; however, most of the objects have either one or the other.
- Processing is the capability to execute instructions through the embedded controller.
- Networking capability involving protocol stack supports the connection of heterogeneous capability.
- Shielding ability is basically the provision of security and privacy to critical readings of smart objects.
- Logging is the capability through which a smart object can register events about itself and the environment.
- Self-awareness capability allows smart objects to keep information about their status, structure, and history.
- Self-management capability refers to the developmental abilities of an object to utilize its collected information to manage the object's *life cycle and its maintenance.*

On the basis of the core, enhanced, and advanced capabilities, smart objects can be classified into three levels, i.e. Essential, Enhanced, and Advanced level smart objects, as shown in Figure 2.8.

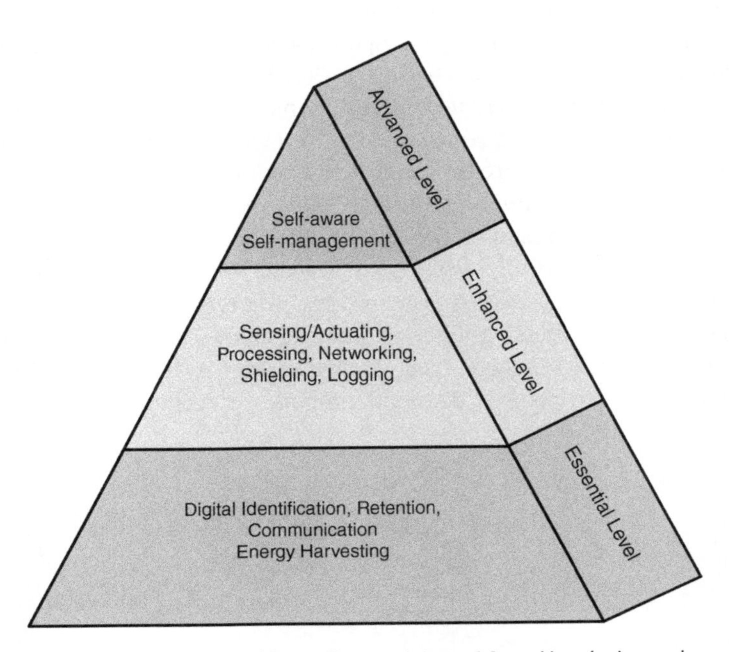


**Figure 2.8** Different levels of smart objects. Source: Adapted from Hernández and Reiff-Marganiec [8].

Considering the three design dimensions (i.e. awareness, representation, and interaction), smart things can be categorized into three types, i.e. Activity-aware smart things, Policy-aware smart things, and Process-aware smart things [3]:

*Activity-aware smart things* have no interactive capabilities and are able to record information and duration of its usage.

*Policy-aware smart things* are essentially an activity-aware thing but are able to interpret activities with reference to organizational policies.

*Process-aware smart things* are able to understand and relate the organizational policies to real-world activities.

## 2.3 The IoT Gateway

IoT gateway can be a dedicated physical device or software that assists connectivity between devices and Cloud [10]. Sensor-acquired data moving toward the IoT Cloud passes through the gateway that pre-processes the sensor data at the edge. Pre-processing on large volumes of sensor data involves the compression of aggregated data to reduce transmission costs. IoT gateway performs the translation of different network protocols to support the interoperability of smart things and connected devices. In addition, IoT gateways provide certain levels of security through advanced encryption networks. Therefore, as a middle layer (between devices and Cloud), IoT gateways protect the IoT system from unauthorized access and malicious attacks. At the abstract level, IoT architecture is shown in Figure 2.9. Figure 2.9 illustrates that the typical IoT gateway is not equipped with any kind of sensors; the software installed on IoT gateway is responsible for the collection, pre-processing, and transmission of received sensor data to the IoT Cloud.

Concerning the anatomy at the generic level, layered IoT gateway technology architecture consists of various hardware components and software modules. The hardware components include MCUs (processors), wireless connectivity modules (i.e. Wi-Fi, ZigBee, Bluetooth, 2G/3G/4G, etc.). Although Linux and Android OSs can be used, however, RTOS is preferred. For communication, one or several (i.e. ZigBee, IPv6 over Low -Power Wireless Personal Area Networks [6LoWPAN], BLE, etc.) is/are available in IoT gateway. Security is implemented in the form of Crypto Authentication chips. In summary, IoT gateway is responsible for the collection, pre-processing, filtering, storage, analysis, and secure transmission of data from sensor nodes to IoT Cloud. This computing (from collection to analysis) at the IoT gateway before transferring to IoT Cloud promotes the concept of Edge computing

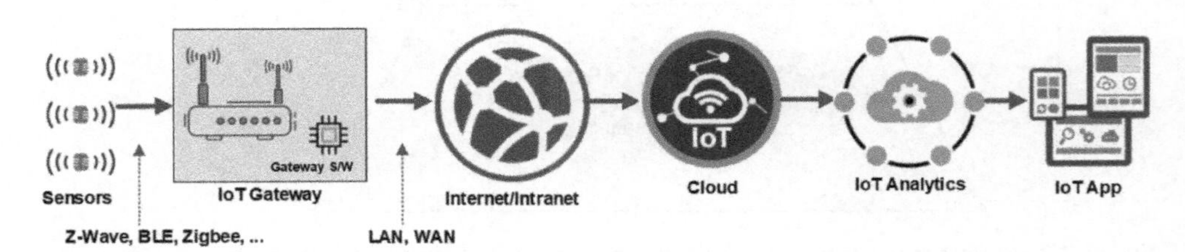


**Figure 2.9** Simple IoT gateway in IoT system.

that complements the IoT Cloud. Edge and Cloud computing are mutually exclusive approaches, and in lager IoT projects, a combination of both is required. Local processing, distributed computation, and quick response for near IoT smart things are the main benefits of Edge/Fog computing over IoT Cloud. Edge/Fog computing is illustrated more in Chapter 6. In addition, the IoT gateway also ensures devices interoperability, security, and privacy of data at the monitoring end of IoT devices [11].

## 2.4 Network Infrastructure

Network infrastructure (i.e. the Internet or internet) enables the processing and transmission of data from smart things to IoT Cloud through several homo-/heterogeneous network devices, i.e. switches, routers, and gateways.

## 2.5 IoT Cloud

Sensor, applications, and/or end users in the IoT system create an enormous amount of data at IoT Cloud. At the architecture level, IoT Cloud is a network of high-performance servers that stores, processes, and manages massive amounts of data for analysis.

IoT Cloud is established through the exploitation of virtualization technology and consists of different components, i.e. Virtual Resource Pool and three major types of VM-based configured servers, i.e. Application Servers, Databases and Load Balancers as shown in Figure 2.10. These servers work independently even running on the same physical machine in the available Virtual Resource Pool. A brief description of these components is given as follows.

### 2.5.1 Virtual Resource Pool

Virtual Resource Pool consists of two components, i.e. [12]:

- Hardware resources that are available in the form of several CPUs, memory, and network connectivity on physical machines.
- Hypervisor software running on these physical machines that provide OS environment (known as VMs) and enable dynamic resource allocation. Regarding the implementation of the hypervisor, virtual OS, i.e. VMWare vSphere, is preferred, which is able to access direct computing and memory resources of physical machines. IoT Cloud services are implemented on VMs to provide high performance at low cost.

### 2.5.2 Application Server

Application Servers in IoT Cloud comprises of both Hypertext Transfer Protocol (HTTP) servers and Message Queuing Telemetry Transport (MQTT) servers. Both are developed using *Node.js*, which is able to provide high concurrency. HTTP servers interact in a request–response manner and require Express (web application framework) to deploy web

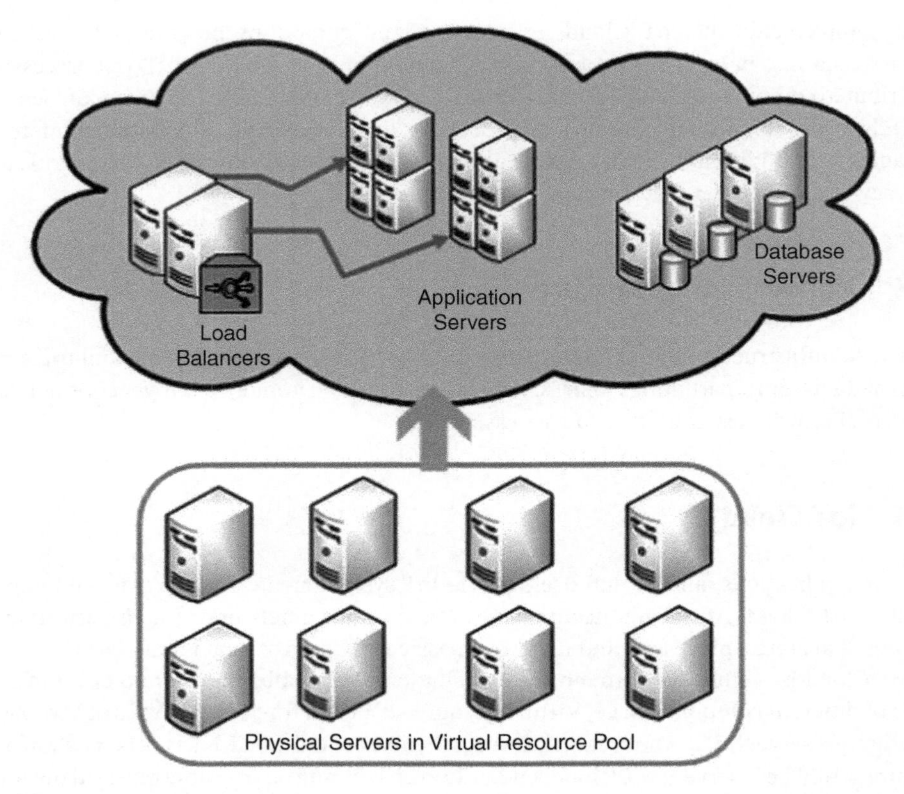


**Figure 2.10** IoT cloud architecture.

and mobile applications. HTTP is not suitable for resource-constrained IoT devices. On the other hand, for business services to the customers, application servers based on MQTT protocol [13, 14] are preferred in contrast to conventional application servers, which are based on HTTP protocol. MQTT is a light-weight publish-subscription-based message transportation protocol and specifically designed for IoT devices that are constrained by computing, communication, and energy resources [15]. MQTT servers are developed using MQTT Connection (an open-source *Node.js* library). Details about the MQTT protocol are available in Chapter 5.

### 2.5.3 Database Servers

Database servers are optional and depending on the nature of IoT application, the data is stored in relational (SQL) and non-relational (NoSQL) databases. However, due to the non-suitability of SQL databases for real-time IoT applications, NoSQL databases are highly recommended in IoT Cloud. Regarding implementation, Redis (a NoSQL database) is used, which is able to store data directly in memory to improve Input/Output speed.

### 2.5.4 Load-balancing Servers

Load-balancing Servers are essential and beneficial in IoT Cloud as these:

- process requests in a scheduled way.
- distribute the workload on application/database servers.
- avoid congestion on application/database servers.
- achieve maximum utilization of available resources.

Regarding implementation, HAProxy (an open-source lightweight load balancer) is preferred that offers HTTP-based and Transmission Control Protocol (TCP)-based load balancing along with the support of high concurrency. For the distribution of requests over HTTP servers, the weighted round-robin strategy is configured in HTTP load balancers. To achieve efficient load balancing, HAProxy first binds requesters' URLs to available ports and then following some pre-defined principle distributes these requests over HTTP servers. In MQTT-based load balancers, a TCP load-balanced mechanism is implemented to distribute the load over MQTT servers.

## 2.6 IoT Analytics

IoT analytics is the process of the formation of useful interpretations for trend forecasting. Predictive analysis of IoT Cloud data can be used for preventive measures as well as for improving products and services. Besides the relationship of IoT analytics with IoT Cloud, there exists a strong association between IoT analytics and BigData analytics because IoT data basically have similar Vs of BigData [16, 17]. For example:

- Volume of data from sensors is too large to store in conventional databases.
- Variety of data (in semantic and format) from different types of sensors.
- Velocity can be considered as high-frequency IoT data collection.
- Veracity is low as noise is required to be removed from collected IoT data.

With these similarities, most of the time the tools and techniques to deal with BigData analytics (e.g. databases, data warehouses, and data mining approaches) are also applicable to IoT analytics. However, IoT analytics/applications have own unique challenges (given in the following), which are distinct from BigData applications (shown in Figure 2.11) and demand new approaches and tools to be proposed and implemented, respectively [16]:

- Acquiring of IoT data through sensors
- Dealing with heterogeneous as well as real-time nature of sensed data
- Refinement/elimination of noise through the implication of statistical and probabilistic techniques on sensed heterogeneous data
- Spatiotemporal dependencies of IoT data streams
- Biasedness of IoT data demands pre-processing (i.e. understanding and scrutiny of training and test datasets)
- Security/privacy requirements necessitate pre-processing of personal data

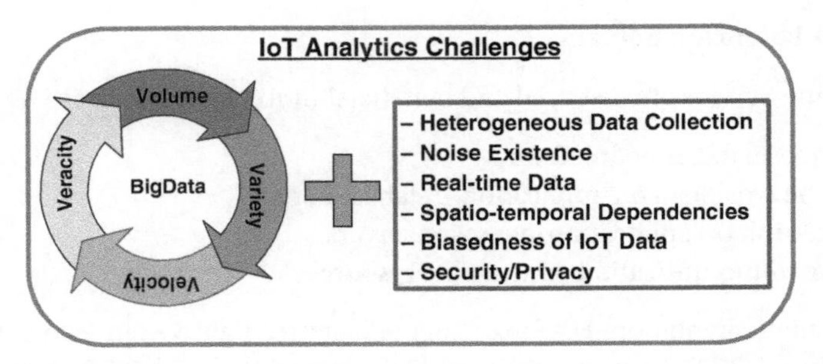


**Figure 2.11** Challenges of IoT data analytics.

### 2.6.1 IoT Analytics – Tools and Techniques

Figure 2.12 shows the disciplines, techniques, and tools associated with IoT analytics. A brief description of these techniques and tools is given in Table 2.1.

**Table 2.1** IoT Analytics - Techniques and tools.

| Technique | Tools |
|---|---|
| Batch Processing (assumes data available in database) | • Hadoop [18–20] is the implementation of the Map Reduce programming model and it allows the parallel processing of large data sets. YARN/ MapReduce 2.0 [21] is the latest version of MapReduce<br>• Apache Spark [22] is an open-source analytical framework that supports various programming languages (i.e. Python, Java, Scala, etc.) and provides a programming interface to execute graphs engine and general-purpose cluster computing system. It is much faster than MapReduce |
| Machine Learning (enables the computing system to learn in an automatic way) | • MLlib [23] is the machine learning library available in Apache Spark. Several common algorithms, i.e. Clustering, Classification, Regression, etc. have been implemented in this MLlib for analysis<br>• Konstanz Information Miner (KNIME) [23] with graphical interface is able to perform analysis using different machine learning algorithms<br>• Mahout [24, 25] is a machine learning library with different implemented classification, clustering, and collaborative filtering algorithms at the top of Hadoop<br>• H20 [26] is an open-source machine-learning-based platform, which is implemented on Hadoop and Spark to train models |
| Stream Processing (dataflow programming) | • Flink [27–29] is an open-source distributed streaming dataflow framework, which is available in Scala and Java and executes dataflow programs in parallel at the top of MapReduce 2.0 or YARN |
| Data Visualization (a graphical representation of data) | • Freeboard with REST API provides a plugin architecture for creating widgets to display information graphically<br>• Elasticsearch (used with Spark and Flink data streams) [30] is an open-source full-text search and analytics engine, which works with JASON documents using web interfaces and capable to search and analyze the huge volume of data in real time<br>• Kibana dashboard [31, 32] is an open-source advanced visualization tool to show data in charts, maps, and tables<br>• Tableau Public [33] is a free software that allows its users to connect spreadsheets for the creation of interactive data visualization |

### 2.6.2 IoT Analytics Life Cycle

Considering all these factors, the lifecycle of IoT Analytics consists of four phases, i.e. data collection, data analysis, and data operational reusability (shown in Figure 2.13).

*IoT Data Collection Phase*: This first phase of the IoT analytics life cycle deals with the collection of IoT data from heterogeneous devices at IoT Cloud.

*IoT Data Unification Phase*: The second phase deals with the validation and refinement of IoT sensed data in terms of integrity, consistency, and accuracy.

*IoT Data Analysis Phase*: This is the third phase of the IoT analytics life cycle and deals with the structuring, storage, and analysis of data through the implications of machine learning and data mining techniques that transform IoT data into actionable knowledge.

*IoT Operational and Reuse Phase*: This fourth phase deals with the actual implementation and operational details of analysis on IoT data. It also supports the visualization and reusability of IoT datasets.

The jobs mentioned in all these phases are supported by several disciplines and tools of computing, i.e. data collection, knowledge discovery, data mining, machine learning, Relational Database Management Systems, Distributed database management systems, NoSQL databases, BigData databases, Hadoop Distributed File System, etc.

## 2.7 IoT Applications

IoT application through the use of mobile and web interfaces promises to reform living, business, and the entire industry. Modern technology related to mobile and web interfaces has high significance as it offers a considerable interactive and user-friendly style to improve customer experience. Details about IoT application domains are available in Chapter 7.

## Review Questions

**2.1** Differentiate among core capabilities, enhanced capabilities, and advanced capabilities of smart things.

**2.2** Briefly explain the self-awareness and self-management capabilities of an object with the help of examples.

**2.3** For smart light bulbs, write the name of the smart object class and the capabilities that it possesses.

**2.4** Write an example of an enhanced level and advanced level of smart object classes.

**2.5** What are the significant features of MCUs? Explain their relationships with development boards.

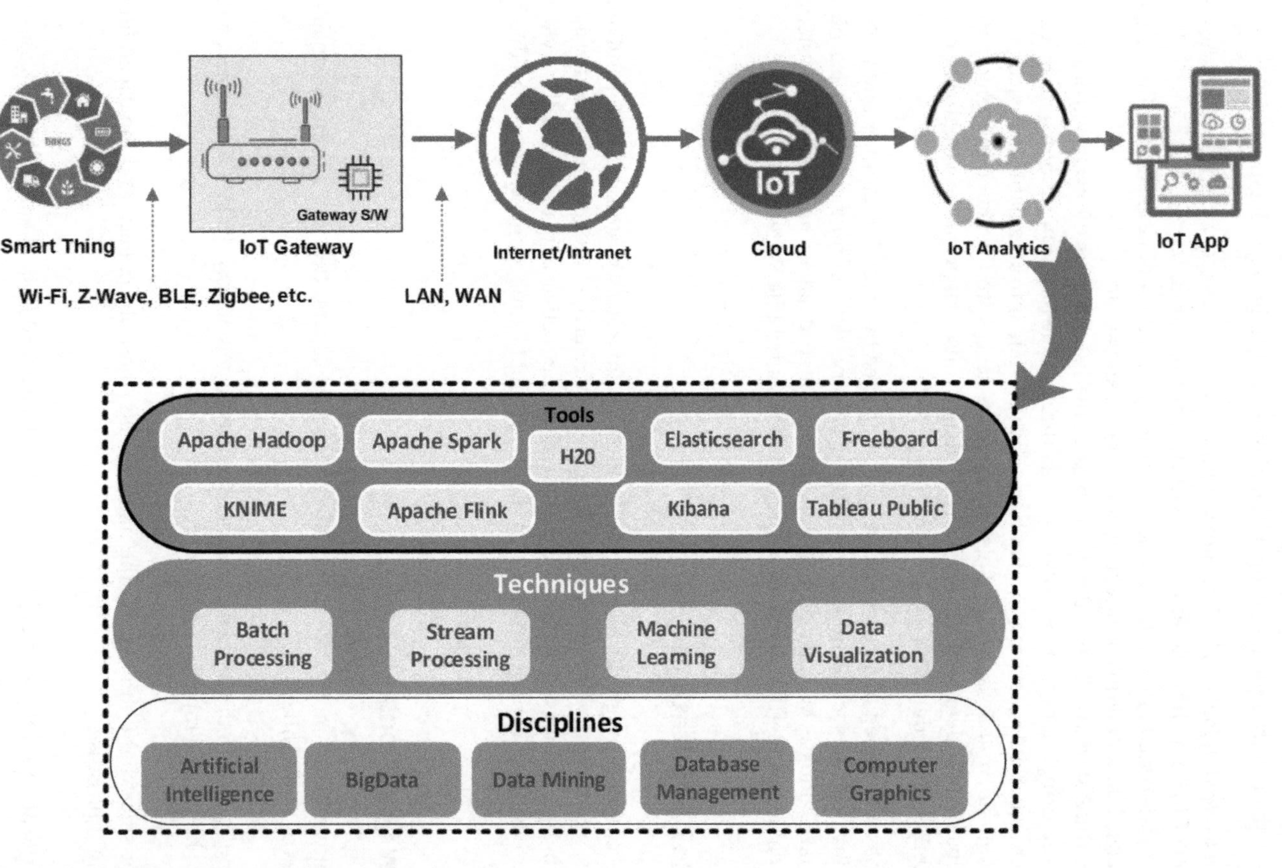


**Figure 2.12** IoT analytics techniques and tools.

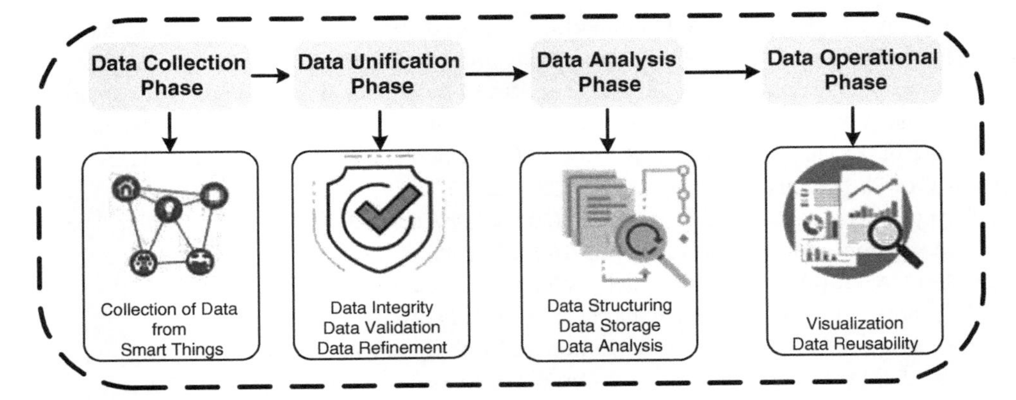


**Figure 2.13** IoT analytics lifecycle phases.

**2.6** In how many ways we can connect IoT components and smart things to MCU?

**2.7** Air pollution is known as "the silent killer" as it is invisible and is very dangerous for human health. Road traffic is one of the main causes of environmental pollution in large cities. An IoT-based system *Trafair* is developed that focuses on developing an air quality forecasting service based on weather forecasts and road-traffic flow. Provide the architecture diagram of the IoT-based Trafair system and explain the functionality of each component in terms of required hardware and software. Trafair system components include sensors, heterogeneous communication technologies, gateways, Cloud, and data analysis (performed at city council and city traffic information center). Moreover, it is required to design this system in Packet Tracer using suitable IoT components for this solution.

**2.8** Two factors, i.e. air temperature and water irrigation amount, affect the production of Strawberries growing in greenhouses. Propose an IoT-based solution that allows to instantly check environmental and production parameters and generate notifications when urgent actions are required to be performed. With the help of an IoT-based smart agriculture system diagram of your deployed solution, discuss:

**A** the role of smart thing controller.
**B** the responsibilities of IoT gateway.
**C** the main challenges of IoT Analytics.

## References

1 Karimi, K. and Atkinson, G. (2013). *What the Internet of Things (IoT) Needs to Become a Reality*. White Paper,, 1–16. FreeScale and ARM.

2 Serbanati, A., Medaglia, C.M., and Ceipidor, U.B. (2011). Building blocks of the internet of things: state of the art and beyond. In: *Deploying RFID-Challenges, Solutions, and Open Issues*. IntechOpen.

3 Kortuem, G., Kawsar, F., Sundramoorthy, V. et al. (2010). Smart objects as building blocks for the internet of things. *IEEE Internet Computing* 14 (1): 44–51.

4 Guoqiang, S., Yanming, C., Chao, Z. et al. (2013). Design and implementation of a smart IoT gateway. in Green Computing and Communications (GreenCom), 2013 IEEE and Internet of Things (iThings/CPSCom). In: *IEEE International Conference on and IEEE Cyber, Physical and Social Computing*, 720–723. IEEE.

5 Anjanappa, M., Datta, K., and Song, T. (2002). *Introduction to sensors and actuators*, The Mechatronics Handbook, vol. 1, 16.1–16.14. CRC Press.

6 Jesin, A. (2014). *Packet Tracer Network Simulator*. Packt Publishing Ltd.

7 Finardi, A. (2018). *IoT Simulations with Cisco Packet Tracer*.

8 Hernández, M.E.P. and Reiff-Marganiec, S. (2014). Classifying smart objects using capabilities. In: *IEEE International Conference on Smart Computing (SMARTCOMP)*, 309–316. IEEE.

9 Püschel, L., Röglinger, M., and Schlott, H. (2016). What's in a Smart Thing? Development of a Multi-layer Taxonomy.

10 Chen, H., Jia, X., and Li, H. (2011). A brief introduction to IoT gateway. In: *IET International Conference on Communication Technology and Application (ICCTA 2011)*, 610–613. IET.

11 Sethi, P. and Sarangi, S.R. (2017). Internet of things: architectures, protocols, and applications. *Journal of Electrical and Computer Engineering*, Hidawi Publisher.

12 Hou, L., Zhao, S., Xiong, X. et al. (2016). Internet of things cloud: architecture and implementation. *IEEE Communications Magazine* 54 (12): 32–39.

13 Yassein, M.B., Shatnawi, M.Q., Aljwarneh, S. et al. (2017). Internet of Things: survey and open issues of MQTT protocol. In: *IEEE 2017 International Conference on Engineering & MIS (ICEMIS)*, 1–6. IEEE.

14 Tang, K., Wang, Y., Liu, H. et al. (2013). Design and implementation of push notification system based on the MQTT protocol. In: *International Conference on Information Science and Computer Applications (ISCA 2013)*. Atlantis Press.

15 Karagiannis, V., Chatzimisios, P., Vazquez-Gallego, F. et al. (2015). A survey on application layer protocols for the internet of things. *Transaction on IoT and Cloud computing* 3 (1): 11–17.

16 Soldatos, J. (2016). *Building Blocks for IoT Analytics*. River Publishers.

17 Marjani, M., Nasaruddin, F., Gani, A. et al. (2017). Big IoT data analytics: architecture, opportunities, and open research challenges. *IEEE Access* 5: 5247–5261.

18 Lam, C. (2010). *Hadoop in Action*. Manning Publications Co.

19 Shvachko, K., Kuang, H., Radia, S. et al. (2010). The hadoop distributed file system. In: *IEEE 26th symposium on mass storage systems and technologies (MSST)*, 1–10. IEEE.

20 White, T. (2012). *Hadoop: The Definitive Guide*. O'Reilly Media, Inc.

21 Suryawanshi, S. and Wadne, V. (2014). Big data mining using map reduce: a survey paper. *IOSR Journals (IOSR Journal of Computer Engineering)* 16 (6): 37–40.

22 Zaharia, M., Xin, R.S., Wendell, P. et al. (2016). Apache spark: a unified engine for big data processing. *Communications of the ACM* 59 (11): 56–65.

23 Meng, X., Bradley, J., Yavuz, B. et al. (2016). Mllib: machine learning in apache spark. *The Journal of Machine Learning Research* 17 (1): 1235–1241.

24 Ingersoll, G. (2009). Introducing apache mahout. IBM developerWorks Technical Library.

25 Sean, O., Robin, A., Ted, D., and Ellen, F. (2012). *Mahout in Action*. Manning Publishers.

**26** Aiello, S., Click, C., Roark, H. et al. (2016). *Machine Learning with Python and H2O*. H2O. ai Inc.

**27** Carbone, P., Katsifodimos, A., Ewen, S. et al. (2015). Apache flink: stream and batch processing in a single engine. *Bulletin of the IEEE Computer Society Technical Committee on Data Engineering* 36: 4.

**28** Katsifodimos, A. and Schelter, S. (2016). Apache flink: stream analytics at scale. In: *IEEE International Conference on Cloud Engineering Workshop (IC2EW)*, 193. IEEE.

**29** Deshpande, T. (2017). *Learning Apache Flink*. Packt Publishing Ltd.

**30** Gormley, C. and Tong, Z. (2015). *Elasticsearch: The Definitive Guide: A Distributed Real-Time Search and Analytics Engine*. O'Reilly Media, Inc.

**31** Gupta, Y. (2015). *Kibana Essentials*. Packt Publishing Ltd.

**32** Bajer, M. (2017). Building an IoT data hub with Elasticsearch, Logstash and Kibana. In: *IEEE 5th International Conference on Future Internet of Things and Cloud Workshops (FiCloudW)*, 63–68. IEEE.

**33** Morton, K., Balazinska, M., Grossman, D. et al. (2014). Public data and visualizations: how are many eyes and tableau public used for collaborative analytics? *ACM SIGMOD Record* 43 (2): 17–22.

# 3

# Sensing Principles and Wireless Sensor Network

**LEARNING OBJECTIVES**

After studying this chapter, students will be able to:

- explain sensor fundamentals and classify sensors.
- understand physical principles of some common sensors.
- describe basics of WSNs.
- determine WSN architecture and types.
- explain the layer-level functionality of WSN protocol stack.
- identify operating systems features of WSNs.

## 3.1 Sensor Fundamentals

The sensor is one of the essential building blocks of IoT which induces the sensing capability to the smart things for perceiving a change in ambient conditions occurring around their environment. In general, a sensor is a device that is able to receive and respond to a stimulus (for example, variation in any natural phenomenon, i.e. temperature, pressure, humidity, motion, position, displacement, sound, force, flow, light, chemical presence, etc.). The following is the classification of a few basic types of stimuli that can be measured by the use of a sensor.

- *Electric Stimuli*: Charge, Electric Field, Current, Voltage, etc.
- *Magnetic Stimuli*: Magnetic Field, Magnetic Flux, Magnetic Flux Density, etc.
- *Thermal Stimuli*: Temperature, Thermal Conductivity, etc.
- *Mechanical Stimuli*: Velocity, Position, Acceleration, Force, Density, Pressure, etc.

In particular, a sensor is a device that translates the received stimulus (a quantity, property, or condition/state of a physical object) into an electrical signal. For example, for temperature and pressure sensors, heat and atmospheric pressure are converted to electrical signals, respectively. Therefore, the input of a sensor is some kind of observation related to the change in the physical property of an object, and output is an electrical signal in the terms of variation in charge, voltage, current, etc., which can be described by frequency,

*Enabling the Internet of Things: Fundamentals, Design, and Applications*, First Edition.
Muhammad Azhar Iqbal, Sajjad Hussain, Huanlai Xing, and Muhammad Ali Imran.

amplitude, and phase. The sensor output is ultimately required to be compatible with electronic circuits. A sensor can be considered as an energy converter, which actually measures the transfer of energy from and into an object under observation. Concerning energy conversion, the sensor must be differentiated from the term *transducer*, which is merely used to convert one form of energy to any other form of energy. Hence, the transducer is more than a sensor; however, these terms are interchangeably used in the literature [1]. Another term, the actuator can be considered a complement to a sensor as it converts the electrical signal into nonelectrical energy. The electric motor is one of the examples of the actuator, which is responsible to produce mechanical action from an electrical signal. Therefore, one may interpret an actuator as a transducer. The key differences between the sensor and transducer are given in Table 3.1.

Comparing to the senses of humans, sensors and actuators are the most fundamental components of a computing device. A comparison between the functionality of the human body and the working of the intelligent machine is shown in Figure 3.1.

**Table 3.1** Difference between sensor, actuators, and transducer.

| Comparison criteria | Sensor | Actuators | Transducer |
|---|---|---|---|
| Functionality | Senses environmental change(s) and converts in an electrical signal | Converts electrical signal to a physical output | Converts one form of energy to another |
| Components | Sensor itself | Mostly motion devices | Sensor and signal conditioning |
| Examples | Proximity sensor, light sensor, motion sensor, accelerometer sensor, etc. | LED, Motor controllers, Robotic arm | Thermocouple, Loudspeaker, Potentiometer, Thermistor, etc. |

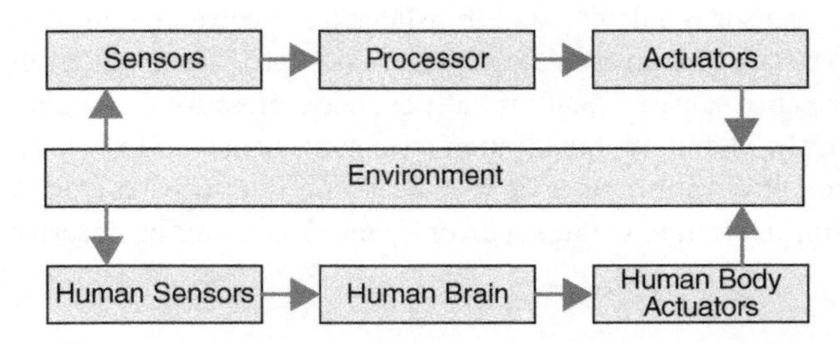


**Figure 3.1** Comparison between human body functionality and working of intelligent machine.

## 3.2 Sensor Classification

Sensors can be classified in the following ways [1]:

- Simple (Direct) Sensors Versus Complex Sensors
- Active Sensors Versus Passive Sensors
- Contact Sensors Versus Noncontact Sensors
- Absolute Sensors and Relative Sensors
- Digital Sensors Versus Analog Sensors (based on output)
- Scalar Sensors Versus Vector Sensors (based on data types)

### 3.2.1 Simple (Direct) Sensor Versus Complex Sensor

The simple or direct sensor is able to transform physical change or stimulus from the environment to an electrical signal (shown in Figure 3.2). On the other hand, complex sensors may require one or more energy transducers to produce an electrical signal (shown in Figure 3.3). Examples of simple sensors include temperature sensor, pressure sensor, light sensor, etc., and accelerometer is an example complex sensors.

### 3.2.2 Active Sensors Versus Passive Sensors

Active (also known as parametric) sensors require external power to be operated. Capacitive and inductive sensors are examples of Active sensors. On the other hand, passive (also known as self-generating) sensors are able to generate electrical signals by itself and are not dependent on the external power supply. A piezoelectric sensor is one of the examples of self-generating sensors.

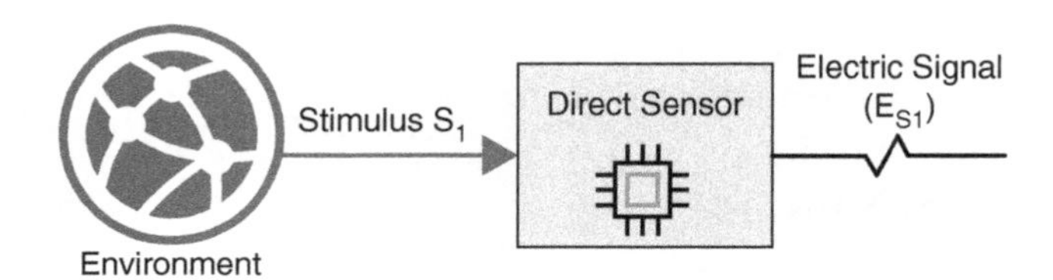


**Figure 3.2** Simple sensor.

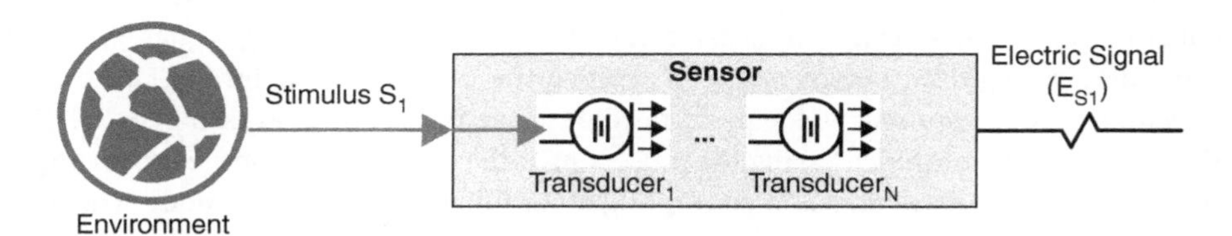


**Figure 3.3** Complex sensor.

### 3.2.3 Contact Sensors Versus Noncontact Sensors

Physical contact to stimulus is essential in Contact sensors (e.g. temperature sensors); however, in Noncontact sensors, physical contact to stimulus is not required (e.g. Infrared thermometers).

### 3.2.4 Absolute Sensors and Relative Sensors

The Absolute sensor reacts to a stimulus using an absolute scale but Relative sensors sense stimulus relative to some reference. Strain gauges and Thermocouple are examples of absolute and relative sensors, respectively.

### 3.2.5 Digital Sensors Versus Analog Sensors (Based on Output)

The output signal of digital nature is produced by the digital sensors. Nevertheless, continuous signals are generated at the output of analog sensors. Pressure sensor, light sensors, and analog accelerometers are examples of Analog Sensors. Digital Accelerometers and digital temperature sensor are examples of digital sensors.

### 3.2.6 Scalar Sensor Versus Vector Sensors (Based on Data Types)

Scalar sensors are used to measure scalar quantities and the output electrical signal is proportional to the magnitude of the quantity, for example, a room temperature sensor. Vector sensors, on the other hand, are used to measure vector quantities, i.e. accelerometer, which measures not only the quantity but also the direction of moving objects.

Various other forms of sensor classifications, i.e. based on Physical Law (e.g. Electromagnetic, Thermoelectric, Photoelectric, etc.), based on Specification (e.g. response time, accuracy, sensitivity, size, weight, etc.), and based on Application (e.g. Military, Meteorology, Agriculture, Transportation, Automotive, Medicine, etc.), have also been proposed.

## 3.3 Anatomy of Sensors

In general, the basic components of a sensor include the sensing unit, processing unit, Analog to Digital Converter (ADC) unit, power unit, storage, and transceiver as shown in Figure 3.4 [2]. Sensing elements are hardware devices, which are responsible to measure any physical stimulus (i.e. light, temperature, sound, etc.) in the environment to collect concerned data. On sensing, the sensor generates a continual analog signal, which is required to be digitized before transmitting it to the controllers for further processing. Therefore, ADC is required, which performs the conversion of an analog signal to the digital signal. A microcontroller having the processing unit is responsible for the processing of received digitized sensed data and controlling other functions of a sensor node. The most common controller is a microcontroller that is used in sensor nodes because of its low power consumption, low cost, and flexibility to connect other devices. Field Programmable

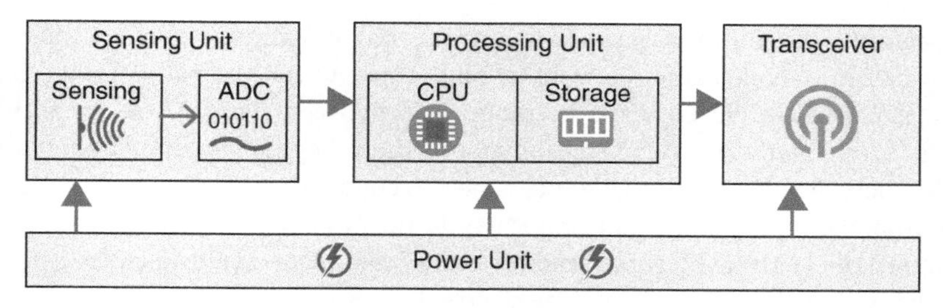


**Figure 3.4** Anatomy of a sensor node.

Gate Arrays (FPGAs) and Application-Specific Integrated Circuits (ASICs) are other alternatives, which can be used as a controller in sensor nodes. The processing unit is generally associated with storage memory. Flash memories are preferred to use because of their low cost and storage capacity. A transceiver is required to connect the sensor nodes to other nodes in the network for the transmission and reception of required data. Mostly, the Industrial, Scientific and Medical (ISM) band is preferred in sensors to use free radio and three common communication schemes, i.e. optical communication (laser), Infrared (IR), and radio frequency (RF) have been used as wireless transmission media. Energy consumption is the most critical issue in a sensor node. Therefore, sensor nodes are more focused on power conservation maintaining QoS in a sensor network. Sensor nodes consume high energy in communication comparing to sensing and any other processing. Natural energy sources and phenomena (e.g. solar and temperature variations) have been preferred to regenerate energy in sensors. In addition, energy consumption issues must be taken into account by operating sensor nodes in idle mode. Dynamic Power Management (DPM) and Dynamic Voltage Scaling (DVS) techniques are in use to address the problem of energy conservation in sensors. DPM works on the principle of shutting down the unused parts of sensor nodes. Contrary, to save energy, DVS chooses the least possible voltage levels for different sensor components to maintain their proper working.

## 3.4 Physical Principles of Sensing

The conversion of physical effects into an electrical signal is based on various basic principles of Physics, i.e. capacitance, magnetism, piezoelectric effect, etc. In the following subsections, these principles have been described from the sensor point of view [1].

### 3.4.1 Capacitance

Capacitance is an electrical phenomenon to store electric charge, and the device that stores charge following this phenomenon is known as a Capacitor. To understand the working of Capacitive sensors, it is first required to comprehend the basics principles of the capacitor world.

A capacitor is an electronic device to store electric energy in the form of an electric charge. In its basic form, a Capacitor comprises two conducting plates (also known as electrodes), which are separated by an insulator/dielectric (a nonconducting matter, i.e. air, plastic, ceramic, etc.) ultimately producing a potential difference. Figure 3.5 illustrates the simplest design of a capacitor, i.e. parallel-plate capacitor. In the parallel-plate capacitor, two metal plates are placed while keeping a gap between them. These metal plates of Capacitor start to store electric charges as soon as the voltage is applied. At a certain potential difference, the electric charge remains on the conducting plates even after its disconnection from the voltage source. In simple words, the placement of electrons onto one plate and removal of electrons (in equal amount) from other plate create an electric field across the gap between conducting plates. The electric field is because of the difference of stored charges on the surface of each plate. These conducting plates are able to hold the electric charge until it is consumed or leakage occurs because of imperfect insulators.

In general, Capacitance is defined as the amount of charge that a capacitor can hold at a given voltage level. Mathematically, it can be represented as

$$C = Q/V \tag{3.1}$$

where

$C$ is the capacitance in Farad (F)
$Q$ is the magnitude of the charge stored on each plate in Coulomb
$V$ is the voltage applied to the plates in Volts (V)

The unit of Capacitance is farad and Eq. (3.1) states that a capacitor having a capacitance of 1 F can store 1 Coulomb of charge when the voltage across its terminals is 1 V [1, 3].

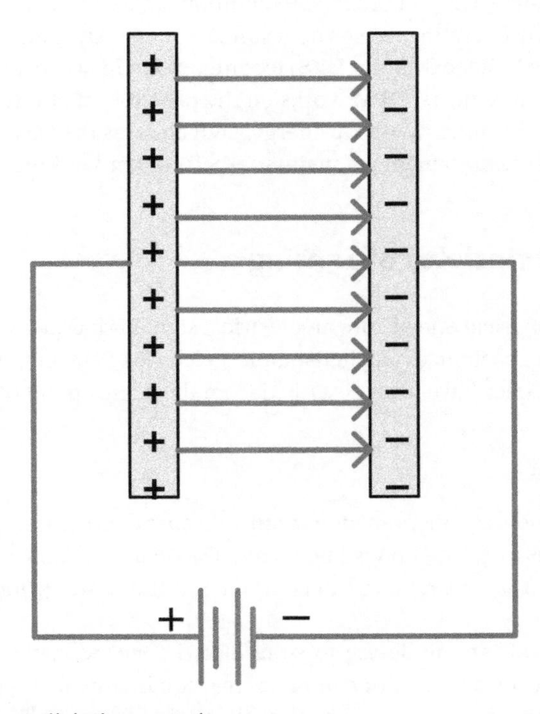


**Figure 3.5** A typical parallel-plate capacitor.

The capacitance depends on the shape/area of each plate, dielectric material, and distance between the plates. Mathematically, this dependency can be represented as

$$C = Q / V = (\varepsilon_0 \times A) / d \tag{3.2}$$

where

$\epsilon_0$ represents the permittivity of free space
$A$ is the area of plates
$d$ is the distance between two plates

From Eq. (3.2), it becomes evident that capacitance is directly proportional to the surface area of conducting plates and inversely proportional to the distance between the conducting plates. This means that lower surface area and higher distance ultimately decrease the capacitance and vice versa.

The nature of the dielectric between the capacitor plates also affects the capacitance. Replacement of vacuum (or air) with any other material increases/decreases the capacitance by a factor, which is known as dielectric constant ($K$). In other words, $K$ is used to represent the material's effect on the electric field or its permittivity. Different materials have different permittivity that in actual represents the material's capacity to transmit electric field. Considering the importance of $K$, Eq. (3.2) is evolved as

$$C = K \times Q / V = K \times (\varepsilon_0 \times A) / d \tag{3.3}$$

Equation (3.3) shows that the capacitance is directly proportional to the magnitude of dielectric constant. The increased or decreased value of dielectric constant causes an increase or decrease in capacitance, respectively.

From Eq. (3.3), it becomes evident that change in any of the three parameter's values (i.e. $A$, $d$, and $K$) can be used to sense environmental changes. Therefore, a number of capacitive sensors have been developed having the ability to convert changing characteristics of dielectric into an electric signal.

#### 3.4.1.1 Examples of Capacitive Sensors

Various environmental factors affect the innate nature of dielectric constant. General principles of few examples of capacitive sensors have been discussed in the following.

*Temperature sensing:* Any change in temperature is subject to change the distance/area of conducting plates and the magnitude of dielectric constant, which eventually affects the capacitance value and used for sensing purposes.

*Humidity sensing:* An increase in humidity affects the permittivity of dielectric, i.e. gas and solid materials. Most of the capacitive humidity sensing devices consist of various layers of specific dielectric, i.e. inorganic hydrophilic oxides [4]. In addition, a large surface area with more absorption of water molecules significantly affects the magnitude of capacitance.

*Proximity sensing:* A capacitive proximity sensor is able to detect the presence of a nearby object. Proximity sensors are based on the simple principle of object detection through the measuring of change in an electric field (capacitance), which is established by two small plates as shown in Figure 3.6.

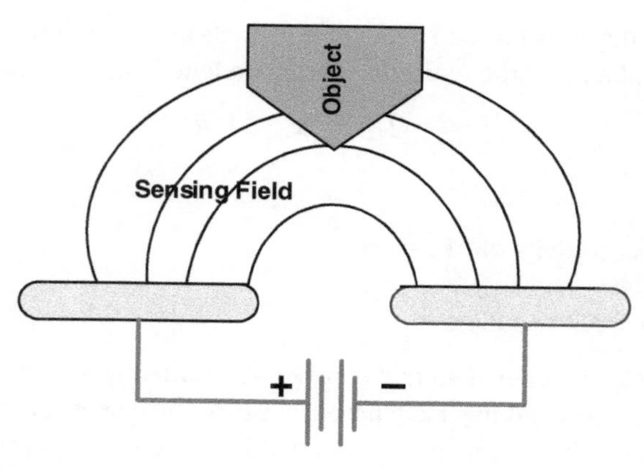


Figure 3.6 Proximity sensor.

Capacitive proximity sensors are able to detect metal objects, wood, paper, plastics [5], human tremors, and object thickness [6].

*Position sensing:* Capacitive position sensors are used to sense the position based on the physical parameters of the capacitor, i.e. capacitive plate area, dielectric constant, and distance between the plates. One of the simplest applications of the capacitive position sensor is liquid level sensing in a storage tank where two fixed space conducted metal rods are immersed in liquid as shown in Figure 3.7. The rise of liquid increases the capacitance between parallel rods and eventually is used to sense fluid levels in a container. Other applications of capacitive position sensors include gear position sensing and touch screen coordinate systems.

The touch screen sensor widely used in our daily life (i.e. in mobile phones) is a type of capacitive sensor. The working of the touch screen capacitive sensor is not complicated and

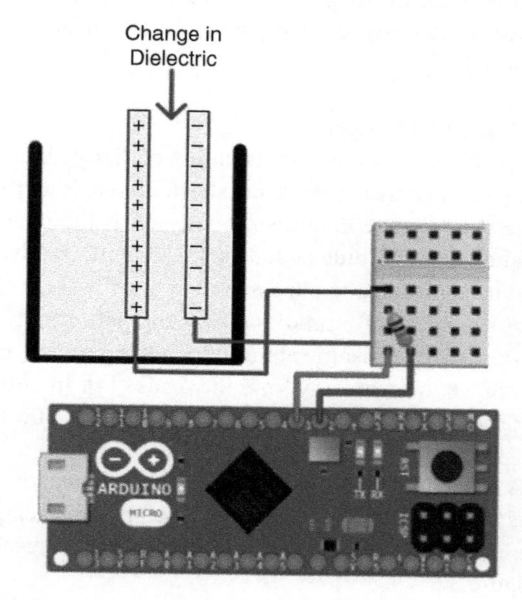


Figure 3.7 Simple capacitive sensor to measure water level.

is based on the principle of capacitive coupling through which devices are able to detect anything having dielectric different from air. It is through the use of this technology that touch screen mobiles are able to detect the place of the finger through the change in the electrostatic field. A capacitive touch screen consists of multiple glass layers where the inner and outer most layers (just representing two parallel rods of a typical capacitor) conduct electricity. Therefore, the whole screen behaves like a capacitor, and upon touching with the finger, the electrical field (existing between the inner and outer glass layer) changes. An image processing controller is required to continuously monitor and detect the coordinates of change of electrostatic field as shown in Figure 3.8.

### 3.4.2 Magnetism and Induction

Concerning fundamental characteristics, electricity and magnetism share strange similarities. For example, the existence of attraction/repulsion forces between two electrically charged rods is very much similar to forces that occur between two (south and north) poles of a magnet. Identical poles repel and opposite poles attract each other comparable to forces that exist between alike and different electric charges. Theoretically, specifically to magnets, electrons in atoms at one end move in one direction, and atoms at the other end move in the opposite direction and eventually create a magnetic field. By definition, the direction of energy forces in a magnetic field is from north to south pole. Magnetism and electricity are associated with each other in such a way that moving charges are able to produce a magnetic field and from the magnetic field it is possible to generate electricity. Simplified models of

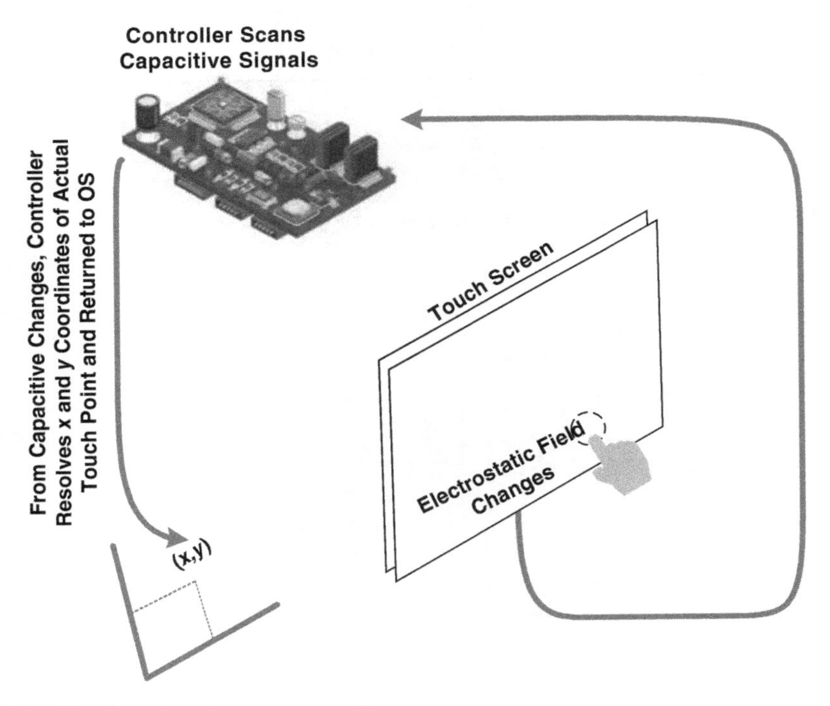


**Figure 3.8** Working of touch screen capacitive sensor.

both phenomena, i.e. production of the magnetic field through electricity and production of electricity through magnetism, are shown in Figures 3.9 and 3.10, respectively.

From Figure 3.9 it becomes evident that when current passes through a simple wire then due to the movement of electrons, a magnetic field is produced around the wire. On the other hand, a varying magnetic field within the vicinity of a wire is able to induce a voltage in that wire or coil (also known as magnetic induction) as shown in Figure 3.10. Michael Faraday in 1831 found that the produced voltage (also known as electromotive force [EMF]) is equal to the rate of change of magnetic flux [1]. Mathematically, it can be represented as

$$\varepsilon = \mathrm{d}\varphi_\mathrm{b} / \mathrm{d}t \tag{3.4}$$

where

$\Phi_\mathrm{b}$ is the magnetic flux
$t$ is the time
$\varepsilon$ is the EMF induced

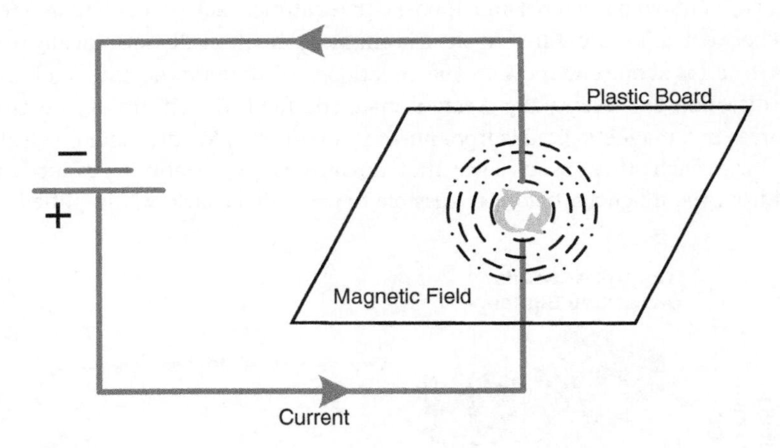


**Figure 3.9** Production of magnetism through electricity.

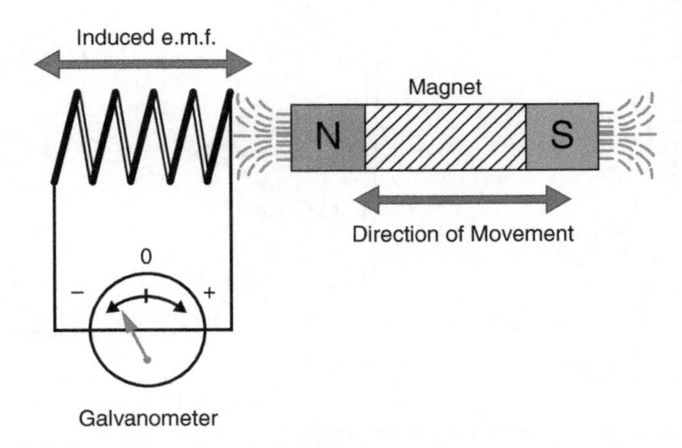


**Figure 3.10** Production of electricity through magnetism.

EMF can be increased through the use of wounded wire coil (having $N$ identical turns). In this way, the resultant EMF becomes N times that of one single wire:

$$\varepsilon = -N \times \left(\mathrm{d}\varphi_{\mathrm{b}}\right) / \mathrm{d}t \tag{3.5}$$

Related to sensor design, Eq. (3.5) can be

$$V = -N \times \left[\left(\mathrm{d}BA\right)\right] / \mathrm{d}t \tag{3.6}$$

Equation (3.6) states that the voltage is directly proportional to the strength of the magnetic field ($B$) or circuit area ($A$). In other words, the induced voltage is dependent on:

- Number of turns in the coil
- Current variation in coil
- Movement of the magnetic field source
- Changing the magnetic source area

#### 3.4.2.1 Magnetic Sensing Examples

Electromagnetic induction is used to sense the position and displacement of an object.

*Eddy Current Sensors:* Eddy current is a phenomenon that explains the induction of electric current in a conductor through varying the intensity of the magnetic field. Through the implication of eddy current in a dual-coil sensor as shown in Figure 3.11, it is possible to sense the proximity of conductive materials. In dual-coil eddy current sensor, one coil is used as a sensing coil and another one is reference coil. Whenever an object approaches the sensing coil, it produces eddy currents that oppose the sensing coil and ultimately affects the inductance, which can be measured in calibrated units.

In addition to proximity sensing, eddy current sensors are also used to check the material conductivity as well as its thickness [1].

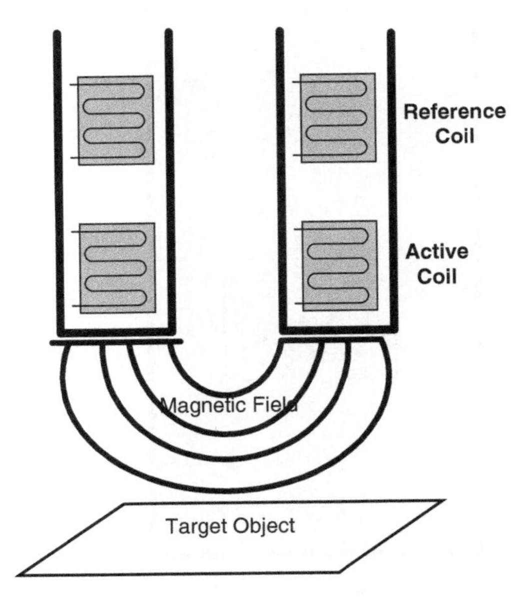


**Figure 3.11** Eddy current-based proximity sensor.

*Hall Effect Sensors:* The production of potential difference across an electric current-carrying conductor in the presence of a magnetic field (applied perpendicular to the direction of current flow) is known as Hall Effect. Hall effect sensor devices are used to measure the voltage that is directly proportional to the magnitude of applied magnetic field strength.

Liquid-level detection or automotive fuel-level detection is one of the applications of Hall effect sensors. The design of the liquid-level Hall sensor is shown in Figure 3.12. It consists of two crucial components, i.e. the permanent magnet inside float (which can slide in vertical and able to point surface level of liquid in a container) and a hall sensor attached at the top of the pole. Upon the rise of liquid level in the container and when the float approaches detection mark, Hall effect is produced and sends the signal to the monitoring device.

In addition to fuel-level detection, the Hall effect phenomenon can also be used to measure DC currents in transformers, position sensing, keyboard switches, etc.

### 3.4.3 Electric Resistance and Resistivity

Electric resistance is a characteristic of all materials and it basically refers to the force of friction that resists the flow or movement of electron.

According to Ohm's law, "the current in a conductor is proportional to the applied voltage while keeping other physical conditions as constant." Therefore, it can be represented as

$$V \alpha I$$

$$V = R \times I$$

where

$V$ is the voltage measured in Volt
$I$ is the current measured in Ampere
$R$ represents the resistance measured in Ohm

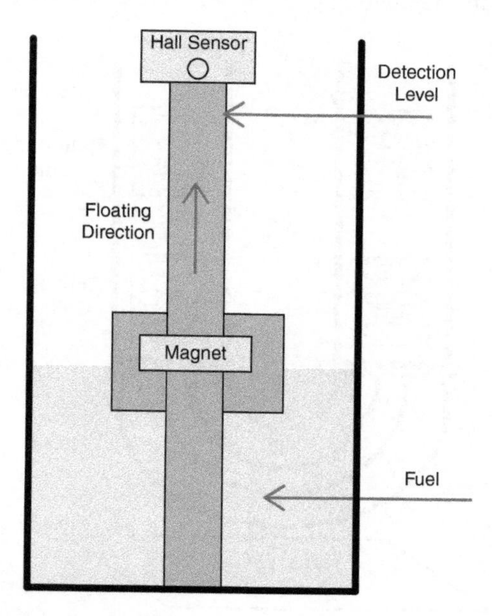


**Figure 3.12** Hall sensor-based fuel detection.

The resistance of metal (with the uniform cross-sectional area) is driven by various factors (i.e. nature, shape, and temperature of the material). For example, in the case of a cylindrical piece of metal, $R$ is directly proportional to the length ($L$) and inversely proportional to the cross-sectional area ($A$) of the cylinder and can be represented as

$$R \alpha L / A$$

$$R = \rho \times (L / A)$$

Here, $\rho$ is the resistivity of the material and is defined as the tendency (or measure) of a material to resist the flow of electrons. Different materials have different resistivity and can be expressed as

$$\rho = R \times (A / L)$$

The unit of resistivity is Ohmmeter.

#### 3.4.3.1 Resistive Sensor Applications

The potentiometer is based on the principle of varying resistance based on variation in the conductor's length. Strain gauge is based on the principle of changing resistance due to strain occurrences and is used to measure the pressure. Resistance varying property at different temperatures is used to make thermometers. Moisture or humidity sensitive resistor can be made up by using hygroscopic material whose resistivity is dependent upon the absorption of water molecules.

### 3.4.4 Piezoelectric Effect

The piezoelectric effect is related to the production electric charge in crystalline material through the consequence of applied mechanical stress. Piezoelectric sensors are based on the principle of transforming a physical dimension into force and act on the opposite faces of the sensing element. The microphone is the simplest example of a piezoelectric effect, which senses pressure variation in the form of sound and converts it into an electric signal.

Other than Capacitance, Induction, Resistance, and Piezoelectric sensor, a number of natural phenomena, i.e. thermal conduction, light polarization, and light scattering, have been used in sensing technology.

## 3.5 Use of Basic Sensing Principles in RFID Technology

All these fundamental sensing principles have been used in RFID technology. The working of these basic units has been explained in the following as shown in Figure 3.13:

*Step 1*: Integrated Circuit (IC) is responsible to send the signal to Oscillator.

*Step 2*: Upon receiving the signal, Oscillator produces Alternating Current (AC) in the reader's coil.

*Step 3*: The generated current in the coil creates a magnetic field, which in turn serves as the tag's power source.

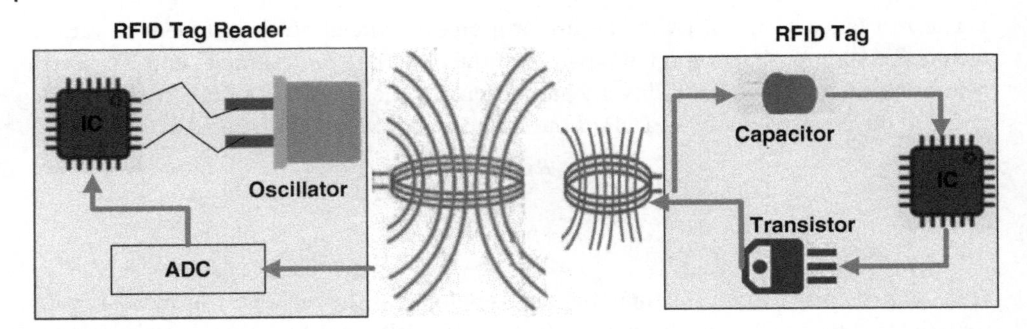


**Figure 3.13** Basic working of RFID technology.

*Step 4*: The magnetic field is powerful enough to interact with the tag's coil, which in turn induces a current and causes the charge to store in a capacitor.
*Step 5*: Charge stored in the capacitor increases the voltage across the circuit and activates the tag's IC, which eventually transmits the digital identifier code.
*Step 6*: High/low levels of digital signal (corresponding to ones and zeros in encoded digital identified code) turn the transistor ON/OFF.
*Step 7*: Upon receiving current (in high/low levels), a varying magnetic field is generated, which ultimately interacts with the reader's magnetic field.
*Step 8*: Following the pattern (zeros and ones) of the tag's transmission, induced magnetic fluctuations cause changes in current flow from the reader's coil to its circuitry.
*Step 9*: ADC device is required to convert the sensed current pattern into a digital signal.
*Step 10*: Finally, the reader's IC is able to discern the tag's identifier code.

## 3.6 Actuators

Actuators as a complement to sensors are required to take actions, which are based on some sensor readings. In simple words, actuators are electronic/mechanical components or devices of an IoT system, which perform direct/indirect actions on the environment to control a certain mechanism. Actuators fundamentally require two things (i.e. energy and control signal) for their proper functioning. Upon the reception of control signals, the actuator converts the received energy and is able to respond through mechanical means. In general, actuators can be categorized into various types, i.e. electric-based actuators, magnetic-based actuators, mechanical-based actuators, hydraulic actuators, and pneumatic actuators.

Electric actuators use electric power motors to actuate the equipment in terms of mechanical torque. For example, a Solenoid valve is an electric actuator, which is used to control water flow in pipes.

Thermal/Magnetic Actuators are actuated by the implication of thermal or magnetic energy. One of the uses of these actuators is the shaping of memory alloys (SMAs).

Mechanical actuators are used in gears, pulleys, chains, etc. and are based on the principle of converting rotary motion into linear motion. Rack and Pinion actuator is an example of a mechanical actuator.

Hydraulic actuators use hydraulic power to facilitate mechanical motion by converting the mechanical motion into linear, oscillatory, or rotary motion. The radial engine is an example of a hydraulic actuator.

Pneumatic actuators are based on the principle of converting energy produced by compressed air into linear/rotary motion. Pneumatic rack and pinion actuators are example of these types of actuators, which are used to control valves of water pipes.

## 3.7 Wireless Sensor Networks (WSNs)

Sensors and actuators are absolutely vital for the realization of an IoT system. But, in contrast to actuators (stand-alone devices), sensors are required to collaborate with one another to achieve an assigned task. WSN is defined as an infrastructure-less and self-configured network consisting of thousands of tiny low power sensor nodes, which are able to monitor physical conditions and communicate wirelessly to transfer collected data to the required destination [7].

### 3.7.1 WSN Architecture

A typical WSN architecture is shown in Figure 3.14.

Wireless sensor nodes (also known as motes) are the devices that directly interact with the environment. All data received from sensor nodes are mainly aggregated at the Sink node before forwarding to the application. Sink nodes are also known as Gateway because

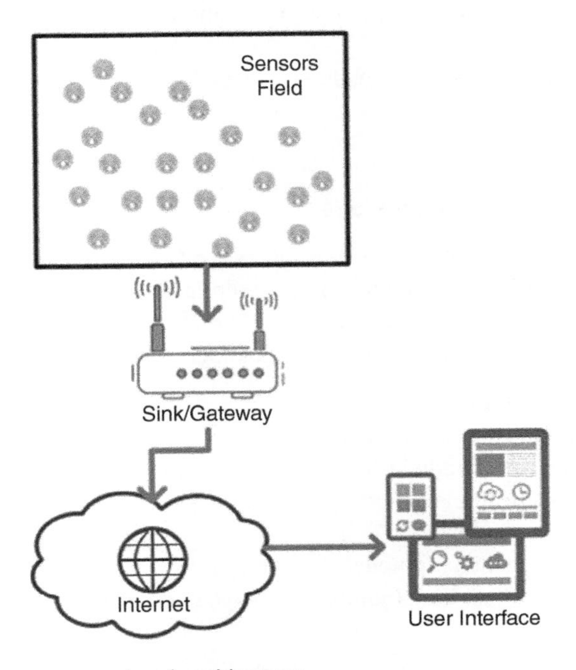


**Figure 3.14** Wireless sensor network architecture.

these nodes connect the WSN to the Internet, Cellular (satellite) network, or Public Switched Telephone Network (PSTN). Sink nodes perform protocol conversions to enable interoperability of sensor networks with non-sensor wired/wireless protocols. It is observed that the energy bottleneck occurs as sensor nodes adjacent to Sink bear high relay loads and eventually limit the lifetime of sensor networks. So, it is suggested to provide more energy to sensor nodes, which are adjacent to the Sink. In WSNs, it is important to note that each node besides sensing is also capable to perform the duties of a router (and also known as relay node).

### 3.7.2 Types of WSNs

There are several types of WSNs including:

- *Wireless Body Area (Sensor) Networks (WBAN/WBASN)*: Networks of implanted and on-body sensors to monitor vital signs of the human body.
- *Terrestrial WSN*: A typical WSN in which a number of sensor nodes are capable to transmit data to the sink or base station.
- *Multimedia WSNs*: These sensor networks are able to monitor real-time events in the form of multimedia, i.e. images, audio, video, etc.
- *Underground WSN*: WSN consists of sensor nodes to monitor changes beneath the earth and demands additional above ground sink nodes to transmit acquired data to the destination.
- *Underwater WSNs*: Sensor networks consist of sensor nodes undersea and dynamic anchor nodes at sea surface to collect gathered data. Underwater WSNs face the challenging environment of underwater communication with long propagation delay and low bandwidth.
- *Mobile WSNs*: WSN consists of mobile sensor nodes to collect data from the physical environment.

### 3.7.3 General Characteristics of WSNs

WSN is a type of wireless ad hoc networks, which supports multi-hop communication. In multi-hop communication, intermediate nodes (acting as routers) are able to forward data packets, which are addressed to other nodes within the network in the absence of a centralized infrastructure. However, WSN has its own distinct characteristics [8, 9] which include:

- Severe power (energy) constraints
- Limited memory and computing capabilities
- Narrow communication bandwidth and short communication range
- Lack of global identifications
- Limited mobility
- Heterogeneity/Homogeneity of nodes
- Self-organizing and self-healing (ability to manage node failures)
- Scalability
- Ability to withstand extreme environmental conditions

- Application dependent
- Distinct (Random as well as Manual) deployment strategies

### 3.7.4 Protocol Stack of WSNs

The protocol stack of WSNs is required to support typical and distinct characteristics of ad hoc networks and sensor networks, respectively [10]. The protocol stack of WSN has been shown in Figure 3.15. Thus, similar to any other ad hoc network, WSN typically consists of the following five layers [7].

#### 3.7.4.1 Physical Layer

Physical layer is responsible for frequency selection, signal detection, signal modulation/demodulation, and encryption/decryption of transmission and reception of data. Network designers must consider the optimum values of three parameters, i.e. transmit power, hop distance, and modulation for the channel modeling of the WSN physical layer. Upon reception of a signal from an adjacent node, it is required to be decoded with certain error probability, and this error probability decreases with an increase in signal power. Therefore, the fundamental challenge to design the physical layer of WSN is related to the adjustment of tradeoff, which exists between signal transmission power and error probability.

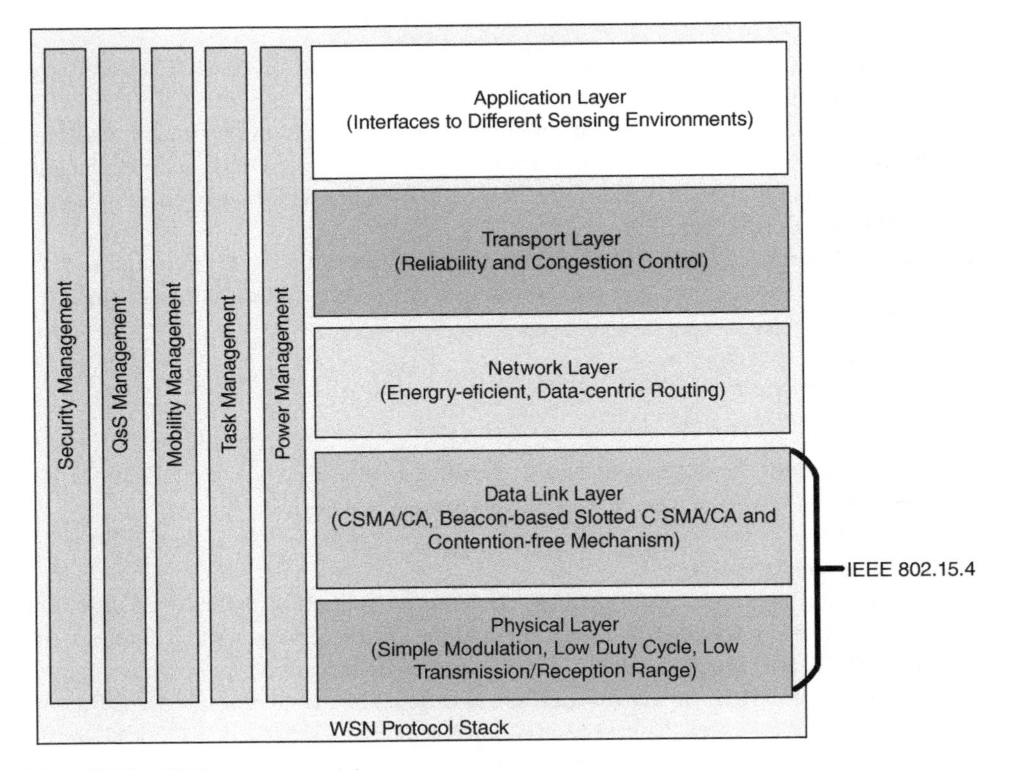


**Figure 3.15** WSN protocol stack layers.

Considering the requirements of different scenarios, IR, RF, and Optical (Laser) are three popular communication schemes used in WSNs.

#### Physical Layer WSN Standard

In general, WSNs use ISM frequency bands, and the following parameters are significantly required to be considered for the designing of Physical layer of WSNs:

- Low transmission and reception range
- Low duty cycle
- Low complexity with simple modulation schemes
- Low power consumption

Considering the aforementioned parameters, a number of standard sensor protocols have been implemented, which can be interfaced with different wireless network standards. The 802.15 working group of IEEE is directing standards for Wireless Personal Area Networks (WPANs) PHY and Medium Access Control (MAC) layers. Considering the demands of WSN applications (i.e. low complex, low power, and low deployment cost), this working group has proposed the IEEE 802.15.4 standard for low rate WPAN devices [11]. IEEE 802.15.4 is also the basis of different wireless technologies, which have been used for IoT, i.e. Zigbee, 6LoWPAN, and WirelessHART.

In 802.15.4, two PHY layer options (i.e. low band and high band) have some fundamental differences and similarities as described in the following [12, 13]:

- Both PHY layers differ in the frequency band (i.e. low band [868/915 MHz] and High band [2.4 GHz]).
- Both PHY layers support different modulation schemes (i.e. Binary Phase Shift Key [BPSK] with Low band and Offset Quadratic Phase Shift Key [O-QPSK] with High band)
- Both PHY layers support different data rates (i.e. 20–40 kbps with Low band and data rate of 250 kbps with High band)
- Both PHY layers are based on DSSS to accomplish low-cost digital IC realizations.
- Both PHY layers share the same packet structure and it consists of following fields:
  - 32-bit preamble
  - 8-bit packet start delimiter
  - 8-bit packet length
  - Variable payload of 2–127 bytes

In summary, both the differences and similarities of two PHY layers of IEEE 802.15.4 have been shown in Figure 3.16. Related details are available in [13, 14].

### 3.7.4.2 Data Link Layer (DLL)

In general, Data Link Layer (DLL) is responsible for reliable transmission (through frame error detection and/or frame error correction) and medium access. According to the specifications of IEEE 802, the DLL layer has been further divided into two sublayers named Logical Link Control (LLC) and MAC. LLC specifications (related to multiplexing, flow control, and Automatic Repeat Request [ARQ]) have been standardized in 802.2 and are the same for all IEEE 802 standards. However, the MAC layer, on the other hand, has different standardizations for each 802-project related to different responsibilities (i.e. frame delivery,

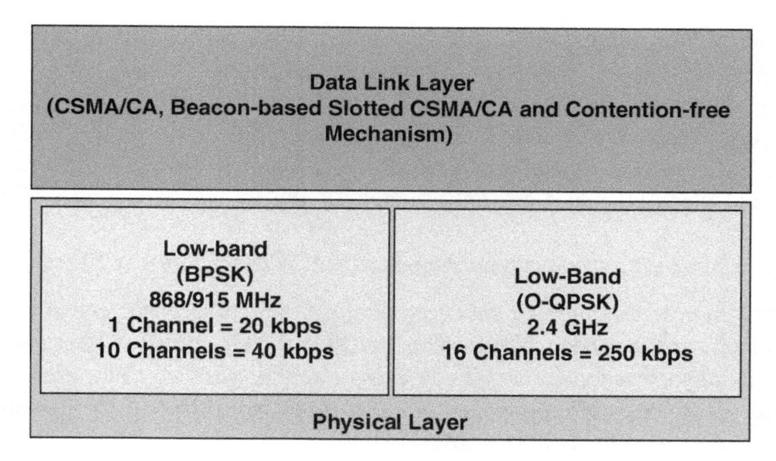


**Figure 3.16** PHY and MAC layer details.

frame validation, device association, and radio access based on the principle of Carrier Sense Multiple Access Collision Avoidance [CSMA/CA]). IEEE 802.15.4 MAC has been designed to support two types of Low Rate WPAN topologies (i.e. Start and Peer).

For three types of data traffic classes (i.e. periodic, intermittent, and repetitive), three types of channel access mechanisms (i.e. CSMA/CA, Slotted CSMA/CA, and Contention-Free [Guaranteed Time Slots]) have been implemented in IEEE 802.15.4 as shown in Figure 3.17.

Slotted CSMA/CA and Contention-Free mechanisms are based on Beacons, which are required to be periodically transmitted by the coordinator node in start topology. Related details are available in [13, 14].

IEEE 802.15.4 standard has been used in a diverse range of wireless sensing applications. However, it is not suitable for most of the WSN applications due to the following reasons [15, 16]:

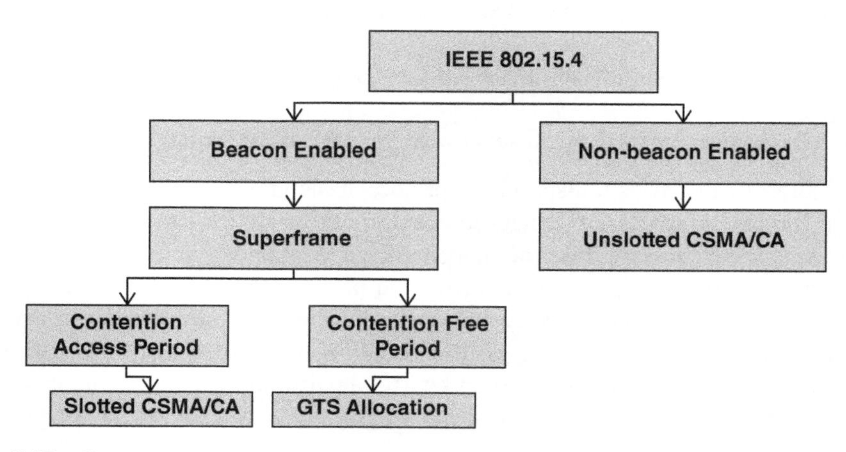


**Figure 3.17** Channel access mechanisms in IEEE 802.15.4.

- High energy consuming due to high probability of collision and frame retransmission (inherent to CSMA/CA), which is used to achieve reliability.
- Idle listening.
- Use of control packet to avoid hidden/exposed node problem
- Operations are not clearly defined for peer-to-peer topology.
- In scalable WSN, all nodes are not available in the transmission range of coordinator node.
- Fair access of all sensor nodes is not required in all WSNs but energy efficiency.

Therefore, in order to optimize the energy consumption and sensor network lifetime, a number of efforts have been proposed in the literature to enhance the existing MAC standard. Detailed discussions about the evolution and taxonomy of various proposed WSN MAC on the basis of different criteria (i.e. traffic prioritization, QoS aware, energy harvesting, etc.) are available in [17–20].

#### 3.7.4.3 Network Layer

Network layer is responsible for the efficient (data-centric) routing from source to the destination node. By taking into account the typical requirement of energy efficiency in WSN, a number of routing protocols have been proposed in the literature, which avoid message broadcast flooding. Mostly data-centric routing approaches are preferred than address-centric routing approaches. Detailed taxonomy and relevant details of WSN routing are available in [21–23].

#### 3.7.4.4 Transport Layer

End-to-end reliability, congestion control, and fair bandwidth allocation are the main concerns that are required to be implemented at the transport layer. The traditional User Datagram Protocol (UDP) and Transmission Control Protocol (TCP) protocols cannot be directly implemented to the WSN due to the following reasons [24]:

- UDP is unreliable.
- TCP connection-oriented overhead is not suitable to WSNs.
- TCP degrades the throughput in WSNs because it adjusts data rate to the lowest when packet loss occurs.
- TCP is reliable but due to end-to-end retransmission mechanism, energy consumption is high.
- TCP reliable mechanism through acknowledgments is not essential in WSNs.

Therefore, efficient WSN transport layer protocol design must address the constraints of fairness and energy efficiency. Congestion control and lost packet recovery are essential components, which directly affect the energy-efficiency, reliability, and QoS, and are implemented separately. The separate implementation provides flexibility to different applications. For example, applications demanding reliability are only concerned with the invocation of lost packet recovery algorithms, and applications related to the congestion control invoke corresponding congestion control algorithms. The taxonomy of the proposed WSN transport layer protocols is shown in Figure 3.18 and the rest of the details are available in [25, 26].

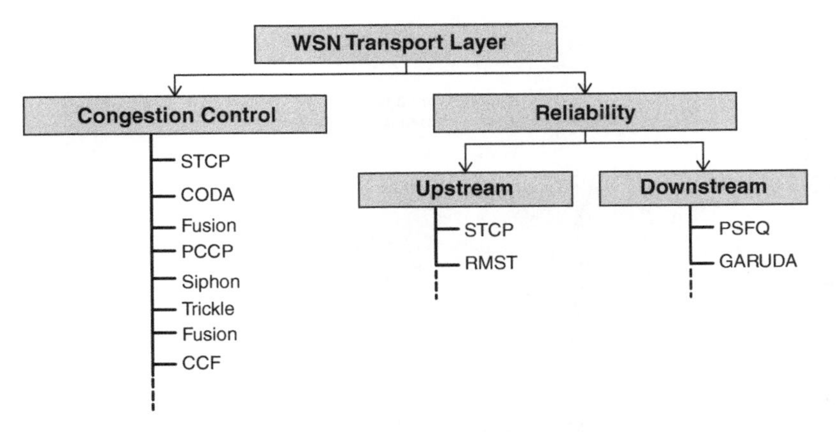


**Figure 3.18** Few proposed WSN transport layer protocols.

#### 3.7.4.5 Application Layer

Considering different types of sensors and corresponding environments, various sensor applications have been developed, which demand data fusion (techniques to combine data to realize inferences), data management (organization of data), clock synchronization (to remove effects of random delay), and positioning (location information of sensor nodes) [10].

#### 3.7.4.6 Cross-layer WSN Protocols

The strict energy, storage, and processing capabilities of sensor nodes, as well as the requirements of application-aware communication protocols for event-centric WSN, demand the implementation of cross-layer approaches. Through interactions or transferring information from one layer to another (Figure 3.19), the cross-layer approaches decrease the significant overhead associated with each layer for WSN approaches and ultimately improves the energy efficiency. Details about cross-layer WSN protocols are available in [27–29].

Besides traditional layers of the networking protocol stack, the following management planes are also the part of WSN protocol stack [7, 30]:

- *Power Management* is related to minimizing power consumption at each layer of the typical protocol stack.
- *Mobility Management* is related to the detection and registration of data route at each layer of a typical protocol stack.
- *Task Management* is related to the scheduling of sensing tasks among nodes.
- *QoS Management* is related to performance optimization in terms of specific QoS performance metrics.
- *Security Management* is related to handle security issues at each layer of the typical protocol stack.

### 3.7.5 WSN Operating Systems

WSNs are basically communication-centric and network protocol stack satisfies the communication needs of WSN applications. However, sensor nodes are resource-constrained

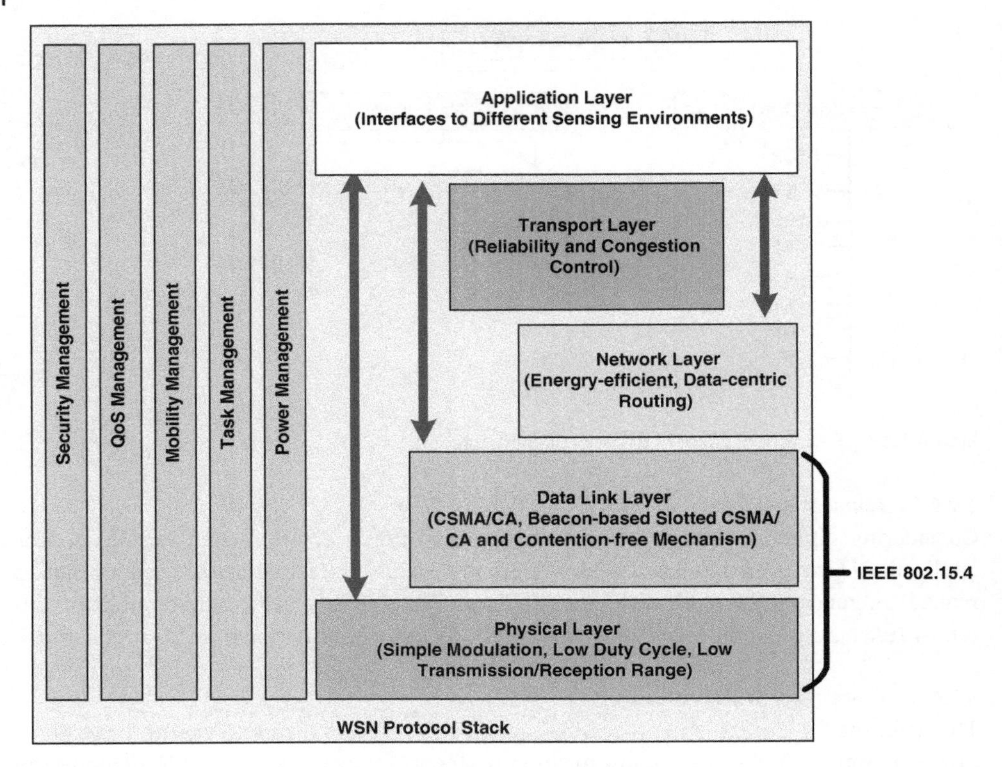


**Figure 3.19** Cross-layer design of WSN communication protocol stack.

in terms of computing, storage, and energy; therefore, a well-organized operating system is required for the efficient implementation and working of network protocols. The design and development of WSN operating system (WSN OS) are challenging because of the following fundamental problems [31, 32]:

- Nodes in WSN are resource-constrained and perform different functionalities in a highly concurrent environment, for example, reception and forwarding of packets from neighboring nodes. WSN OS face the challenge of accommodating concurrency of multiple ongoing transmissions and must be able to manage microcontroller processing and storage of transmissions in resource-efficient manner. Moreover, for a particular application, selection of WSN OS execution model should be appropriate enough between event-driven and multithreaded.
- Nodes in WSN face the problem of memory fragmentation issue (noncontiguous or scattered regions of unused memory) in memory allocation mechanism. Memory fragmentation can cause failure of memory allocation. Therefore, in general, dynamic memory is avoided in WSN OS. However, if dynamic memory requires, it must be provided with pre-allocation of few static buffers to reduce risks related to issues of memory fragmentation.
- Tracking of energy consumption is required in WSNs. WSN OS must monitor and be able to switch off unused components.

- To support real-time processing capability, WSN OS is required to act as real-time OS (RTOS).
- WSN OS must be able to support multi-interface heterogeneity for the provisioning of multi-homing.
- Security and Privacy issues are required to be addressed in a resource-constrained environment.
- WSN OS framework must be able to support the efficient implementation of various network protocol stacks for efficient nodes' communication. Contrary to traditional layered architecture, WSN OS prefers cross-layer implementation to save resources.

#### 3.7.5.1 WSN OS Design Issues

Several concerns, i.e. architecture, programming model, scheduling, memory management, communication protocol stack implementation, and resource sharing, affect the design of WSN OS.

- *Architecture:* WSN OS is required to adopt appropriate architecture (among monolithic, micro-kernel, and virtual-machine architectures), which must be of small kernel size, and supports the loading of application-required services to the system.
- *Programming Model:* Concerning the nature of IoT application, event-driven or multi-threading programming models are required to be selected.
- *Scheduling Algorithm:* Depending on demands of IoT application, WSN OS must be efficient in terms of energy conservation and are required to support scheduling of real-time and non-real-time environments.
- *Memory Management:* WSN OS is required to support hybrid (static + dynamic) memory allocation mechanism with the provisioning of static buffers to reduce risks related to issues of memory fragmentation.
- *Networking Protocol Support:* WSN OS must be able to support energy-efficient implementation of MAC, Network, and Transport layers.

Considering these design issues, the comparison of several WSN OS has been shown in Table 3.2 [33].

**Table 3.2** Comparison of WSN OSs.

| WSN OS | Architecture | Programming Model | Scheduling | Memory Management | Communication Protocol Support |
|---|---|---|---|---|---|
| Tiny OS | Monolithic | Both event-driven and multithreaded | FIFO | Static | Active Message |
| Contiki | Modular | Both event-driven and multithreaded | Priority interrupts | Dynamic | *u*IP |
| MANTIS | Layered | Multithreading | Priority based | Dynamic | TCP/IP |
| LiteOS | Modular | Both event driven and multithreaded | Round-robin | Dynamic | File based |

## Review Questions

**3.1** Explain the basic principles of sensing.

**3.2** Explain the difference between Capacitance and Induction.

**3.3** Provides a few examples of Capacitive and Magnetic Sensing.

**3.4** Explain the use of basic sensing principles in RFID technology.

**3.5** Which parameters significantly affect the design of the Physical layer of WSNs?

**3.6** Why are traditional transport layer protocols not suitable for WSN scenarios?

**3.7** How do cross-layer approaches improve energy efficiency?

**3.8** Describe the role of management planes that are part of the WSN protocol stack.

## References

1 Fraden, J. (2004). *Handbook of Modern Sensors: Physics, Designs, and Applications*. Springer Science & Business Media.
2 Wadaa, A., Olariu, S., Wilson, L., Eltoweissy, M., and Jones, K. (2005). Training a wireless sensor network. *Mobile Networks and Applications,* 10 (1–2): 151–168.
3 Terzic, E., Terzic, J., Nagarajah, R., and Alamgir, M. (2012). A neural network approach to fluid quantity measurement in dynamic environments. *Springer Science & Business Media.*
4 Gründler, P. (2007). *Conductivity sensors and capacitive sensors. In Chemical sensors: An introduction for scientists and engineers* (pp. 123–132). Berlin: Springer.
5 Pallás-Areny, R., and Webster, J.G. (2001). *Reactance variation and electromagnetic sensors. In Sensors and signal conditioning* (pp. 207–273). New York: Wiley.
6 Baxter, L.K. (1997). *Capacitive sensors—design and applications. In R. J. Herrick (Ed.) IEEE Press.*
7 Akyildiz, I.F., Su, W., Sankarasubramaniam, Y. et al. (2002). A survey on sensor networks. *IEEE Communications Magazine* 40 (8): 102–114.
8 Yong-Min, L., Shu-Ci, W., and Xiao-Hong, N. (2009). The architecture and characteristics of wireless sensor network. In: *IEEE International Conference on Computer Technology and Development*, 561–565. IEEE.
9 Ahmed, M.R., Huang, X., Sharma, D. et al. (2012). Wireless sensor network: characteristics and architectures. *International Journal of Information and Communication Engineering* 6 (12): 1398–1401.
10 Fahmy, H.M.A. (2016). *Wireless Sensor Networks. Concepts, Applications, Experimentation and Analysis.*
11 Baronti, P., Pillai, P., Chook, V.W. et al. (2007). Wireless sensor networks: a survey on the state of the art and the 802.15. 4 and ZigBee standards. *Computer Communications* 30 (7): 1655–1695.

12 Howitt, I. and Gutierrez, J.A. (2003). IEEE 802.15. 4 low rate-wireless personal area network coexistence issues. In: *IEEE Wireless Communications and Networking (WCNC)*, 148–1486. IEEE.
13 Gutierrez, J.A., Naeve, M., Callaway, E. et al. (2001). IEEE 802.15. 4: a developing standard for low-power low-cost wireless personal area networks. *IEEE Network* 15 (5): 12–19.
14 Gutierrez, J.A., Callaway, E.H., and Barrett, R.L. *Low-rate Wireless Personal Area Networks: Enabling Wireless Sensors with IEEE 802.15. 4. 2004*. IEEE Standards Association.
15 Anastasi, G., Conti, M., and Di Francesco, M. (2010). A comprehensive analysis of the MAC unreliability problem in IEEE 802.15. 4 wireless sensor networks. *IEEE Transactions on Industrial Informatics* 7 (1): 52–65.
16 Bhar, J. (2015). A mac protocol implementation for wireless sensor network. *Journal of Computer Networks and Communications* 2015: 1.
17 Huang, P., Xiao, L., Soltani, S. et al. (2012). The evolution of MAC protocols in wireless sensor networks: a survey. *IEEE Communication Surveys and Tutorials* 15 (1): 101–120.
18 Kosunalp, S. (2015). MAC protocols for energy harvesting wireless sensor networks: survey. *ETRI Journal* 37 (4): 804–812.
19 Yigitel, M.A., Incel, O.D., and Ersoy, C. (2011). QoS-aware MAC protocols for wireless sensor networks: a survey. *Computer Networks* 55 (8): 1982–2004.
20 Khanafer, M., Guennoun, M., and Mouftah, H.T. (2013). A survey of beacon-enabled IEEE 802.15. 4 MAC protocols in wireless sensor networks. *IEEE Communication Surveys and Tutorials* 16 (2): 856–876.
21 Bhattacharyya, D., Kim, T.-H., and Pal, S. (2010). A comparative study of wireless sensor networks and their routing protocols. *Sensors* 10 (12): 10506–10523.
22 Akkaya, K. and Younis, M. (2005). A survey on routing protocols for wireless sensor networks. *Ad Hoc Networks* 3 (3): 325–349.
23 Pantazis, N.A., Nikolidakis, S.A., and Vergados, D.D. (2012). Energy-efficient routing protocols in wireless sensor networks: a survey. *IEEE Communication Surveys and Tutorials* 15 (2): 551–591.
24 Wang, C., Sohraby, K., Hu, Y. et al. (2005). Issues of transport control protocols for wireless sensor networks. In: *IEEE International Conference on Communications, Circuits and Systems*, 422–426. IEEE.
25 Wang, C., Sohraby, K., Li, B. et al. (2006). A survey of transport protocols for wireless sensor networks. *IEEE Network* 20 (3): 34–40.
26 Rahman, M.A., El Saddik, A., and Gueaieb, W. (2008). *Wireless Sensor Network Transport Layer: State of the Art, Sensors*, 221–245. Springer.
27 Melodia, T., Vuran, M.C., and Pompili, D. (2005). The state of the art in cross-layer design for wireless sensor networks. In: *International workshop of the EuroNGI network of excellence*. Springer.
28 Chilamkurti, N., Zeadally, S., Vasilakos, A. et al. (2009). Cross-layer support for energy efficient routing in wireless sensor networks. *Journal of Sensors* 2009.
29 Jagadeesan, S. and Parthasarathy, V. (2012). Cross-layer design in wireless sensor networks. In: *Advances in Computer Science, Engineering & Applications* (eds. D. Wyld, J. Zizka and D. Nagamalai), 283–295. Springer.

30 Wang, Q. and Balasingham, I. (2010). Wireless sensor networks-an introduction. In: *Wireless Sensor Networks: Application-Centric Design*, 1–14. IntechOpen.
31 Dutta, P. and Dunkels, A. (2012). Operating systems and network protocols for wireless sensor networks. *Philosophical Transactions of the Royal Society A: Mathematical, Physical and Engineering Sciences* 370 (1958): 68–84.
32 Zikria, Y.B., Kim, S.W., Hahm, O. et al. (2019). *Internet of Things (IoT) Operating Systems Management: Opportunities, Challenges, and Solution*. Multidisciplinary Digital Publishing Institute.
33 Farooq, M.O. and Kunz, T. (2011). Operating systems for wireless sensor networks: a survey. *Sensors* 11 (6): 5900–5930.

# 4

# IoT Gateway

| LEARNING OBJECTIVES |
| --- |
| After studying this chapter, students will be able to:<br>• differentiate IoT architecture domains.<br>• elaborate IoT gateway architecture.<br>• define IoT gateway functionalities.<br>• evaluate IoT gateway selection criteria.<br>• associate IoT gateway and edge computing.<br>• construct edge computing-based solution for specific IoT applications. |

## 4.1 The IoT Gateway

In general, IoT architecture can be divided into three domains, i.e. sensing domain, network domain, and application domain [1] as shown in Figure 4.1.

The sensing domain (consisting of sensors and sensor networks with smart things) enables the physical things for getting and transmission of environment information to the required applications. Details about the sensing domain have already been discussed in Chapter 3. Network domain (consisting of communication infrastructure) is responsible for the collection, processing, and relaying of sensed data toward its destination. Application domain provides services to the user of particular IoT application. Generally, at the network domain of any system, a network gateway device is used to transfer data or information from one network to another. Simple gateways are sometimes referred to routers because of the similarity in regulating the flow of network traffic. However, the subtle difference exists in the regulation of traffic flow between similar (in the case of routers) and dissimilar networks (in the case of gateways). Therefore, due to the provisioning of interoperability (between heterogeneous networks through protocol conversion), network gateways are also known as protocol translation gateways or mapping gateways. In enterprise networks, network gateways are able to support the functionality of proxy server and firewall. In IoT systems, the IoT gateway [2] can be a dedicated physical device or software (installed on any core network device), which assists the connectivity between sensing

*Enabling the Internet of Things: Fundamentals, Design, and Applications*, First Edition.
Muhammad Azhar Iqbal, Sajjad Hussain, Huanlai Xing, and Muhammad Ali Imran.

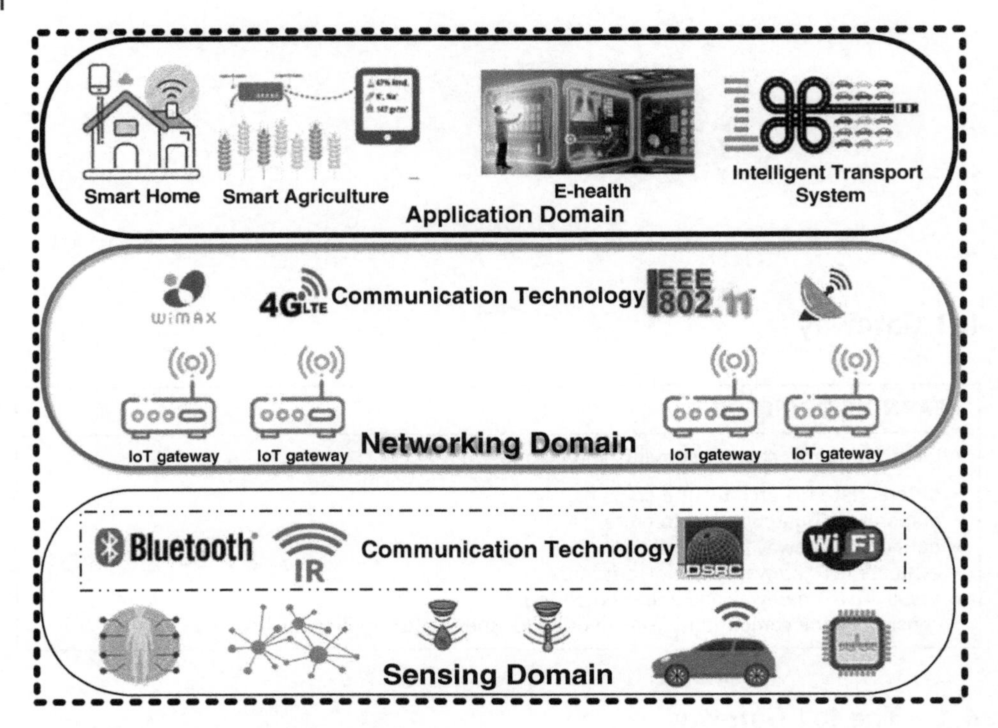


**Figure 4.1** Three domains of IoT architecture.

devices at sensing domain and services at application domains [1]. IoT gateways act as communication link among IoT edge devices (i.e. smart things or sensors), IoT Cloud, and end-user equipment (i.e. desktop computers, laptops, smartphones, etc.). Sensor acquired data moving toward the IoT Cloud passes through the IoT gateway. IoT gateway performs preprocessing on large volumes of sensor data, which involves the compression of aggregated data to reduce transmission costs. It also performs the translation of different network protocols to support interoperability of smart things and various connected devices. In addition, IoT gateways provide certain levels of security through advanced encryption algorithms. Therefore, as a middle layer (between sensor and IoT Cloud) in IoT architecture, IoT gateways protect the IoT system from unauthorized access and malicious attacks.

At the abstract level, two types of IoT gateways exist, as shown in Figures 4.2 and 4.3. Figure 4.2 illustrates that in the simplest IoT gateway design, the IoT gateway is not

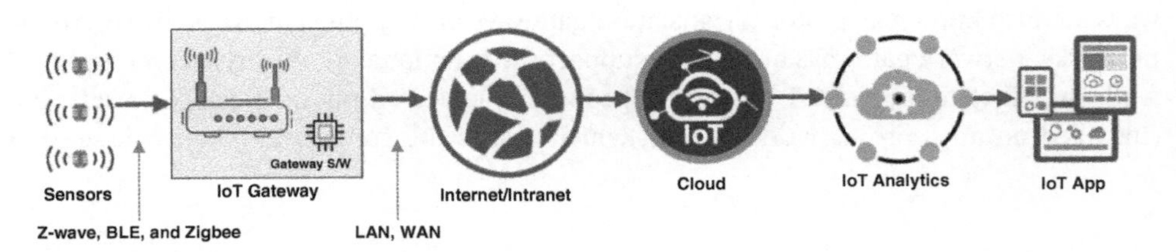


**Figure 4.2** Simple IoT gateway in the IoT system.

**Figure 4.3** IoT gateway with embedded sensors in the IoT system.

equipped with any kind of sensors; the software installed on IoT gateway is responsible for the collection, preprocessing, and transmission of received sensor data to the IoT Cloud. On the other hand, Figure 4.3 shows that few sensors such as, (i.e. Global Positioning Systems [GPSs] unit, temperature sensor, humidity sensors, etc.) can be placed on the gateway device.

In summary, IoT gateways are typically specialized or core network hardware devices with gateway software, which facilitate device-to-device communication or device-to-Cloud communication in IoT systems, and have been evolved to perform many tasks [3], for example:

- Data reception, buffering, preprocessing, aggregation, and forwarding
- Communication link between WSN and IoT Cloud
- Data encryption and Security Features
- Information Visualization
- Provisioning of Edge Computing
- Several Communication Interfaces (e.g. 2G/3G/4G/5G, Local Area Networks [LANs], WLAN, ZigBee, Bluetooth, LTE, PSTN)
- Protocol Conversion to Support Heterogeneous Communication
- Controllability of associated smart things

Figure 4.4 provides an overview of IoT gateway functionalities.

## 4.2 Sensing Domain and IoT Gateways

Depending on the IoT gateway use-case scenario, the nature of the network (i.e. WBAN, WPAN, WSN, Vehicular Ad hoc Network [VANET]) varies in the sensing domain. Ultimately the nature of the network affects the choice of IoT gateway (Figure 4.5). For example, in the case of:

- WBAN - smart phone, single board computer system (i.e. Raspberry Pi, Arduino, AdaFruit Flora, BeagleBone Black, etc.), or customized router can be used as gateways in order to get patient's vital signs information.
- WPAN - with different communication technologies (i.e. Bluetooth, ZigBee, Wireless Universal Serial Bus [USB], etc.), laptop, smart phone, or any core network device can be configured as gateways to process received information.

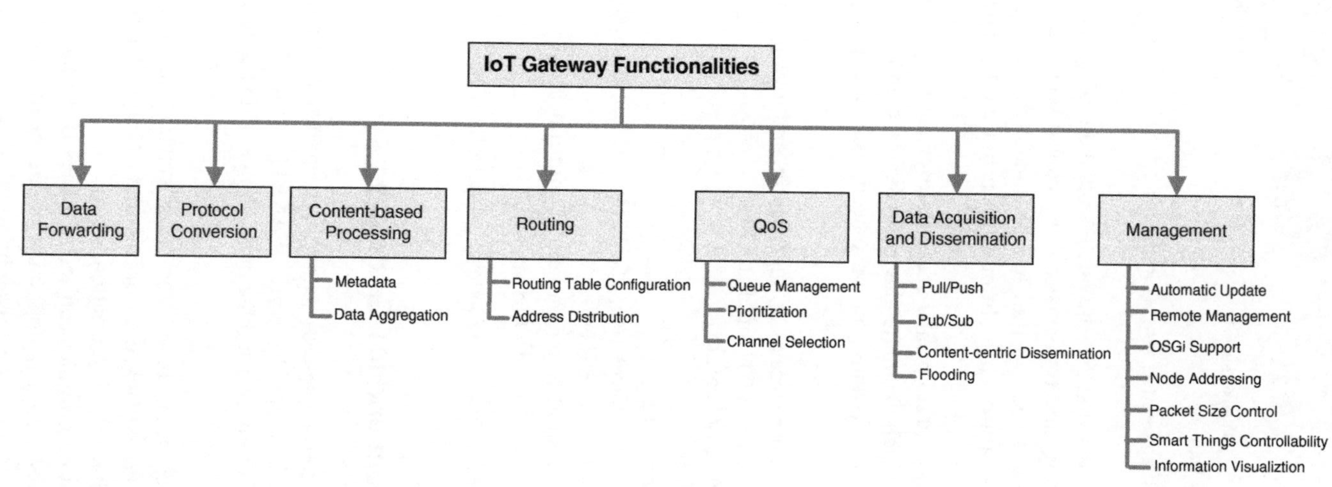


**Figure 4.4** Overview of the IoT gateway functionalities.

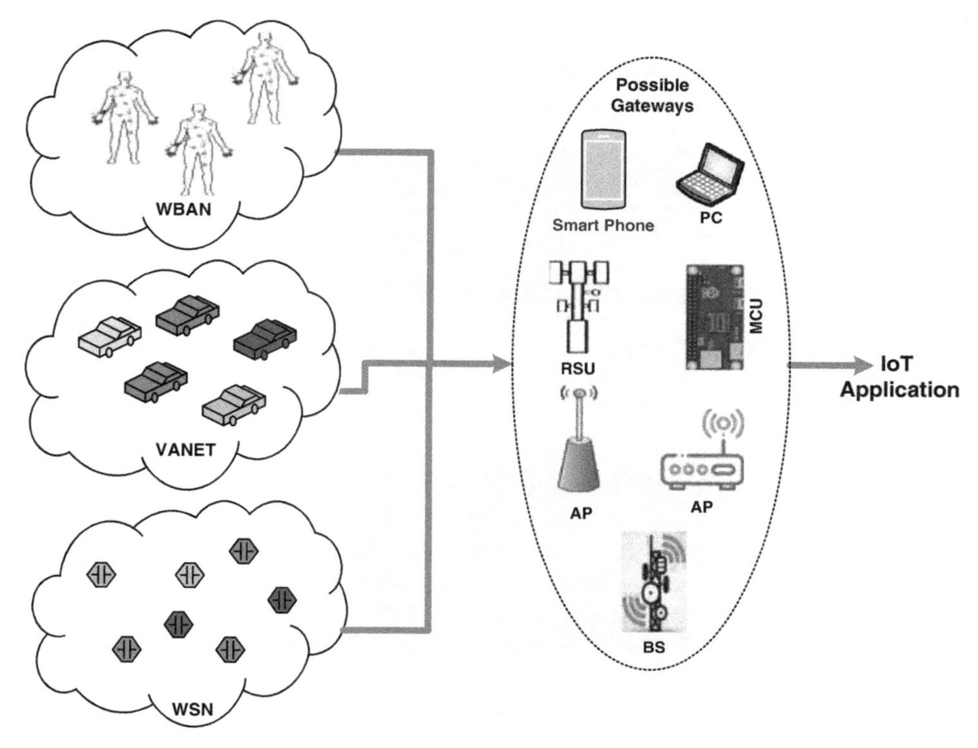


**Figure 4.5** Different types of IoT gateways.

- WSN - an Access Point, Base Station (BS), or specialized core network device can be used as gateways to collect and process sensor data.
- VANETs - vehicular gateway connected to Road-Side Units (RSUs), to make services available to on-road vehicles with short delay.

## 4.3 The Architecture of IoT Gateway

The design of the IoT gateway is reliant on the requirements of IoT application (i.e., health safety, connected car, agriculture, industrial automation, etc.). However, concerning the anatomy at the generic level, layered IoT gateway technology architecture consists of various hardware/software modules as shown in Figure 4.6.

### 4.3.1 Hardware Layer of IoT Gateway

The components at the hardware layer of the IoT gateway include sensors (optional), microcontrollers (processor), and communication/networking modules (i.e. Bluetooth, LAN, Wi-Fi, ZigBee, etc.). The type of microcontroller, processing capability, and memory

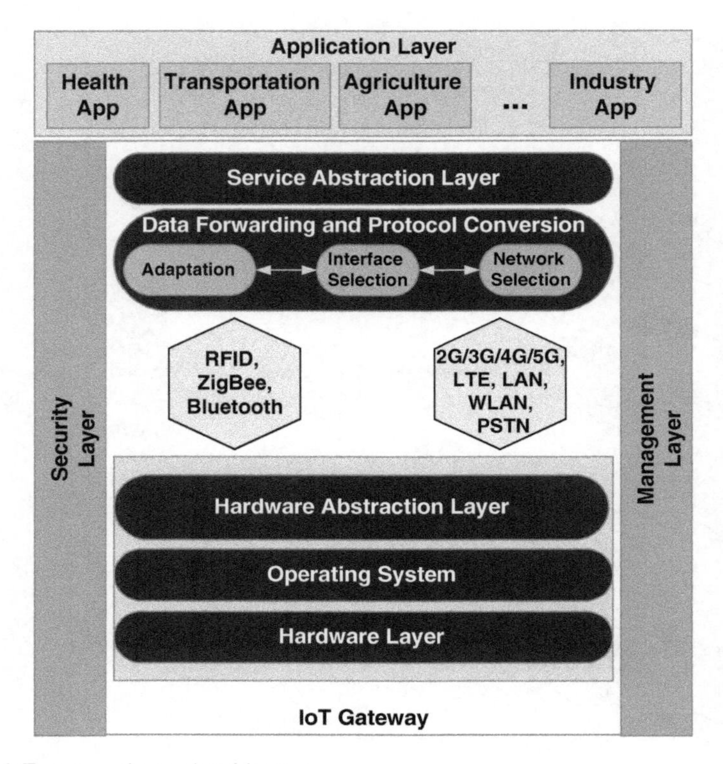


**Figure 4.6** IoT gateway layered architecture.

size of IoT gateway is fundamentally based on the underlying operating system (OS) that ultimately depends upon the complexity of IoT applications.

### 4.3.2 OS Layer of IoT Gateway

IoT gateway OS is designed to fulfill the requirements of IoT applications running on it. For small-scale to medium-scale IoT applications, RTOS is preferred. However, Linux OS and Android OS are also designated to support complex and Graphical User Interface (GUI)-enriched IoT applications, respectively.

### 4.3.3 Hardware Abstraction Layer

This layer supports software design independence over diverse hardware platforms and ultimately reduces the hardware cost to redesign IoT applications.

### 4.3.4 Data Forwarding Layer

The Data Forwarding and Protocol conversion layer consists of network selection, interface selection, and adaptation submodules. The network selection module is responsible for the provisioning of underlying network level information to interface selection module, which ultimately takes decision for the selection of appropriate interface for communication. The adaptation module is responsible for the conversion or adjustment of packet sizes of two different technologies.

### 4.3.5 Service Abstraction Layer

Service abstraction layer offers standard interfaces for application developers to access underlying network technologies of IoT gateway.

### 4.3.6 Manageability Layer

Concerning device manageability, an IoT gateway must be able to keep track of all connected sensors. IoT gateway device should have a trivial configuration for storing sensor and data access settings at IoT gateway. Other than device management, IoT gateways' data management module is responsible for data streaming, filtering, and storing. Data is temporarily stored on the IoT gateway to support disconnectivity problems with IoT Cloud.

### 4.3.7 Security Layer

The security layer ensures the security of things, data, and networks through the implementation of crypto authentication chips. In addition, IoT gateway is responsible to encrypt all transmissions prior to forwarding it to the IoT Cloud.

### 4.3.8 Application Layer

The customized design of IoT gateway application allows communication and interaction among different types of applications.

## 4.4 Selection of IoT Gateway

It is already mentioned that all data moving to and from sensing devices to IoT application server is to pass through IoT gateways. Therefore, the selection of right IoT device for successful deployment of an IoT system is very crucial and depends on certain requirements and factors, which are given as follows.

### 4.4.1 Nature of IoT System Architecture

IoT gateway selection is ultimately dependent on the selection of any one of the IoT system architectures, i.e. IoT systems having end devices with built-in gateway and IoT systems with central gateway as shown in Figure 4.7a,b.

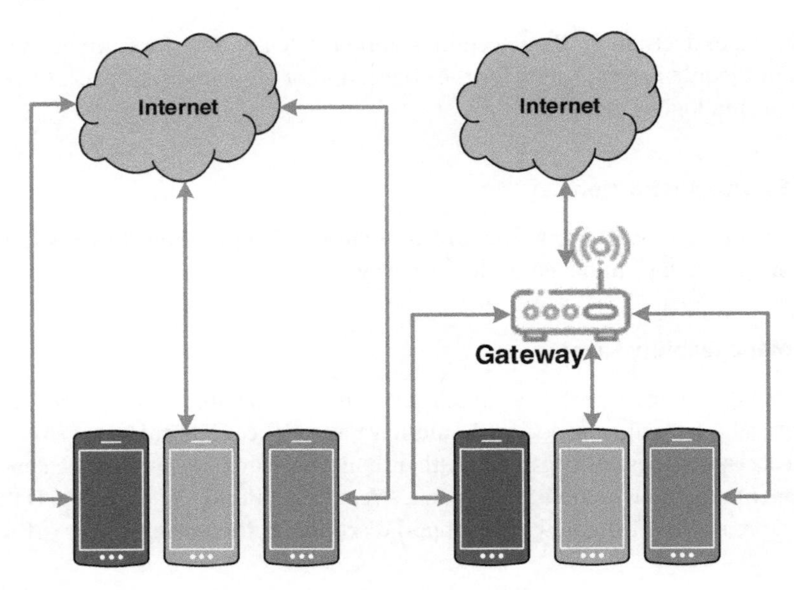


**Figure 4.7** (a) End-devices with built-in gateway, (b) central gateway.

### 4.4.2 Multiple Network Connectivity Support

IoT gateways acquire data from smart things and (or sensing devices) and send to the IoT Cloud. This transmission of data to and from smart things (or field sensors and end-user devices) and IoT Cloud in IoT system is relied on network infrastructure. Therefore, keeping in mind the energy constraints of an IoT system, IoT gateway must be able to support different kinds of networks (i.e. Wi-Fi, ZigBee, General Packet Radio Service [GPRS], etc.) to support ubiquity and data traffic demands. In short, IoT gateways are required dynamic accommodation of short-range and wide-range connectivity options.

### 4.4.3 Data Storage Capacity

Sensor data is stored in IoT Cloud and real-time retrieval of this data is relied on network connectivity. However, to support all-time availability of critical information in case of poor network connectivity, IoT gateway memory is required to be enough to store critical information.

### 4.4.4 Development Environment

IoT gateway should be able to support the multiple development environments, i.e. Python, C, C++, etc.

### 4.4.5 Robust Security Mechanism

IoT gateway security is important to secure the demands of the entire information flow in IoT systems. Therefore, IoT gateways with built-in security mechanism having advanced encryption standards, automatic discovery, and authentication mechanisms should be preferred.

### 4.4.6 External Hardware Watchdog Timer

Most of the time microcontrollers have integrated watchdog timer to detect and recover the hanging of edge application through restarting of microcontroller. External hardware watchdog timer saves gateway site visits through restarting the hanged microcontroller, which is not possible through internal watchdog.

### 4.4.7 Time Synchronization

The IoT gateway must uphold the Real-Time Clock synchronization through Network Time Protocol (NTP) or other sources (e.g. Global System for Mobile Communications [GSMs]) to handle clock drift and clock error.

### 4.4.8 Firmware Update

On The Air (OTA) programming feature saves the visit(s) of IoT gateway sites for the upgradation of the firmware.

### 4.4.9 LED Indication and Remote Reboot

IoT gateways with LED indications are helpful to detect connectivity issues. Moreover, remote rebooting of IoT gateways is helpful to recover from faults without visiting the physical site of deployed IoT gateways.

### 4.4.10 Support for Legacy Equipment

The chosen IoT gateway must support legacy equipment to ensure its data transmission in IoT system with existing devices.

### 4.4.11 Standard Protocol Support

Standard Transmission Control Protocol (TCP)/IP protocols (i.e. HTTP, TCP, User Datagram Protocol [UDP], etc.) and typical IoT protocols (i.e. Message Queuing Telemetry Transport [MQTT], Constrained Application Protocol [CoAP], IPv6 Low WPAN [6LoWPAN] Protocol, etc.) are required to be supported by IoT gateways.

### 4.4.12 Gateway Certification

The gateway model should be certified by different communication commissions (i.e. Federal Communications Commission [FCC] or Industry Canada [IC]). In addition, E-Mark, Society of Automotive Engineers (SAE), and J1455 certifications verify the durability of IoT gateway devices.

### 4.4.13 Control of Low Power Footprint

IoT gateway design should ensure the hardware operation on low power footprints and have capabilities to adopt power saving options.

### 4.4.14 Support for Edge Computing

IoT gateways should be able to support Edge computing with advanced features of data filtering and analytics.

## 4.5 IoT Gateways and Edge Computing

Considering swift processing demands of real-time IoT applications, edge computing (compared to the Cloud computing) provides local processing at the edge of the network. This local processing near the end users mitigates the computational stress of Cloud and reduces latency of response time in IoT systems. Despite the independent evolution of edge computing and IoT, there exists many similarities as shown in Table 4.1.

In IoT systems, the IoT gateway delivers edge computing power for preprocessing, aggregation, and analysis of data near the end user. IoT edge gateway receives data from edge device in IoT system and takes decision to send data back either to Cloud or to edge device while considering the computation needs of certain IoT application. Therefore, in edge computing paradigm, smart things at the edge of the system are data consumers and data producers at the same time. The layered architecture of edge computing-based IoT is shown in Figure 4.8.

### 4.5.1 Benefits of Edge Computing

The following are some of the potential benefits of edge computing in IoT systems:

- *Better Performance of IoT Applications*: Storage and processing of data near mobile edge devices reduces response time and energy consumption, which ultimately improves the overall performance of IoT systems.
- *Real-time Data Analysis*: Edge computing leverages the analysis of data at local level that facilitates real-time analysis of data.

**Table 4.1** Similarities of IoT systems and edge computing.

| | IoT | Edge |
|---|---|---|
| Components | Smart Things/Sensors | Edge Nodes |
| Deployment | Distributed | Distributed |
| Computation | Limited | Limited |
| Storage | Small | Limited (greater than IoT) |
| Response Time | Application Dependent | Fast |

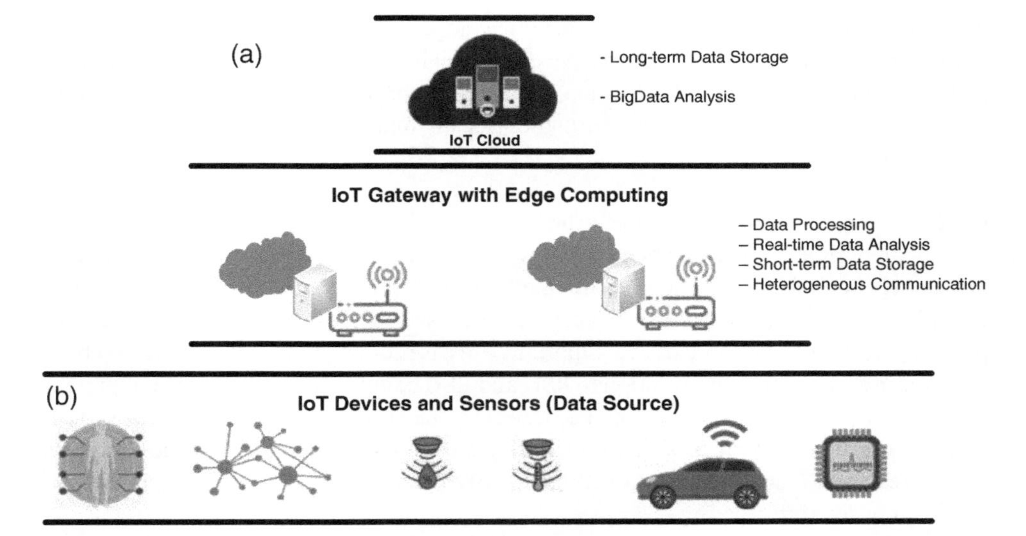


**Figure 4.8** Edge computing-based IoT architecture.

- *Scalability*: Edge computing permits organizations the expansion of computing capacity through the usage of cost-effective edge computing devices, which can process data without imposing extra network demands.
- *Reduced Operational Cost*: Storing and processing of data at the edge eventually reduces the cost of back end Cloud infrastructure.
- *Security*: Edge computing assists to achieve important security benefits in IoT systems, e.g.:
  - Distributed nature of edge computing helps to avoid single point overall network disruption.
  - Less data transmission to Cloud decreases the risk of data forgery.
  - Local data analysis within enterprise premises protects it from potential security threats.
  - Secures network from cyberattacks and improves data privacy with implementation of supplementary security measures.
- *Reliability*: Distributed implementation of edge computing ensures data transmission reliability with the provisioning of multiple pathways from edge to Cloud infrastructure.

### 4.5.2 Use Cases of Edge Computing

The followings are few important IoT edge computing use cases, which are helpful to understand the concept of edge computing [4].

#### 4.5.2.1 Smart Home

Smart things in home produce a huge amount of data, which is mostly required to be consumed in home. Therefore, edge gateways with specialized edge Operating System (edgeOS) in smart homes is the most suitable choice. The edgeOS collects data from available smart home appliances and residents' mobile devices through heterogeneous wireless technologies and apply data filtration to take decision about storage of data locally and/or on Cloud. IoT edge gateways in homes process information locally and lessens the burden on network bandwidth.

#### 4.5.2.2 Cooperative Safety Smart Vehicles

It is not an optimal choice for smart vehicles on road to access information, which is stored on remote Cloud for instant decision-making. Through edge computing at RSUs, a vehicle can utilize certain type of stored information to get assistance about safe driving. For example, cooperative safety is one of the important use cases of edge computing where vehicles send speed and position information to RSU and RSU computes updated map and provides various types of recommendations, i.e. trajectory recommendation, merging speed and merging time recommendations, lane changing recommendation, etc. IoT edge computing at RSUs can also help autonomous vehicles to take instant decisions, for example, decision to stop at pedestrian crossings.

#### 4.5.2.3 Provisioning of Infotainment Services for Smart Vehicles

Edge computing also provides infotainment services (i.e. toll collection, locations of nearby gasoline stations or restaurants, and notifications about movie/song availability, etc.) to facilitate drivers and passengers in their travel journeys. Mostly, these types of information are required immediately by drivers and passengers.

#### 4.5.2.4 Online Shopping Service

In online shopping services, customers' shopping cart manipulations are mainly stored on Cloud. Therefore, the updated view of shopping cart takes long time to download on end-user mobile devices. Cloud data offloading of necessary workload (e.g. shopping cart data with operations of item addition, updation, and deletion) at IoT gateways (equipped with appropriate computing resources) addresses the problems of high latency. Synchronization with Cloud data and collaboration of multiple edge gateways is done in the background.

#### 4.5.2.5 Healthcare and Collaborative Edge

Edge computing at healthcare IoT gateways is very important in different scenarios. For example:

- In close-loop systems, sensors demand instant analysis of data to maintain physiologic homeostasis, insulin levels, cardiac rhythms, respiration, etc. of seriously ill chronic disease patients.
- Real-time collection, analysis, and transmission of critical vital sign data from patients in Intensive Care Unit (ICU) to the concerned medical personnel require edge computing to be implemented.
- Quick data analysis to assist remote surgery through robots demands edge computing to be implemented.
- Critical patient data can be transmitted from ambulances to remote hospital in real time.

#### 4.5.2.6 Video Monitoring and Analysis

IoT edge computing technology is suitable to perform searching or some sort of detection in collection of available videos captured by personal mobile devices and urban security cameras. For example, finding a lost child using video-captured data is difficult if all the large quantity of video data is available on Cloud. However, with the use of edge computing technology, this task becomes easier *to implement in a targeted area.*

#### 4.5.2.7 Smart City

Edge computing technology for smart city is very important and most suitable for the collection, storage and forwarding of filtered data to Cloud from different public and private institutions, i.e. transportation, health, education, utility, etc. Centralized storage at Cloud is not realistic because of high data traffic load. Public safety, transportation, and healthcare applications demand storage of critical data at edge in a distributed way. This distributed data storage at IoT edge gateways improves the speed of diagnosis and real-time decision-making and transmission of required information to the end users' devices.

Sometimes, collaboration among various geographically restricted applications of government institutes and private enterprises located at different locations of a metropolitan is required to support a single data distributed application. Data distributed applications exploit collaborative edge to compose complex services and respective interfaces. Potential advantages of collaborative edge computing in a metropolitan can be comprehensible with the help of an example of connected healthcare working in case of a disease outbreak [4]. Concerning disease outbreak, patients flow to hospitals is increased in a specific region of a metropolitan. These hospitals share summarized information of that disease outbreak related to symptoms, precautions, prescribed medicines, etc. to the infected people. Infected patients theoretically purchase prescribed medicine from pharmacies in that area. Pharmacies share drug inventory information to pharmaceutical companies and government healthcare departments. Ultimately, the government healthcare agencies ask pharmaceutical companies to update production plan to meet certain requirements of particular drug supplies to that infected region.

#### 4.5.2.8 Security Surveillance

Edge-enabled IoT security surveillance systems are able to send real-time alerts to end users about any unusual activity, e.g. trespassing of a restricted geographical boundary.

#### 4.5.2.9 Retail Advertising

Demographic information collection by retail organizations is very important for the advertisement of specific products. In this case, edge computing through the implication of data encryption schemes can protect user privacy by sending encrypted data to the Cloud. Moreover, advertisement data stored at edge can be transferred to end user devices with low latency.

### 4.5.3 Challenges of Edge Computing-based IoT Systems

System integration, resource management, heterogeneous communication, security and privacy, and smart system support are main challenges of IoT systems, which are based on edge technology [5, 6].

#### 4.5.3.1 System Integration

Heterogenous nature of edge devices complicates the development of applications as the discovery of the devices with different communication technologies, unsuitability of traditional Domain Name System (DNS) and IP-based name schemes, huge programming overhead on server-side are main problems, which are required to be addressed. Therefore, there is a need to introduce new approaches for discovery, naming, and programmability of heterogeneous devices.

#### 4.5.3.2 Resource Management

Edge computing promises to provide nearby computation and resource management to resource-constrained IoT devices. However, heterogeneous nature of edge devices, available services, and IoT applications make the management of resources a difficult task. Allocation, sharing, and service pricing of IoT systems are required to be optimized through adapting different strategies.

#### 4.5.3.3 Security and Privacy

In edge computing-based IoT systems, involvement of heterogeneous nodes in unique scenarios, migration of services, and distributed structure raise new challenges for security and privacy issues.

For example:

- Authentication of gateways at different levels
- Implementation of efficient privacy-preserving schemes to protect end-user data collection at the edge gateways
- Secure transmission of data from IoT gateways to Cloud and edge devices
- Deployment of technologies to monitor network traffic and intrusion detection system
- Securing the uploading of computational tasks to edge computing nodes

#### 4.5.3.4 Heterogenous Communication

Integration of advanced communication technologies (including Multiple-Input and Multiple Output [MIMO], 5G cellular networks, and millimeter-wave) is a big challenge for edge computing-based IoT gateways.

#### 4.5.3.5 Data Analysis Support for Smart Systems

Data analysis at edge is challenging for IoT systems, which belong to different domains of real life, e.g. smart grid, smart transportation, smart city, etc. For example:

- In smart grid, the utilization of multiple edge servers to process and filter data streams from smart meters to Cloud and timely transmission of energy management decisions to edge devices
- In smart cities, non-interoperability of heterogeneous technologies involved at different edge devices, i.e. video analysis for public safety at public places, traffic monitoring at roads, and patient monitoring in hospitals
- In smart transportation, real-time management of vehicles on road

## 4.6 IoT Gateway Providers

Concerning IoT edge technology market, two groups of vendors (i.e. Cloud Backend Technology Vendor Group and Edge Device Technology Vendor Group) are involved in the designing and manufacturing of IoT gateways. Cloud Backend Technology vendor group provides software components for edge devices and certified devices. Edge device vendors have their own certified software stack, which works with an IoT Cloud. A plethora of vendor and service providers for IoT edge technologies are available and few names have been mentioned in Table 4.2.

## Review Questions

**4.1** Describe the layered architecture of IoT gateway.

**4.2** Which requirements and factors play important roles in the selection of suitable IoT gateway for an IoT system?

**4.3** What is IoT edge computing and what are its main benefits?

**4.4** Explain the concept of IoT edge computing with the help of three different real-life case studies.

**4.5** Assume that you are requested to create a national-level IoT system consisting of several IoT devices placed on the Internet, combining medical records from hospitals, doctors in general practice, and specialists. Such a system would, for example, allow access to vital information in a patient's medical record from casualty wards, ensuring the availability of the right information for unconscious patients. In this scenario, vital information, such as allergies, medicine intolerances, chronic disease history, etc., can be accessed on any system, while other information may be only accessed at

**Table 4.2** Names of IoT edge technology vendors and service providers.

| | | | |
|---|---|---|---|
| Adlink | Eurotech | Kontron | Relayr |
| Advantech | Emerson | Microsoft | Robotron |
| AWS | FogHorn | NEXIONA | Samsung |
| Bosch | Fujitsu | Nordcloud | Software AG |
| Cisco | Gemalto | OpenHab | Telekom |
| Citrix Systems | Hitachi | Oracle | Telit |
| Davra Networks | Huawei | OSRAM | VMware |
| Dell | IBM | Qualcomm | Vodafone |
| DG Logik | Intel | Jitsu | Wago |
| Digital Concept | | | Zebra |

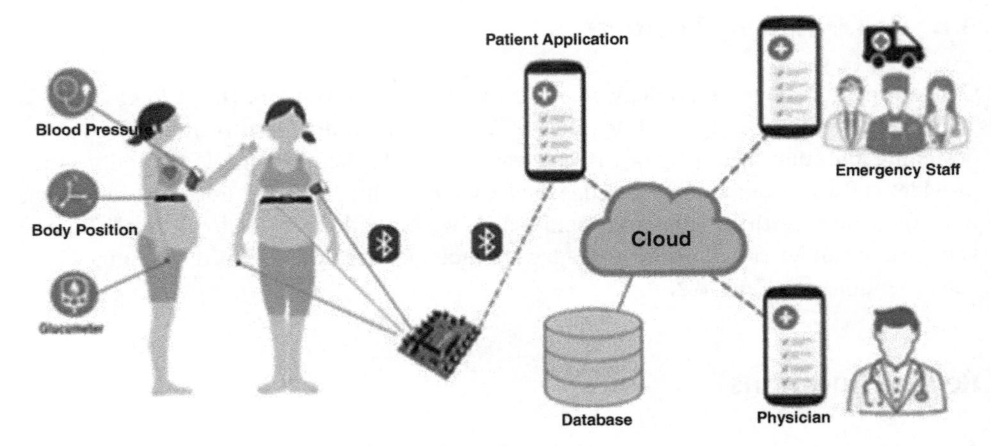


**Figure 4.9** Figure for Question 4.6. Source: Libelium [7].

the location of origin. Only new information is required to be added to the medical records in the Cloud. Considering this scenario, draw the architecture diagram and explain the usage and benefits of edge computing in this scenario.

**4.6** Hypertensive disorders and gestational diabetes are the most common diseases during pregnancy. Remote monitoring of pregnant woman's vital signs can help in the early detection of risk situations. Princess Nora University students developed an IoT system, which uses mobile devices for the remote monitoring of vital signs of pregnant women. Suppose the Figure 4.9 as an architecture diagram of this system, which is not correctly related to the actual realization of this application. As an expert in IoT system, correct this architecture diagram and propose an edge computing-based solution so that patients, physicians, and emergency staff can get information in real time. Explain the overall working of your proposed solution.

## References


1 Chen, H., Jia, X., and Li, H. (2011). A brief introduction to IoT gateway. In: *International Conference on Communication Technology and Application (ICCTA 2011)*, 610–613. IET.

2 Guoqiang, S., Yanming, C., Chao, Z. et al. (2013). Design and implementation of a smart IoT gateway. in Green Computing and Communications (GreenCom), 2013 IEEE and Internet of Things (iThings/CPSCom). In: *IEEE International Conference on and IEEE Cyber, Physical and Social Computing*, 720–723. IEEE.

3 Gazis, V., Görtz, M., Huber, M. et al. (2015). A survey of technologies for the internet of things. In: *International Wireless Communications and Mobile Computing Conference (IWCMC)*, 1090–1095. IEEE.

4 Shi, W., Cao, J., Zhang, Q. et al. (2016). *Edge computing: vision and challenges. IEEE Internet of Things Journal* 3 (5): 637–646.

**5** Salman, O., Elhajj, I., Kayssi, A. et al. (2015). Edge computing enabling the Internet of Things. In: *IEEE 2nd World Forum on Internet of Things (WF-IoT)*, 603–608. IEEE.

**6** Yu, W., Liang, F., He, X. et al. (2017). A survey on the edge computing for the internet of things. *IEEE Access* 6: 6900–6919.

**7** Libelium (2017). *E-Health application developed with MySignals first winner in health competition ISHIC 2017*. Libelium.

# 5

# IoT Protocol Stack

| LEARNING OBJECTIVES |
| --- |
| After studying this chapter, students will be able to:<br>• describe the mapping of IoT protocols to layered IoT architecture.<br>• explain the functionality of infrastructure, service discovery, and application layer protocols of IoT protocol stack.<br>• determine the role of application layer protocols in real-life applications. |

## 5.1 IoT Protocol Stack

IoT systems consisting of hardware and software components help in processing, storage, and analysis of data retrieved from smart things. For successful working of all these tasks of IoT systems, a number of IoT protocols have been proposed by different groups (such as Institute of Electrical and Electronics Engineers [IEEE], Internet Engineering Task Force [IETF], World Wide Web Consortium [W3C], European Telecommunications Standards Institute [ETSI], EPCglobal, etc.) to assist the working of application programmers and service providers. These protocols can be categorized into three types, i.e. Infrastructure Protocols, Service Discovery Protocols, and Application Protocols as shown in Figure 5.1 [1–3].

IoT protocols can be considered as the implementation manifestation of different layers of IoT architecture and offer different functionalities to achieve fast, reliable, and secure implementation of any IoT system. In Chapter 1, it is mentioned that there is no standard IoT architecture. However, various widely agreed reference IoT layered architectures have been proposed in the literature [4]. These layers of IoT architectures encompass the details associated with the basic building blocks (i.e. Smart Thing [Object/Device], IoT Gateway, IoT Cloud, IoT Analytics, and IoT Applications) [5, 6] of IoT systems. In this chapter, for the ease of simple and all-inclusive understanding, following [1] we have also considered five-layer IoT architecture, and the mapping of different categories of IoT protocols to five-layer IoT architecture is shown in Figure 5.2.

A quick recap of five-layered IoT architecture is provided in Table 5.1.

*Enabling the Internet of Things: Fundamentals, Design, and Applications*, First Edition.
Muhammad Azhar Iqbal, Sajjad Hussain, Huanlai Xing, and Muhammad Ali Imran.

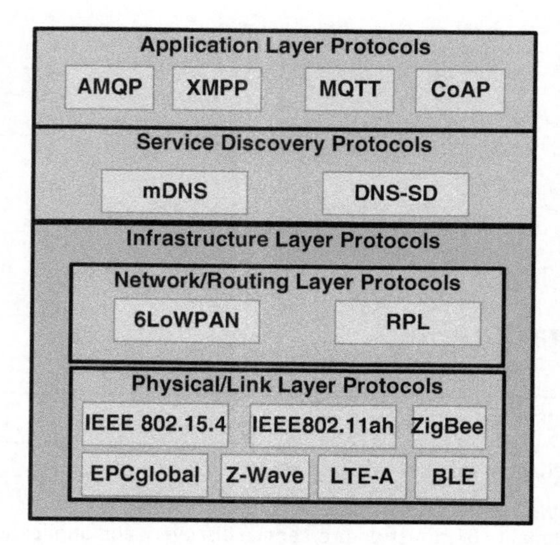


**Figure 5.1** IoT protocol stack. Sources: Raj and Raman [1]; Al-Fuqaha et al. [2]; Hersent et al. [3].

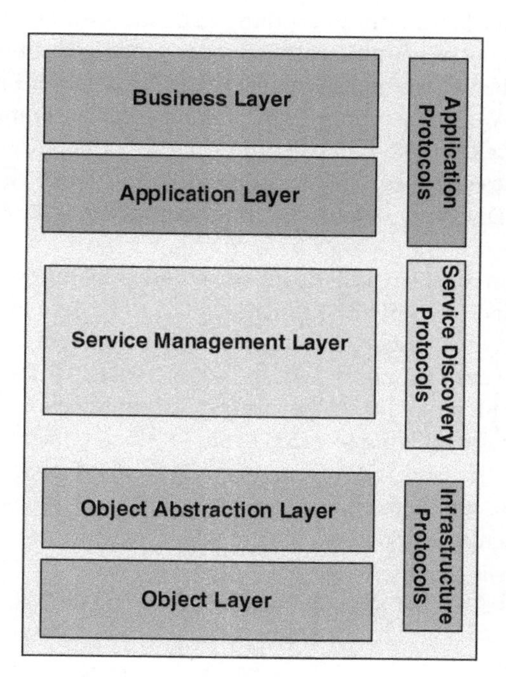


**Figure 5.2** Mapping of IoT protocol categories to five-layer architecture.

**Table 5.1** Functional features of five-layered IoT architecture.

| Layer name | Features |
|---|---|
| Object (Perception) Layer | The object layer primarily deals with the sensing, identification, collection, digitizing, and transmission of real-world information to Object Abstraction Layer |
| Object Abstraction Layer | Object Abstraction Layer or Network layer is responsible for secure data transmission from physical sensors to information processing systems by using various technologies, i.e. Wi-Fi, Infrared, ZigBee, BLE, WiMax, GSM, 3G/4G/5G, etc. |
| Service Management Layer | The service management layer provides pairing of services to requesters' applications (based on name and address) and enables IoT application programmers to deal with heterogeneous data collected by smart things with different hardware specifications. This layer includes the processing of received data before transmitting to the application layer |
| Application Layer | The application layer is responsible for the provisioning of services requested by the users |
| Business Layer | The business layer is responsible to manage overall activities/services of the IoT system through the creation of flow charts, business models, and graphs on received processed data from the application layer |

## 5.2 IoT Protocols

The general overview and core functionalities of few IoT protocols (shown in Figure 5.1) belonging to different categories have been discussed in the following subsections.

### 5.2.1 Infrastructure Protocols

The IoT protocols at this layer can be further categorized into different sublayers, i.e. Physical Layer Protocols, Link Layer Protocols, and Network/Routing Layer Protocols [2].

EPCglobal, Z-Wave, and Long-Term Evolution – Advanced (LTE-A), are a few well-known protocols of Physical Layer. On the other hand, IEEE 802.15.4, IEEE 802.11ah, BLE, and ZigBee are examples of protocols, which implemented both Physical and Link layer functionalities.

#### 5.2.1.1 EPCglobal

RFID technology used in IoT devices supports smart thing identification and service discovery. At the abstract level, an RFID tag consists of two main components, i.e. the electronic circuit to keep device identity and radio signal transponder, which uses radio waves to transfer the signal from tag to the tag-reader system. RFID tag reader forwards this unique electronic tag number to object-name service applications, which ultimately perform look-up operations to obtain related information from the database, i.e. manufacturing time and manufacturing place. The unique tag number on the RFID tag is known as the Electronic Product Code (EPC) number. This EPC number mainly helps in the identification and tracking of smart items. EPCglobal organization is responsible for the development and management of EPC and RFID standards [7, 8]. The basic format of EPC number assigned to smart things consists of four parts (as shown in Figure 5.3) [9], i.e.:

*Header (8 bits)*: Assigned by EPCglobal and used to identify the version, type, structure, length, and generation of EPC
*EPC Manager Number (28 bits)*: Assigned by EPCglobal to maintain partitions
*Object Class (28 bits)*: Assigned by EPC manager owner to identify the class of object
*Serial Number (36 bits)*: Assigned by EPC manager owner to identify a particular instance

Descriptions related to different categories of EPC tags are shown in Table 5.2 [1, 2].

#### 5.2.1.2 Z-wave

Z-Wave is a low-power communication protocol developed by Sigma Designs and improved by Z-Wave Alliance [10] to support point-to-point communication up to 30 m [2]. Z-Wave technology uses the Industrial Scientific and Medical (ISM) band and is used in remote control applications developed for the automation of home network [11], i.e. household heating, ventilation, air conditioning (HVAC) appliances, fire detection, and light control. Z-Wave network consists of controllers (primary and secondary) and slave nodes. Primary and secondary controller devices are responsible for assigning of home/node IDs to Z-wave nodes and maintaining routing tables in the Z-wave network, respectively. On the other hand, slave nodes do not maintain any routing tables but the network map, which contains routes to different nodes and is obtained through source routing. Medium Access Control (MAC) layer of Z-Wave protocol uses Carrier Sense Multiple Access Collision Avoidance (CSMA/CA) and guarantees reliable transmission using optional acknowledgement (ACK) messages [3].

**Table 5.2** Categories of EPC tags.

| EPC No. | RFID tag | Tag category | Functionality |
|---|---|---|---|
| 0 | Read | Passive | Write once – Read multiple |
| 1 | Write Once and Read Only | Passive | Write once – Read multiple |
| 2 | Read or Write | Passive | Read Multiple – Write Multiple |
| 3 | Read or Write | Semi-passive | Attached to Sensor |
| 4 | Read or Write | Active | Attached to Sensor with Radio Wave Field to Communicate User |

Sources: Raj and Raman [1]; Al-Fuqaha et al. [2].

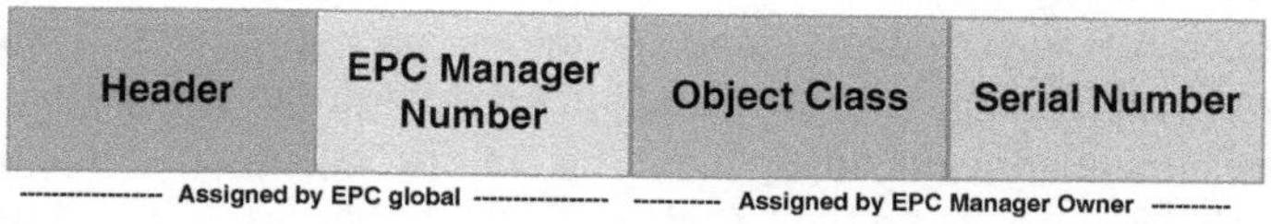

Header | EPC Manager Number | Object Class | Serial Number

---------------- Assigned by EPC global ---------------- ---------- Assigned by EPC Manager Owner ---------

**Figure 5.3** Basic format of EPC code.
Source: Modified from Jones Chung [9].

#### 5.2.1.3 Long-term Evolution – Advanced (LTE-A)

LTE-A is a communication standard for high-speed cellular networks, which is also referred as 4G LTE [12]. LTE-A is suitable for smart city IoT infrastructure in terms of its service cost and scalability. Live streaming for the coverage of ongoing events in the city and real-time streaming of important TV shows are two well-known examples of LTE-A use cases.

The architecture of LTE-A consists of a Core Network (CN) infrastructure and Radio Access Network (RAN) infrastructure. CN deals with the controlling of mobile devices and IP message flows. On the other hand, RAN consisting of base stations deals with the communication of devices using radio waves. The Physical layer of LTE-A is based on Orthogonal Frequency Division Multiplexing (OFDM), which is used to partition channel bandwidth [2, 13].

#### 5.2.1.4 Bluetooth Low Energy (BLE)

BLE standard is an evolution of Bluetooth Core, which utilizes short-range radio waves with minimal power consumption [14, 15]. Communication range of BLE is 100 m and is considered as one of the best protocols for IoT devices as it is available in many modern-day smartphones.

The communication stack of BLE consists of:

*Physical Layer*: For transmission of bits
*Link Layer*: For connection establishment, medium access, error control, flow control
*Logical Link Control and Adaptation Protocol (L2CAP) Layer*: For multiplexing, fragmentation, and reassembly of packets
*Host Control Interface (HCI) Layer*: Interface for the link layer to access data from upper layers
*Attribute Protocol (ATT) Layer*: For GATT server to communicate with GATT client
*Security Manager (SM) Layer*: For authentication and security of devices
*Generic Attribute Protocol (GATT) Layer*: For collection of sensor data
*Generic Access Profile (GAP) Layer*: For scanning and connection management of devices

#### 5.2.1.5 IEEE 802.15.4

IEEE 802.15.4 is a technical standard, which provides the functionalities of Physical and MAC layers for Low-Rate Wireless Private Area Networks (LR-WPAN) [16]. IEEE 802.15.4 offers reliable high data rate communication at low power consumption [17]. Due to its features of node handling at large scale and low power consumption for data transmission, it is widely employed in various IoT systems. Moreover, 802.15.4 also forms the basis of different protocols, i.e. ZigBee, Internet Protocol version 6 over Low-Power Wireless Personal Area Network (6LoWPAN), which are developed to support communication in IoT systems. The Physical layer of 802.15.4 is based on Direct Sequence Spread Spectrum (DSSS), and data transmission is possible on three frequency bands (i.e. 2.4 GHz, 915 MHz, and 868 MHz). Depending on the frequency band, it supports communication at different data rates (ranging from 40 to 250 kbps). Lower frequencies cover large distances and are good in providing better sensitivity. On the other hand, high frequencies support low latency and high throughput. The MAC layer of IEEE 802.1.4 is based on the CSMA/CA protocol, which reduces collisions for ongoing simultaneous transmissions of multiple nodes in a network [18].

IEEE 802.15.4 has been implemented to support two types of network nodes, i.e. Reduced Function Device (RFD) and Full Function Device (FFD). RFD are simple network devices with constrained resource and communication requirements, and FFD devices have more resources. FFD devices can:

- Act as coordinator (in WPAN). The coordinator is a special kind of FFD node, which controls and manages the PAN network.
- Act like a normal node.
- Support full MAC implementation.
- Store routing tables.

Network nodes of the IEEE 802.15.4 standard can communicate with each other in star, peer-to-peer, and cluster-tree topologies as shown in Figure 5.4. In star topology, FFD nodes (at least one) act as coordinator, and RFD nodes can only communicate through that FFD coordinator(s) to transfer data packets to other nodes. FFD coordinators are responsible to control and maintain connectivity of RFD nodes in PAN. In peer-to-peer topologies, a coordinator exists but all other nodes are able to communicate with each other directly. Cluster topology is a kind of peer-to-peer topology, which consists of PAN Coordinator (PC), Cluster Head (CH), and RFD nodes. In a cluster-tree network, a coordinator can designate another FFD to act as a coordinator for a subset of that PAN nodes.

#### 5.2.1.6 IEEE 802.11ah

Although IEEE 802.15.4 is a low power consumption protocol and is suitable for several IoT applications, however 802.15.4 is not suitable for IoT applications consisting of a large number of IoT devices, which are dispersed in larger areas [19]. To fill this gap, IEEE802.11 initiated a task group known as 802.11ah to develop a new standard by considering the advantages of both IEEE802.11 standard and low-power sensor communication technologies. The IEEE 802.11ah-based network structure for large-scale IoT systems is shown in Figure 5.5.

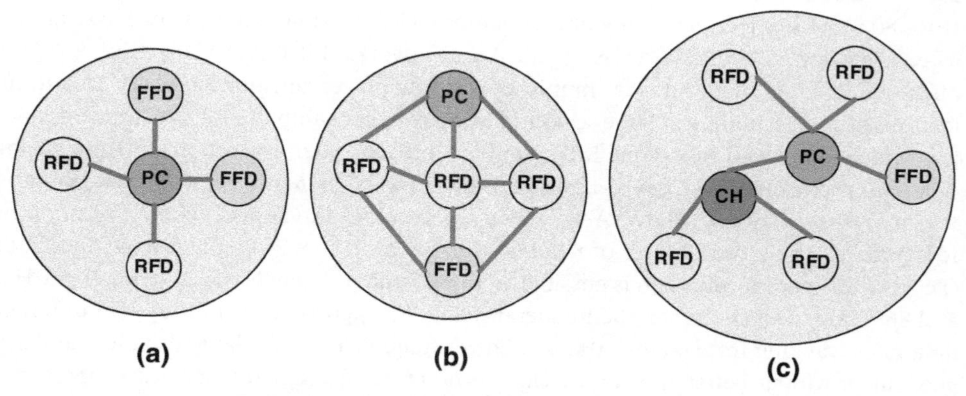


**Figure 5.4** IEEE 802.15.4 Topologies (a) Start, (b) Peer-to-Peer, and (c) Cluster.

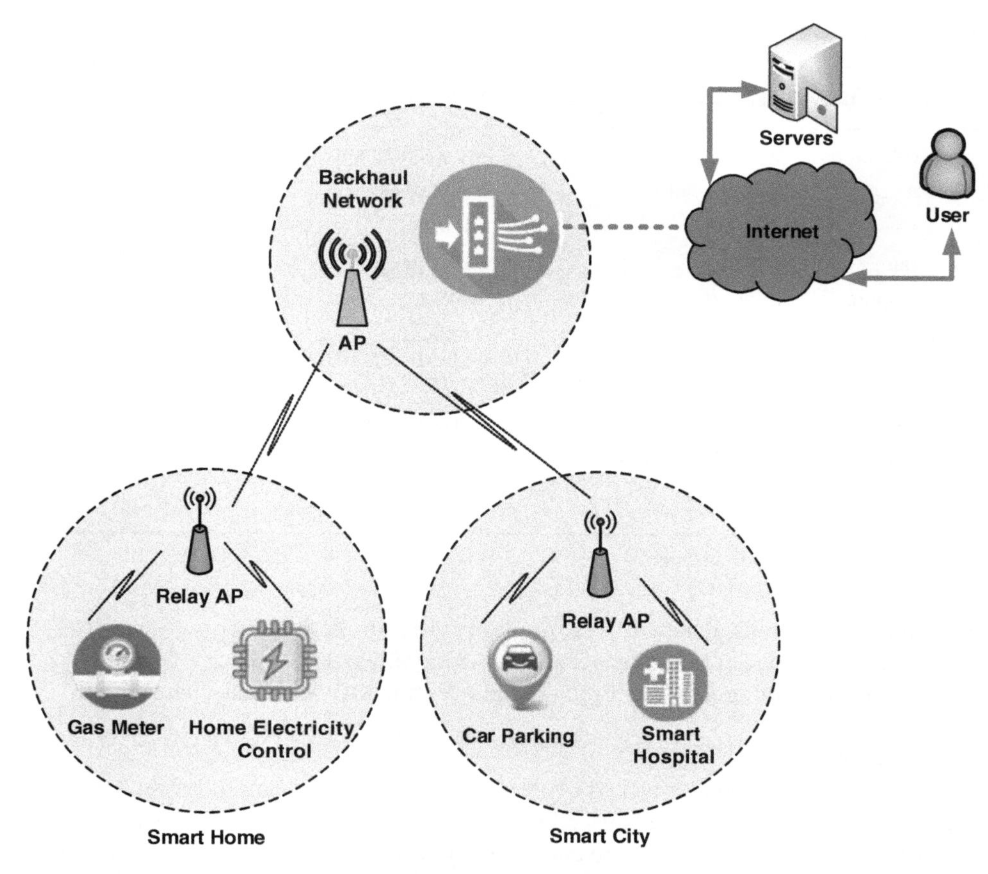


**Figure 5.5** IEEE 802.11ah-based network structure for IoT systems.

Considering the core features of IEEE 802.15.4 and IEEE 802.11ah, the comparison is provided in Table 5.3 [20].

IEEE 802.11ah supports long-distance communication through the implementation of the following concepts at the Physical and MAC layers:

- Hierarchical Association IDentification (AID)
- Short MAC frames
- Increased Sleep Time
- Restricted Access Window (RAW)
- Null Data Packet
- Traffic Indication Map (TIM)
- Delivery Traffic Indication Map (DTIM)
- Target Wake Time (TWT)

The following are the salient features of IEEE 802.11ah:

- It supports single-hop communication over 1000 m.
- It supports the usage of Relay APs to extend connectivity up to two hops.

**Table 5.3** Comparison between IEEE 802.15.4 and IEEE 802.11ah.

| Protocol Feature | IEEE 802.15.4 | IEEE 802.11ah |
|---|---|---|
| Network Support | WSN | WSN, Backhaul |
| Frequency | 2.4 GHz, Sub-1 GHz | Sub-1 GHz |
| Range | ~100 m | ~1000 m |
| Data rate | 250 Kbps | 78 Mbps |
| MAC Support | CSMA/CA | RAW |
| Power Saving | Sleep–Wake Mechanism | Traffic Indication Map (TIM), Delivery Traffic Indication Map (DTIM), Target Wake Time (TWT) |
| Relay Node | FFD | Access Point (AP) |
| Application Support | Smart Environment Monitoring, Smart Agriculture | Smart City, Smart Home |

Source: Based on Ahmed et al. [20].

- It supports hierarchical network organization to improve scalability.
- It extends the range of Wi-Fi networks by utilizing sub-1GHz bands.
- It reduces collision probability through implementing RAW-based mechanism.

#### 5.2.1.7 ZigBee

ZigBee protocol has been developed by the ZigBee alliance [21], and it is based on IEEE 802.15.4 specifications. ZigBee is used in small and low-power digital radios to create ad hoc networks, i.e. WPANs for home energy monitoring in home automation, WBANs for vital sign data collection in medical applications, remote operations in smart metering, etc. ZigBee has certain other features to support various IoT applications, i.e.:

- Support scalability (i.e. a large number of network nodes [≤65K nodes])
- Low power consumption
- Data rate of 25 kbps, which is desirable for intermittent data transmissions from sensor nodes
- Less expensive than Z-Wave and BLE
- Support decentralized topology similar to the Internet
- Robust in terms of finding new routes in case of certain route failure

The layered architecture of ZigBee consists of several layers as shown in Figure 5.6. It consists of two main sections (i.e. foundation layers section and application layers section) and network layer [22].

Foundation layers section consists of two layers, i.e. physical (PHY) and MAC layers, which are based on 802.15.4 standard. These two layers are responsible for the

- Channel selection
- Sending/receiving of packets on the selected channel
- Energy detection within the channel
- Collision avoidance through CSMA/CA

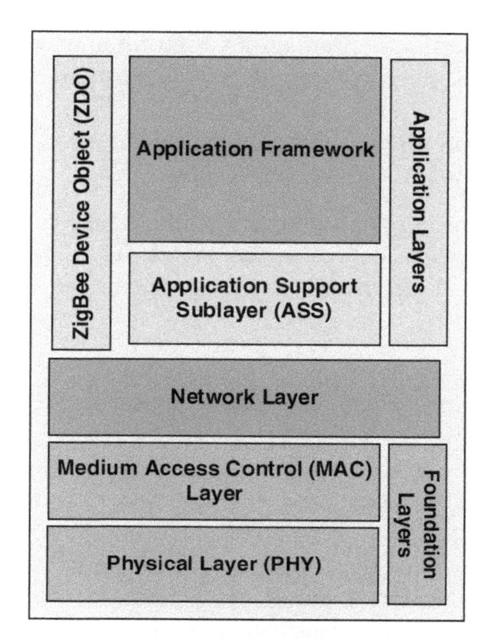


**Figure 5.6** ZigBee layered architecture.

- Beacon generation for data management
- Management of guaranteed time slots
- Data transfer to upper layers

ZigBee defines its own Network layer, which is responsible for the:

- Network initialization
- Address assigning to network nodes
- Device configuration
- Secured transmission
- Adding of routing capabilities to support the multihop transmission of data packets from the source node to destination node

The ZigBee defines its own upper layers, i.e. Application Support Sublayer (ASS), ZigBee Device Object (ZDO) layer, and Application Framework layer in the application layers section.

ASS layer is responsible for the following:

- Defines addressing of various objects along with mapping and management of profiles
- Filtering of packets for non-registered end devices or profiles
- Reassembling of data packets

ZDO layer is responsible for:

- Discovery of devices and available services
- Advanced network management

The Application Framework provides an environment in which application objects are hosted on ZigBee devices. Here, key-value pair service is used to get attributes of application objects.

ZigBee is able to support several network topologies, i.e. star, mesh, cluster, etc. The following are the few main areas of ZigBee applications:

- Home automation (sensing, monitoring, automation, and security)
- Healthcare systems (sensing, data collection, monitoring, and diagnostic)
- Industrial and commercial (sensing, monitoring, automation, and control)

#### 5.2.1.8 6LoWPAN

The 6LoWPAN is an open standard for IoT communication [22]. It was implemented by IETF to support communication with 802.15.4 devices. However, now it supports a number of WPAN devices based on BLE, low-power RD, low-power Wi-Fi, etc. Unique characteristics of WPAN devices, i.e. variable address lengths, limited packet size, and low bandwidth, demand an adaptation layer to accommodate IPv6 specifications. 6LoWPAN protocol implementation addresses this issue with the provisioning of header compression [23]. This header compression feature reduces various overheads (i.e. fragmentation overhead and transmission overhead) and provides very promising support for IP communication of IoT devices. IP packets enveloped by 6LoWPAN protocol have a combination of 2-bit headers, for example [22]:

- No LoWPAN Header (00) indicates that packet is required to be discarded as not following 6LoWPAN specifications.
- Dispatch Header (01) indicates packet has IPv6 header in compressed form.
- Mesh Addressing (10) indicates that the 802.15.4 packet is required to be forwarded to the link layer.
- Fragmentation Header (11) indicates that the IP packet exceeds the length of 802.15.4 frame and fragmentation of this packet is required.

In this way, 6LoWPAN provides support to large mesh network topology with low power consumption. The architecture of the 6LoWPAN mesh network is shown in Figure 5.7. 6LoWPAN network consists of routers (R) and hosts (H). Hosts are the end-point devices and send data to routers, which ultimately route received data to destination nodes. The destination device can be a 6LoWPAN device within the network or any IP-based device outside the WPAN network. The 6LoWPAN network is connected to the IPv6 network through the implementation of a 6LoWPAN edge router. The gateway device to the Internet is IPv6 router to whom different IP-based devices, i.e. PCs and servers, have been connected. Exchange of data between WPAN devices and IP-based devices on the Internet is possible through the implication of the 6LoWPAN edge router [1].

#### 5.2.1.9 Routing Protocol for Low-Power and Lossy Networks (RPL)

WSN and WPAN are examples of low-power and lossy networks in which nodes are mostly resource-constrained. RPL is based on IPv6 protocol and is standardized by IETF's Routing Over Low-power and Lossy Links (ROLL) working group [24]. RPL is based

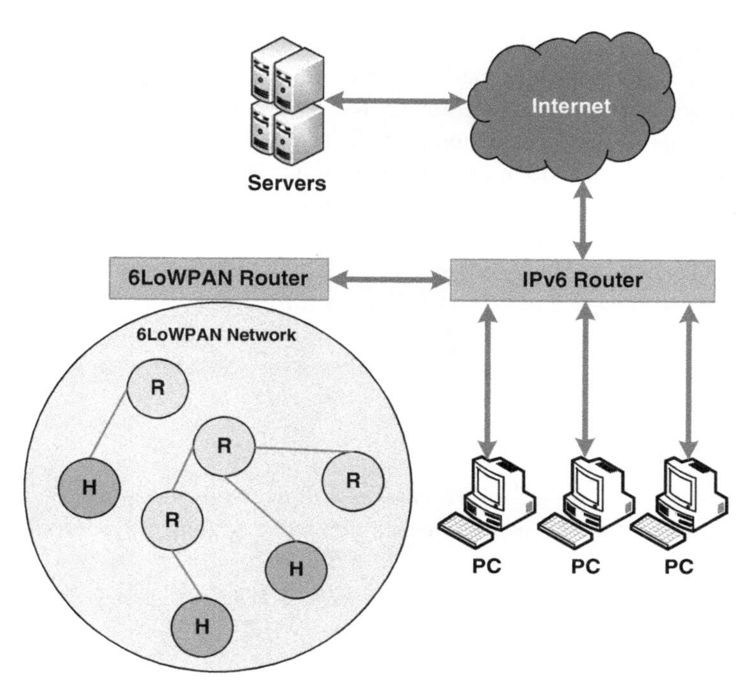


**Figure 5.7** 6LoWPAN network architecture.

on Destination Oriented Directed Acyclic Graph (DODAG) to build robust topology over connected lossy links (without cycles) to support different traffic models (i.e. Point-to-Point, Point-to-Multipoint, and Multipoint-to-Point) [25]. Particular characteristics of DODAG are the following:

- Only one single root node
- Other nodes contain information about parent node
- No node contains information about child nodes

RPL exploiting this information is able to maintain at least one single path from each node to the root node and preferred parent to track the faster path. To keep this information updated, RPL uses four types of control packets, i.e.:

- *DODAG Information Object (DIO)*: This packet is used:
  - To keep information of the current node level
  - To determine node distance to root node
  - To select the preferred parent

  *Destination Advertisement Object (DAO)*: This control packet is used to help RPL to support upward and downward traffic.
- *DODAG Information Solicitation (DIS)*: This control packet is used to acquire DIO messages from other reachable adjacent nodes.
- *Destination Advertisement Object Acknowledgment (DAO-ACK)*: This control packet is a response to the DAO packet from the recipient node.

DODAG formation starts with the transmission of the DIO control packet from the root, which contains its location information to all nodes (routers) at different levels of the low-power lossy network. Upon reception of the DIO packet, each node stores two paths, i.e. parent node path and participation path for each node. This propagation of the DIO control packet to other network nodes gradually builds DODAG. Upon the construction of DODAG, the preferred parent is selected as a default path toward the root. After receiving DIO control packets, the root node stores the upward route to other nodes. Moreover, downward routes build through unicasting of DAO control packets toward the root node. Considering these upward and downward routes, RPL nodes work either in non-storing mode or in storing mode. Non-storing is based on source routing to route packets toward downward routes, and storing mode routing is based on IPv6 addresses.

### 5.2.2 Service Discovery Protocols

Service Discovery is important in IoT systems and demands an automatic and efficient resource management mechanism for registration and discovery of resources and services in the IoT network.

The following are two well-known service discovery protocols, which have been used in IoT systems:

- Multicast Domain Name System (mDNS)
- Domain Name System Service Discovery (DNS-SD)

#### 5.2.2.1 Multicast Domain Name System (mDNS)

The mDNS [26] is developed by joint efforts of IETF Zero Configuration Networking (Zeroconf) and Domain Name System Extensions (DNSEXT) working groups. mDNS service works similarly to that of the unicast DNS server and is very flexible because DNS namespace is used locally without any supplementary configuration. mDNS is a preferred choice for embedded IP-based devices due to different factors, i.e.:

- Automatic configuration.
- Additional infrastructure is not required to manage devices.
- High level of fault tolerance.

An IoT service that requires service discovery using mDNS first sends an IP multicast message to all nodes in the local domain to inquire names. With this message, a client asks different devices with a specific name to respond back. Upon reception, the target device containing its name sends a multicast response packet, which contains its IP address. All devices in the IoT network store target device's name and corresponding IP address in its cache memory to use that service later.

#### 5.2.2.2 DNS Service Discovery (DNS-SD)

In IoT systems, DNS-SD [27] uses standard DNS messages to facilitate users with the discovery of desired services. Moreover, DNS-SD protocol assists to connect devices without additional configuration. Service discovery in DNS-SD is a two-step process, that is, finding of hostnames of required services and pairing of IP addresses with hostnames using mDNS. The finding of the hostname is necessary as IP address has the possibility to be changed, but the hostname will not be changed as DNS-DS supports to keep hostnames constant within the network to be used later.

### 5.2.3 Application Layer Protocols

Considering the diverse aspects of the performance of IoT applications, a number of application-level IoT protocols have been developed. The following are the few well-known IoT application layer protocols developed to implement unique message capabilities of IoT, which are required in IoT applications:

- Data Distribution Service (DDS)
- Message Queue Telemetry Transport (MQTT)
- Constrained Application Protocol (CoAP)
- Advanced Message Queuing Protocol (AMQP)
- eXtensible Messaging and Presence Protocol (XMPP)

#### 5.2.3.1 Data Distribution Service (DDS)

DDS is an application-level protocol, which is developed by Object Management Group (OMG) to support real-time M2M communications in the IoT system [28, 29]. DDS is based on broker-less publish–subscribe architecture and uses multicasting to fulfill the reliability and QoS demands of IoT applications. DDS functionality has been divided into two layers, i.e.:

- *Data-Centric PublishSubscribe (DCPS) Layer*: This layer is responsible in providing information to subscribers. The DCPS layer of the DDS model consists of five components to regulate the flow of data:
  - Publisher to disseminate data.
  - DataWriter component that interacts with the publisher to indicate that application is going to publish data.
  - Subscriber component responsible for the reception of published data and its delivery to the application.
  - DataReader component is employed to receive data.
  - The topic is the fifth component, which is identified by a data type and a name.
- *Data-Local Reconstruction Layer (DLRL) Layer*: This layer is an optional layer above DCPS and assists the sharing of distributed data among distributed objects.

Data transmission within the DDS domain is allowed for connected publishing and subscribing applications. The conceptual architecture of the DDS protocol is shown in Figure 5.8.

#### 5.2.3.2 Message Queue Telemetry Transport (MQTT)

MQTT is a publish–subscribe, lightweight, and message queuing protocol, which was standardized by OASIS for resource-constrained devices having low bandwidth links [30]. The architecture of the MQTT protocol (shown in Figure 5.9) is based on the client-server model where the client acts as a publisher or subscriber and the server acts as a broker [31]:

- The publishers as a generator of data publish messages on a topic.

- The subscribers are interested client devices, which register for specific topics to get notifications from the broker when publishers publish topics of interest.

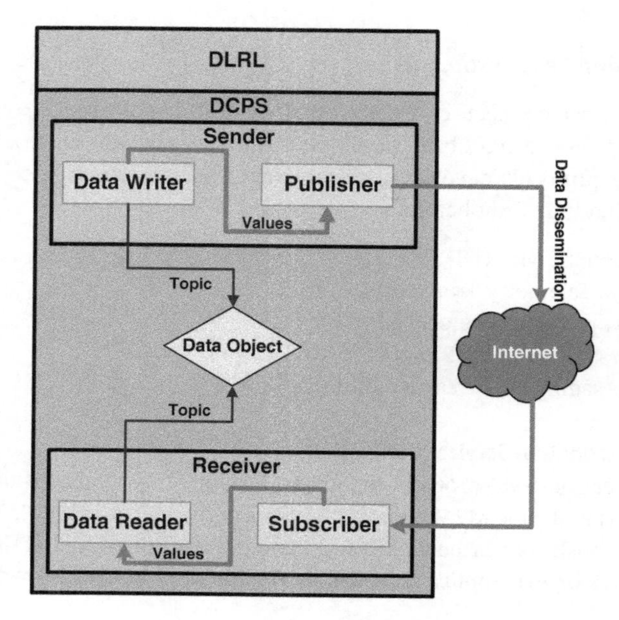


**Figure 5.8** DDS working model.

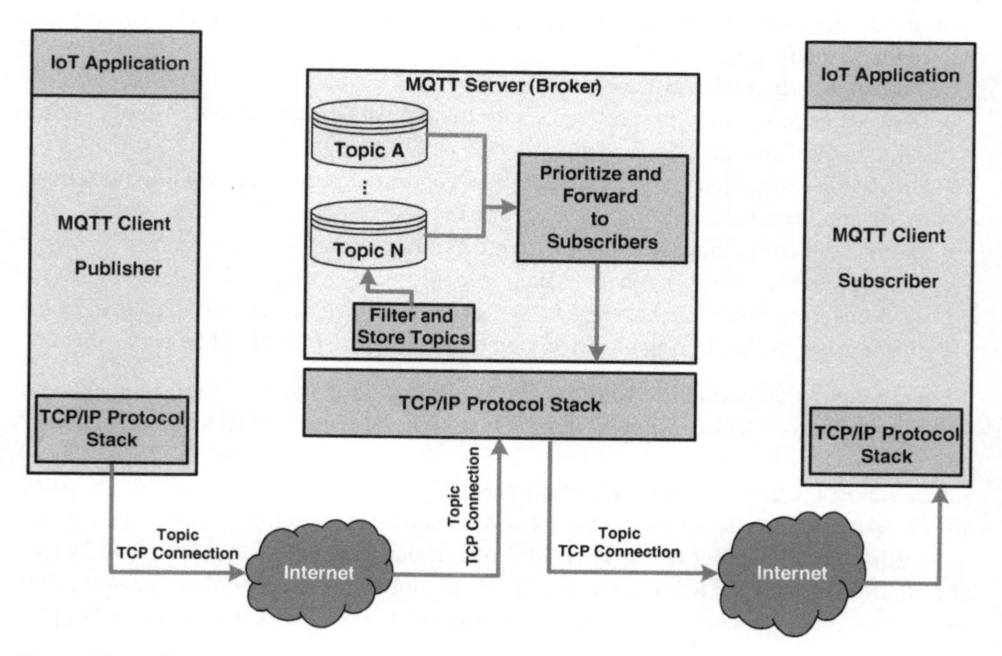


**Figure 5.9** MQTT Architecture.

- The server (acts as a broker) receives published message (topic) from publishers, stores the list of topics, and notifies concerned subscribers. In addition, the broker also performs an authorization check to implement security.

Originally, the MQTT architecture implementation is based on TCP protocol and aims to connect embedded devices and networks (i.e. M2M, WSN, and IoT) where sensor nodes and actuators communicate with each other through MQTT broker. A variant known as MQTT-SN also supports data transmission over UDP and Bluetooth. A number of IoT applications, i.e. smart home, smart healthcare system, and energy meter, utilize MQTT [2, 31]. For example, consider one of the scenarios of a blind control application of smart home:

- Light sensors sense and transfer sensed data to the broker.
- Broker forward sensor data to the application.
- The application sends a blind activation message to the broker.
- The broker transfers activation message to blind actuators.

MQTT message consists of three parts as shown in Figure 5.10:

- 2-byte fixed header (details of header fields are available in Table 5.4)
- Variable-length Option Header
- Variable-length payload

The message format of the MQTT protocol is shown in Figure 5.11.
Message types:

| 0 | Reserved | 4 | PUBACK | 8 | SUBSCRIBE | 12 | PINGREQ |
|---|---|---|---|---|---|---|---|
| 1 | CONNECT | 5 | PUBREC | 9 | SUBACK | 13 | PINGRESP |
| 2 | CONNACK | 6 | PUBREL | 10 | UNSUBSCRIBE | 14 | DISCONNECT |
| 3 | PUBLISH | 7 | PUBCOMP | 11 | UNSUBACK | 15 | Reserved |

**Table 5.4** Details of fixed length packet header field.

| Field name | Description |
|---|---|
| Message Type | 4 bits in length |
| DUP Flag | 1-bit field indicates message already received |
| QoS Level | 2-bit field used to indicate the level of delivery assurance of a PUBLISH message |
| RETAIN | 1-bit field to instruct the broker to retain last received PUBLISH message |
| Remaining Length | 8-bit field indicates the number of remaining bytes in the message including the length of (optional) variable length header and (optional) payload |

| Fixed Length Header (2 bytes), Present in All MQTT Control Packets |
|---|
| Variable Length Header (1–4 bytes), Present in Some MQTT Control Packets |
| Variable Length Packet Payload, Present in some MQTT Control Packets |

**Figure 5.10** Parts of MQTT message.

| Bit | 0 | 1 | 2 | 3 | 4 | 5 | 6 | 7 |
|---|---|---|---|---|---|---|---|---|
| byte 1 | MQTT Control Packet type | | | | DUP | QoS Level | | RETAIN |
| byte 2 | Remaining Length (Length of Optionsand Payload) | | | | | | | |
| byte 3 to byte n | Optional (Variable Length Header) | | | | | | | |
| byte n+1 to byte m | Optional (Variable Length Payload) | | | | | | | |

**Figure 5.11** MQTT Message format.

*CONNECT*: (Client-to-Server) when the client sends connect request to the server
*CONNACK*: (Server-to-Client) for the acknowledgment of connection request
*PUBLISH*: (Bidirectional) to publish a message
*PUBACK*: (Bidirectional) for the acknowledgment of publish request
*PUBREC*: (Bidirectional) indicates that publish request is received
*PUBREL*: (Bidirectional) to release a publish message
*PUBCOMP*: (Bidirectional) to indicate the completion of publish message
*SUBSCRIBE*: (Client-to-Server) when the client sends a subscription request
*SUBACK*: (Server-to-Client) when the server sends an acknowledgment of subscription
*UNSUBSCRIBE*: (Client-to-Server) when the client sends a request to unsubscribe
*UNSUBACK*: (Server-to-Client) when the server sends an acknowledgment for unsubscription
*PINGREQ*: (Client-to-Server) when the client sends a ping request to the server
*PINGRESP*: (Server-to-Client) when the server sends a ping response
*DISCONNECT*: (Client-to-Server) when a client sends disconnection request to the server

**DUP flag**

This 1-bit field indicates that the message with this ID has already been received. This field is set as true in all resend messages.

**MQTT and QoS levels**

Although, the MQTT protocol is based on TCP, which provides guaranteed delivery, data can be lost in case the TCP connection breaks down. To avoid such scenarios, MQTT defines three levels of QoS on top of TCP as shown in Table 5.5.

*QoS Level 1: At-most-once Delivery, no guarantees:* This level does not provide any guarantee. Messages can be dropped, not assured to arrive, and the delivery of messages is based on the delivery guarantees of the underlying network. Publishers send (message to the broker) and delete the message (from its cache) at the same time as shown in Figure 5.12.

**Table 5.5** MQTT QoS Levels.

| QoS value | Bit | Description |
|---|---|---|
| 0 | 00 | At-most-once delivery |
| 1 | 01 | At-least-once delivery |
| 2 | 10 | Exactly once delivery |
| — | 11 | Reserved – Not used |

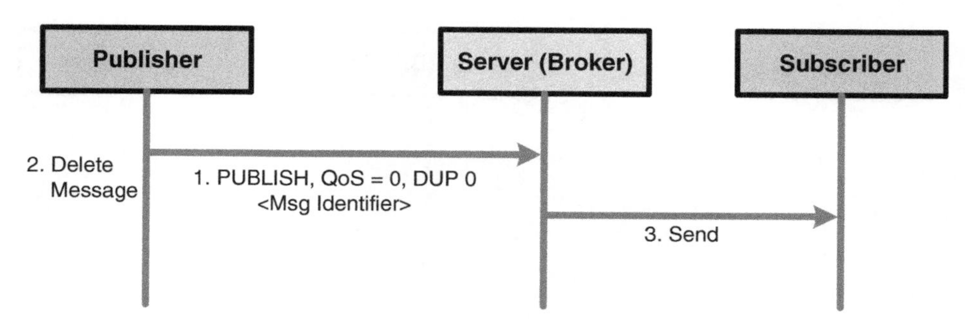


**Figure 5.12** MQTT QoS Level 0.

*QoS Level 2:At-least-once Delivery:* At this QoS level, messages are guaranteed to arrive but duplicate message arrival is possible. Client stores sent message in its cache, and upon reception of PUBACK, it deletes it from cache (as shown in Figure 5.13).

*QoS Level 3: Exactly-once delivery:* This level assures the exactly once delivery of message arrival with no duplication or loss at the cost of additional control packets. Client stores sent message in its cache. The server (broker) stores the message in its memory, sends to the interested subscriber, and sends PUBREC (publication received) message to the client. After a few intervals of time, the client contacts the server about the release of the publication by sending (PUBREL) message. Upon receptions of publication complete (PUBCOMP) message from the server, the client deletes the message from its memory (as shown in Figure 5.14).

**RETAIN**

This is a 1-bit field to instruct the broker to retain last received PUBLISH message and deliver it as a first message to new subscribers. RETAIN field is important as subscribers at the time of subscription are interested only to know the previously stored value about a topic. For example, subscriber Y will receive only last updated values, which are available on the server (as shown in Figure 5.15).

**Remaining length field**

This field provides information about the number of bytes remaining in the current packet (including the length of the variable header field plus the length of the actual

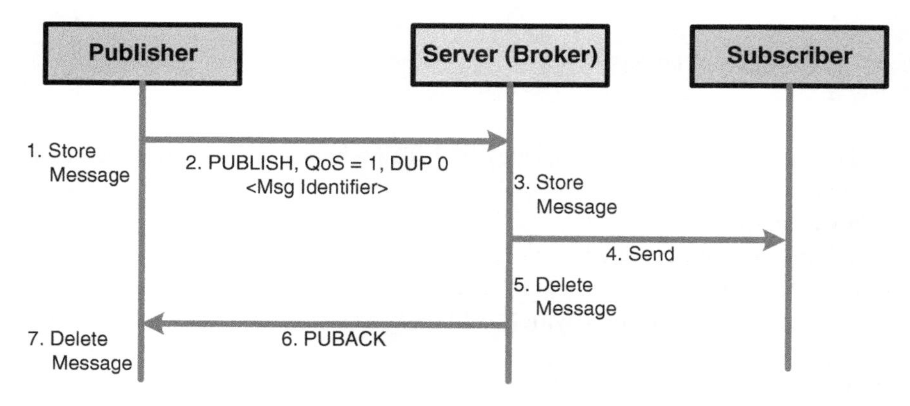


**Figure 5.13** MQTT QoS Level 2.

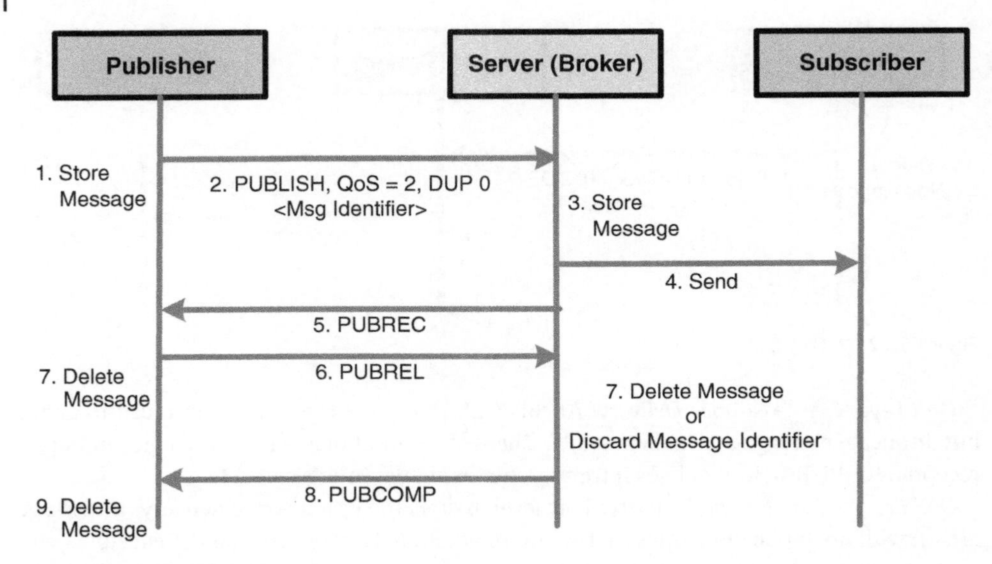


**Figure 5.14** MQTT QoS Level 1.

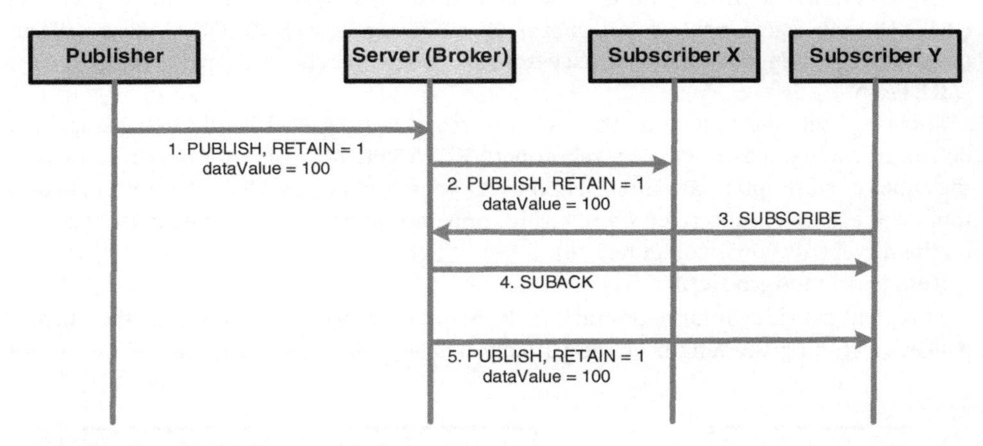


**Figure 5.15** MQTT working with RETAIN = 1.

payload). This field does not include bytes, which have been used to encode the Remaining Length field.

**Variable header**

The variable-length header contains two fields, Topic Name and Packet Identifier. The topic name provides information about the channel to which payload data is published. The Packet Identifier field is only used with PUBLISH Packets considering the QoS level is 1 or 2.

**Payload**

Few MQTT control packets contain payload as shown in Table 5.6.

**Table 5.6** Control packets that contain a payload.

| Control packet | Payload |
|---|---|
| CONNECT | Required |
| CONNACK | None |
| PUBLISH | Optional |
| PUBACK | None |
| PUBREC | None |
| PUBREL | None |
| PUBCOMP | None |
| SUBSCRIBE | Required |
| SUBACK | Required |
| UNSUBSCRIBE | Required |
| UNSUBACK | None |
| PINGREQ | None |
| PINGRESP | None |
| DISCONNECT | None |

#### 5.2.3.3 Constrained Application Protocol (CoAP)

The Constrained RESTful Environments (CoRE) working group of IETF has developed a lightweight application-layer REST-based CoAP over Hypertext Transfer Protocol (HTTP) functionalities [32, 33]. The fundamental goal of CoAP development was to enable web-based services in constrained wireless networks, for example, networks with 8-bit micro-controllers, limited memory, and low power. The design of CoAP addresses the limitations of HTTP communication and REST architecture for constrained devices and constrained networks. However, unlike HTTP and REST, it works on the top of UDP and has the following unique features [34]:

- Uniform Resource Identifier (URI) and content-type support
- Low header overhead
- Low parsing complexity
- Asynchronous message exchange
- Optional reliability over UDP to support unicast and multicast requirements
- Security bindings to Datagram Transport Layer Security (DTLS) protocol

The following are the fundamental Design Requirements (DR) of CoAP:

*DR1*: CoAP must be appropriate for resource-constrained nodes (having low code size and memory).

*DR2*: CoAP protocol must be appropriate for the network with constrained resources. Therefore, protocol overhead is required to be optimized for constrained networks and may possess a high degree of packet loss and low throughput.

*DR3*: CoAP must be able to deal with the power management of resource-constrained nodes with periodic sleeping and waking up of radio transmitters.
*DR4*: Caching is also required to be supported for recent resource requests.
*DR5*: CoAP must support the manipulation of resources through simple push, pull, and notify approach.
*DR6*: CoAP must allow devices to publish values for other devices, which have subscribed for those services.
*DR7*: CoAP must support mapping to standard request, response, and error codes of REST API.
*DR8*: Resource discovery must be based on URI, which may include device name and list of its resources.
*DR9*: CoAP must support non-reliable IP multicast for simultaneous manipulation of resources on different devices.
*DR10*: CoAP is required to be implemented on UDP. (However, optional functionality, i.e. large data chunks delivery, is required to be implemented on TCP.)
*DR11*: CoAP must support reliable transmission of unicast messages over UDP.
*DR12*: CoAP must minimize the latency for home area networks.
*DR13*: CoAP must support a subset of Internet media types.
*DR14*: CoAP must consider operational aspects of the protocol and able to inform that device is powered off or not.

The functionality of CoAP is shown in Figure 5.16. It is shown that resources are located on the server and can be accessed via URI. Low-power client devices (CD) in the IoT system uses CoAP to request resources from the HTTP server either directly or through using proxy servers. Proxy servers perform conversion between CoAP protocol and REST architecture.

**CoAP Layers, Message Models, and Message Format**

CoAP consists of two sublayers [32], i.e.:

*Message Sublayer*: Responsible for the detection of duplicate packet transmission and provides reliable communication over UDP protocol using exponential backoff.
*Request–Response Sublayer*: Responsible to handle REST communication.

To support both reliable and non-reliable services, the CoAP protocol utilizes four types of messages (i.e. Confirmable [CON], Non-confirmable [NON], Acknowledgement [ACK], and Reset [RST]) with different modes of responses.

*Mode 1*: CoAP client sends a CON request to the server and waits for the corresponding ACK message as shown in Figure 5.17. This mode is used to support reliable data transmission. The client retransmits data upon the elapses of ACK waiting time.
*Mode 2*: CoAP client sends a message of NON type to the server and does not require corresponding ACK message as shown in Figure 5.18. This mode is used to support unreliable data transmission as data can be lost or received out of order.
*Mode 3*: The client sends CON type request and receives piggybacked ACK, i.e. ACK with response data as shown in Figure 5.19. Clients retransmit if the piggybacked response is lost.
*Mode 4*: The client sends a request to the server and the server responds empty ACK as it is not able to respond immediately with requested data as shown in Figure 5.20. In this way, the client does not retransmit and wait for a response. When the requested response is ready, then the server sends data in the CON message, which is required to be acknowledged by the client.

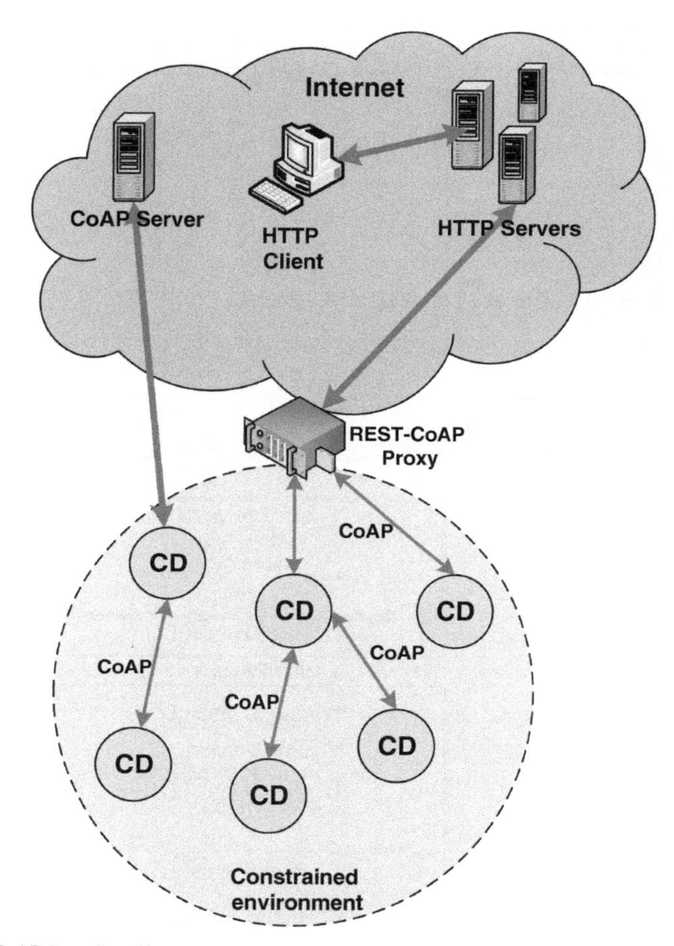


**Figure 5.16** CoAP functionality.

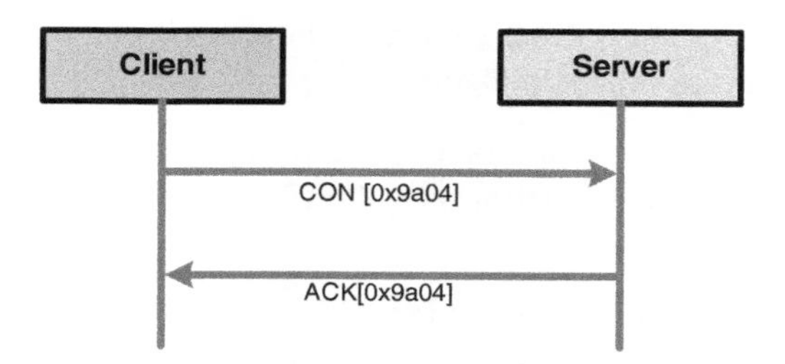


**Figure 5.17** Reliable message transmission.

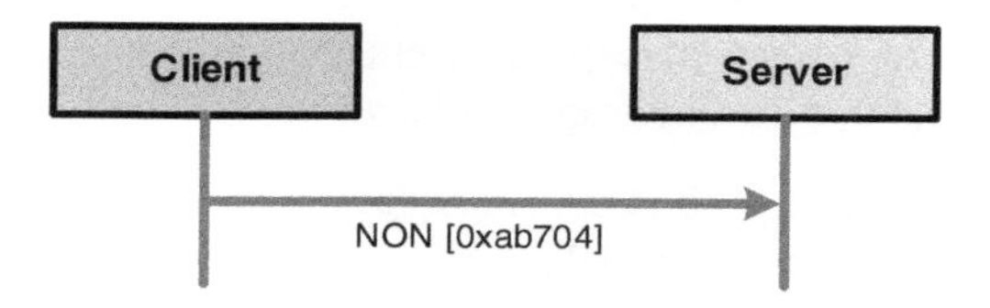


**Figure 5.18** Unreliable message transmission.

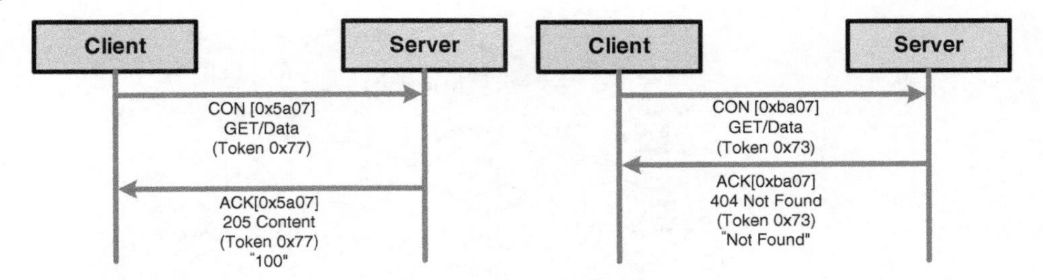


**Figure 5.19** Requests and piggybacked responses.

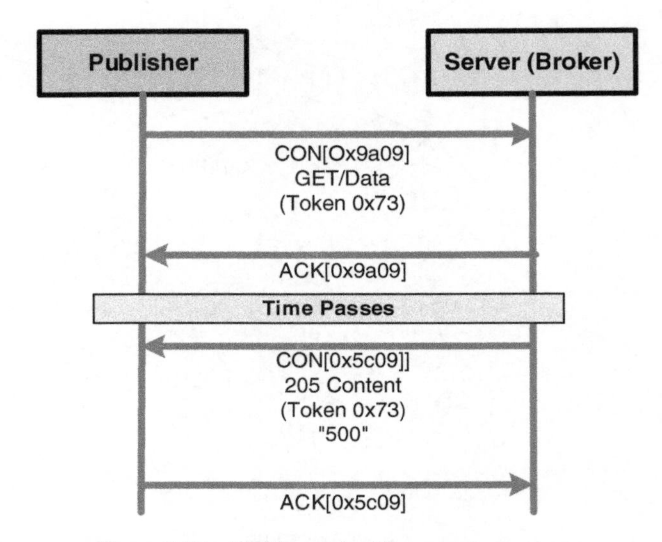


**Figure 5.20** GET Request with separate response.

*Mode 5*: RST (instead of ACK) message is sent to the client if the server is unable to handle the incoming request as shown in Figure 5.21.

**CoAP message format** CoAP message communication is based on compact messages, which utilize the data section of UDP packets. CoAP message consists of a four-byte header with five fields, i.e.:

Version (2 bit) field to indicate CoAP version
Type (2 bit) field to indicate message type, for example, 0 for CON, 1 for NON, 2 for ACK, and 3 for RST.
Token Length (4 bit) field to indicate token field length
Code (8 bit) field indicate request or response method
Message ID (16 bit) field to detect message duplication

Following the message header, the CoAP message may consist of the Token field, Option field, and Payload as shown in Figure 5.22.

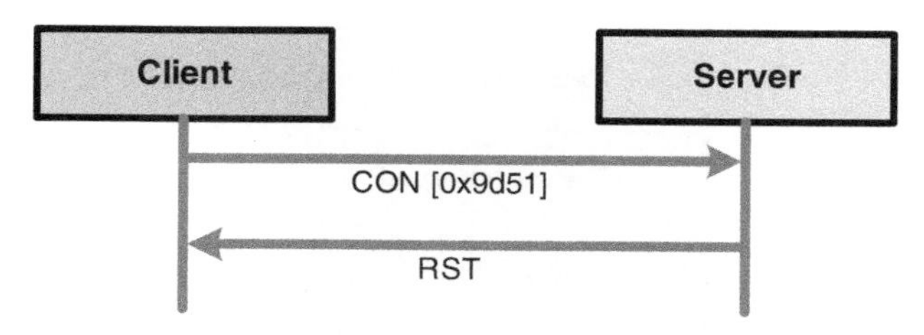


**Figure 5.21** RST response from server.

| Version (2-bit) | Type (2-bit) | Token length (4-bit) | Token length (8-bit) | Token length (16-bit) |
|---|---|---|---|---|
| Token (0–8 byte) | | | | |
| Options (If Any) | | | | |
| Payload Marker (1 1 1 1 1 1 1 1) | | | Payload (If Any) | |

**Figure 5.22** CoAP message format.

Concerning communication over resource-constrained IoT devices, the advantages of CoAP can be summarized as follows.

- CoAP is lightweight compared to HTTP REST and enables resource-constrained sensor nodes to utilize a minimal subset of REST-based requests using GET, POST, PUT, and DELETE methods.
- Works over UDP at the transport layer.
- Exploit data section of UDP protocol.
- Exponential backoff provides reliability over UDP.
- Compact header reduces communication overhead and reduces power consumption.
- Support of asynchronous data push is used to transmit data only if a change is observed, and it allows resource-constrained IoT devices to conserve power with more sleeping time.

To implement security, CoAP relies on DTLS on the top of UDP as shown in Figure 5.23. DTLS provides integrity, authentication, and confidentiality. Considering DTLS, four types of security modes have been supported by the CoAP:

- *No Security Mode*: DTLS is disabled.
- *PreSharedKey Mode*: DTLS is enabled, and the connection is encrypted by a pair of pre-shared key and support cipher suite TLS-PSK-WITH-AES-128-CCM-8.
- *RawPublicKey Mode*: DTLS is enabled, and the connection is encrypted by a pair of pre-shared key and support cipher suite TLS-ECDHE-ECDSA-WITH-AES-128-CCM-8.
- *Certificate Mode*: DTLS is enabled, and the connection is encrypted by a pair of asymmetric keys with an X.509 Certificate.

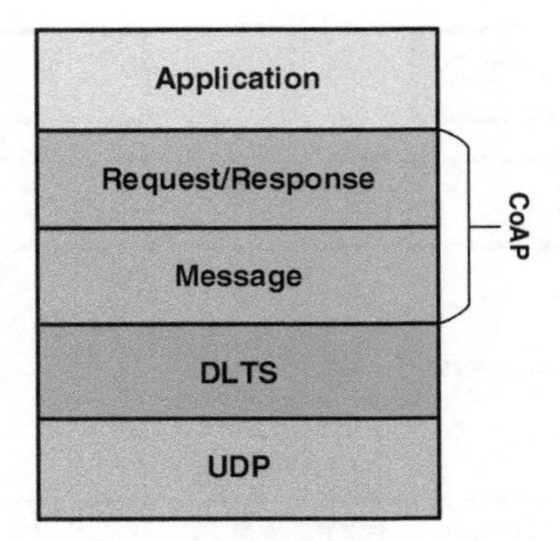


Figure 5.23 CoAP layers over DLTS.

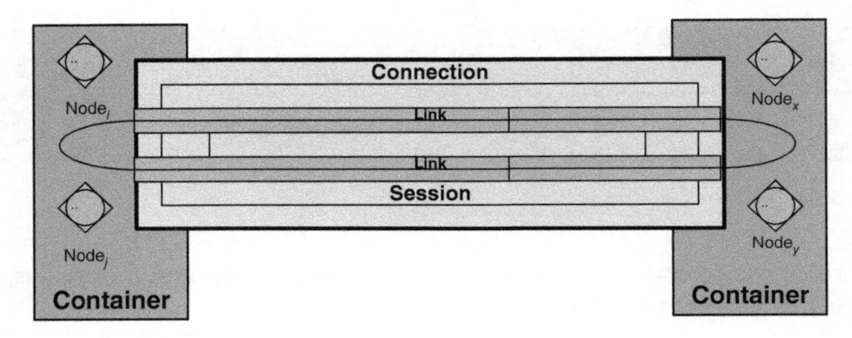


Figure 5.24 AMQP communication between nodes.

#### 5.2.3.4 Advanced Message Queuing Protocol (AMQP)

AMQP [35, 36] is an application layer protocol, which:

- Is open standard
- Is a message-oriented protocol
- Provides reliable communication
- Provides secure communication
- Supports at-most-once, at-least-once, and exactly once delivery
- Is suitable for both point-to-point and publish–subscribe models

There are three participants and one element of AMQP protocol:

- The Producer to create messages.
- The Consumer to access/edit a message after taking it from message queues.
- The Broker to distribute messages to appropriate queues by following certain defined rules.
- The Message is the central element of AMQP communication.

Nodes and Containers are the two main entities that are part of an AMQP network.

Nodes are the named (addressable) entities, which are organized in a flat/hierarchical way and responsible to store and deliver messages. In the messaging space, a producer/

consumer/queue can be considered as nodes. The container is the application that contains nodes. At the architectural level, there are two layers of AMQP, i.e. Transport layer and Function layer.

*Transport Layer*: At this layer, communication is frame-oriented and AMQP has implemented its own link protocol to define the connection behavior and security layer between peers.

*Function Layer*: Provides messaging capabilities at the application level.

**Transport layer**

AMQP communication at this layer is known as the session, and each session is further attached to several links, which are responsible for the actual exchange of frames between nodes as shown in Figure 5.24.

AMQP frames consist of Frame Header, Extended Header, and Frame Body as shown in Figure 5.25 [37].Frame Header (of size 8 bytes) has three fields, i.e.:

*Size*: Contains a value representing the frame size

*DOFF*: Data offset value that indicates the position of frame body within the frame

*TYPE*: Value here indicates frame type, i.e.:

- *OPEN*: To open connection
- *BEGIN*: To begin channel session
- *ATTACH*: To attach the link to a session
- *FLOW*: To update link state
- *TRANSFER*: To send message
- *DISPOSITION*: To inform remote peers about the delivery state change
- *DETACH*: To detach link from session
- *END*: To end the session
- *CLOSE*: To close connection

*Extended Header* is of variable length and represents an extension point defined for future expansion.

*Frame Body* is of variable length and contains the actual payload.

In summary, the Transport layer is responsible to perform the following roles:

- Channel multiplexing
- Encoding
- Framing

| Byte | 0 | 1 | 2 | 3 | |
|---|---|---|---|---|---|
| Byte 0 | Size | | | | Frame Header |
| Byte 4 | DOFF | Type | Type-specific | | |
| Byte 8 | Type-specific | | | | Extended Frame |
| Byte DOFF*4 | Type-specific | | | | Frame Body |

**Figure 5.25** AMQP frame format.

- Data representation
- Error handling

**Functional layer**

The Function layer provides message capabilities. At functional-level, it is shown that the broker (as shown in Figure 5.26) in AMQP has three main components, i.e. Exchange, Queue, and Binding. The Exchange component receives messages from the publishers, stores these messages in the appropriate Queue, and forwards to intended subscribers. Binding designates the connectivity of Exchange to the appropriate Queue. Routing between Exchange and Queue components is based on predefined rules.

AMQP defines two types of messages, i.e. Bare messages (with payload sent by sender) and Annotated messages (received at the receiver) as shown in Figure 5.27. The Bare message consists of three sections, i.e. system properties, application properties, and body. The Annotated message includes three sections, i.e. head annotations (optional), bare message (mandatory), and footer (annotations).

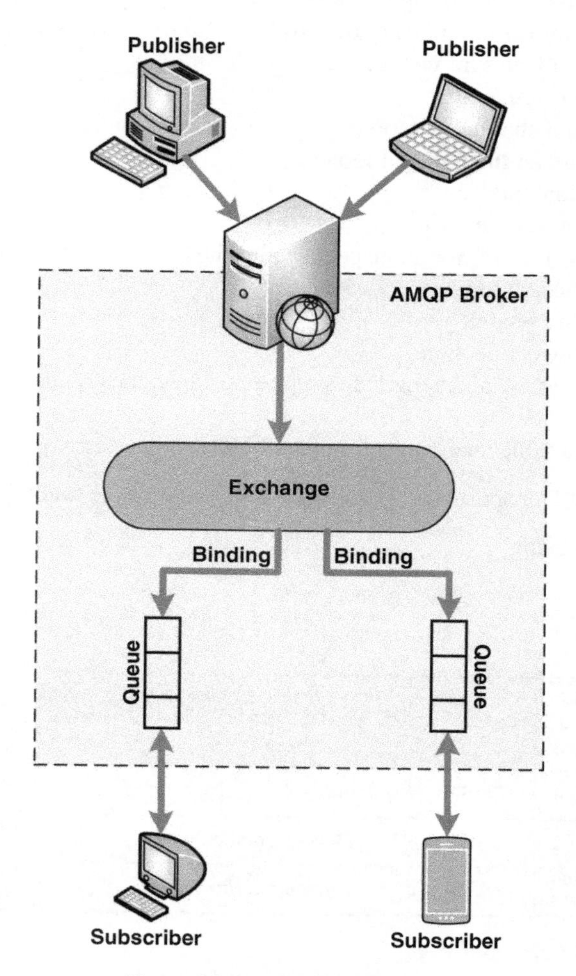


**Figure 5.26** AMQP architecture.

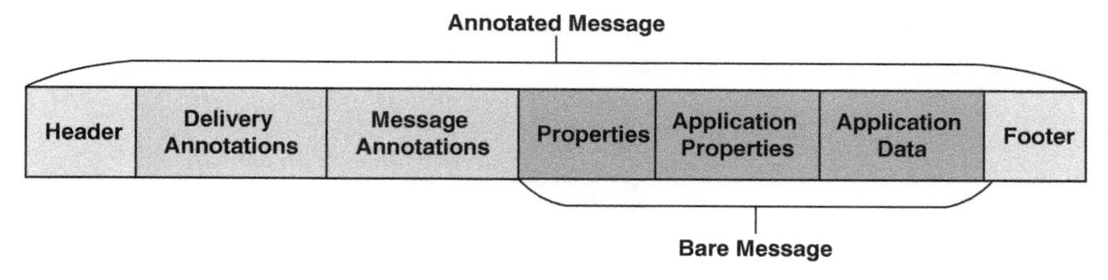


**Figure 5.27** AMQP message format.

*Header*: Contains information about priority, time to live (TTL), and message delivery count
*Delivery Annotation:* Contains message delivery information (that is not included in Properties section)
*Message Annotation*: Contains message information (that is not included in Properties section)
*Properties*: Contains information about message ID, Source, Target, Subject, content type, message creation time, and message expiry time
*Application Properties*: Contains structured application data, which is used by intermediaries for filtering and routing
*Footer*: Contains encryption details (message hashes and signatures)

#### 5.2.3.5 eXtensible Messaging and Presence Protocol (XMPP)

eXtensible Messaging and Presence Protocol (XMPP) [38, 39] is XML-based, open, secure, spam-free, and application layer protocol, which:

- Allows users to communicate with each other through platform-independent instant messaging (i.e. multiparty chatting, voice, and video calling)
- Enables instant messaging applications to accomplish authentication, encryption, access control, and privacy
- Permits file transfer, gaming, and social network services in IoT system

The generic XMPP framework has been shown in Figure 5.28 where the XMPP gateway backbone can bridge multiple foreign messaging networks and support Internet-based instant messaging.

XMPP clients use stream of XML stanzas to connect to XMPP-enabled server. The XML stanza [40] (as shown in Figure 5.29) consists of three components (i.e. `<presence/>`, `<message/>`, and `<iq/>`), where:

- `<message/>` stanza is used to identify the source, destination, types, and IDs of XMPP entities (that ultimately utilize PUSH method to retrieve data). An example of `<message/>` stanza is shown in the following.
- `<presence/>` stanza shows and notify authorized customers Presence Information status. An example of `<presence/>` stanza is shown as follows.

```
<presence xml:lang='en'>
<show>WS</show>
<status>William Stallings</status>
</presence>
```

- `<iq/>` (stands for info/query) is used to establish sessions and pairs message senders and receivers in a roster (contact list). Examples of `<iq/>` stanza are shown in the following.

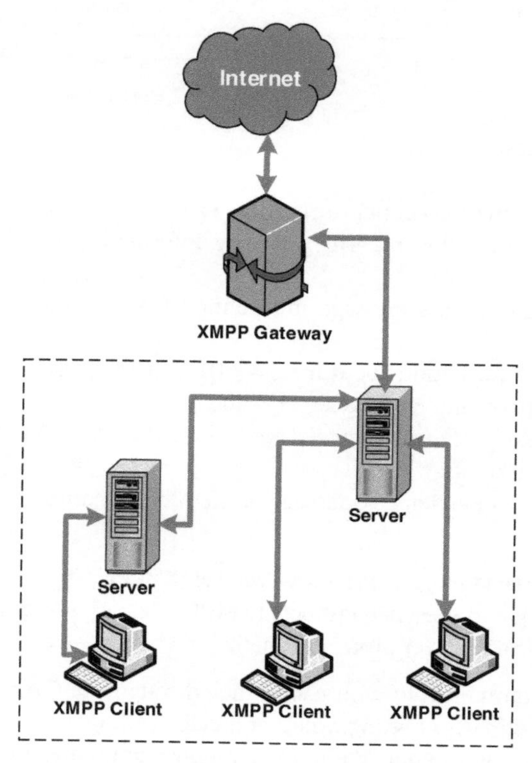


**Figure 5.28** Generic framework of XMPP.

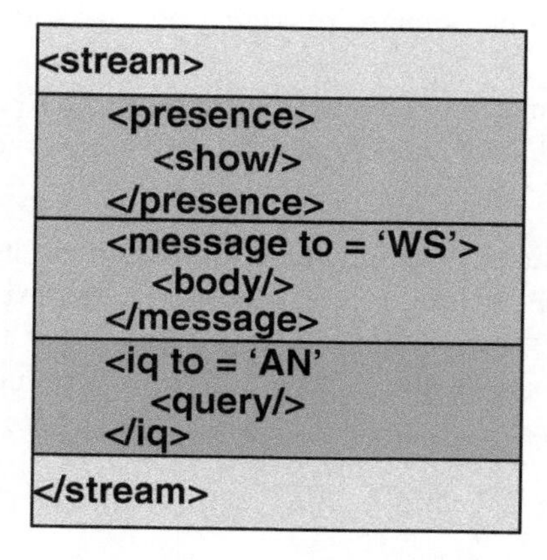


**Figure 5.29** Generic structure of XMPP Stanza.

**Session Establishment Example Using <iq> Stanza**

For session establishment, client requests session from server using <iq/> stanza.

```
<iq to='http://example.com'
               type='set'
               id='sess_1'>
<session xmlns='urn:ietf:params:xml:ns:xmpp-session'/>
</iq>
        Server informs client that session has been created
using <iq/> stanza.
<iq from='http://example.com'
            type='result'
            id='sess_1'/>
```

**Roster Request Example Using <iq> Stanza**

Client requests current roster from server using following <iq> stanza.

```
<iq from='john@example.com/balcony' type='get' id='roster_1'>
<query xmlns='stell:iq:roster'/>
</iq>
Client receives roster result from server using following <iq> stanza
<iq to='john@example.com/balcony' type='result' id='roster_1'>
<query xmlns='stell:iq:roster'>
<item jid='peter@example.net'
          name='Peter'
          subscription='both'>
<group>Friends</group>
</item>
<item jid='xiaou@example.org'
          name='Xiaou'
          subscription='from'>
<group>Friends</group>
</item>
<item jid='rockey@example.org'
          name='Rock'
          subscription='both'>
<group>Friends</group>
</item>
</query>
</iq>
```

The comparison of the aforementioned IoT application layer protocols is given in Table 5.7.

**Table 5.7** Comparison of IoT application layer protocols.

| Protocol name | Request/response | Publish/subscribe | Broker-based | Transport protocol | Security protocol | Header size (bytes) |
|---|---|---|---|---|---|---|
| DDS | No | Yes | No | TCP/UDP | SSL/DTLS | Varies |
| MQTT | No | Yes | Yes | TCP | SSL | 2 |
| CoAP | Yes | Yes | Yes | UDP | DTLS | 4 |
| AMQP | No | Yes | Yes | TCP | SSL | 8 |
| XMPP | Yes | Yes | Yes | TCP | SSL | Varies |

## Review Questions

**5.1** Explain the working of MQTT.

**5.2** What are the requirements of the CoAP Protocol?

**5.3** Explain the difference between IEEE 802.15.4 and 802.11ah standards.

**5.4** Kiwi fruit is one of the most sensitive fruits in terms of quality, which is given by size, sweetness, and dry matter. To reach the best quality, farmers need to develop a good irrigation strategy to attain marketable products and to reduce product losses. To develop accurate irrigation strategies for farmers, the company has installed wireless sensors to monitor soil water status to plan irrigation in a kiwi orchard. These sensors transmit soil water level information to the Cloud. Explain how 6LoWPAN provides support to large network topology to transmit data to Cloud (Figure 5.30).

**5.5** To help milk producers, different companies developed dairy production management services known as Dairy Production Analytics (DPA). Besides collecting data about cow activities, DPA collects livestock farm environmental information, e.g. temperature and humidity through sensors deployed at different locations of livestock farms. Different kinds of analysis are performed on data aggregated in Cloud,

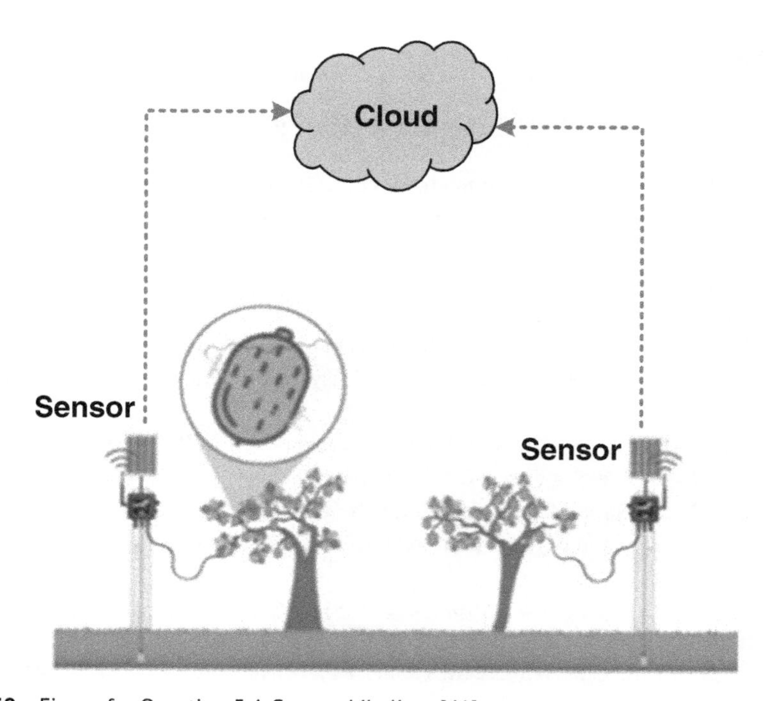


**Figure 5.30** Figure for Question 5.4. Source: Libelium [41].

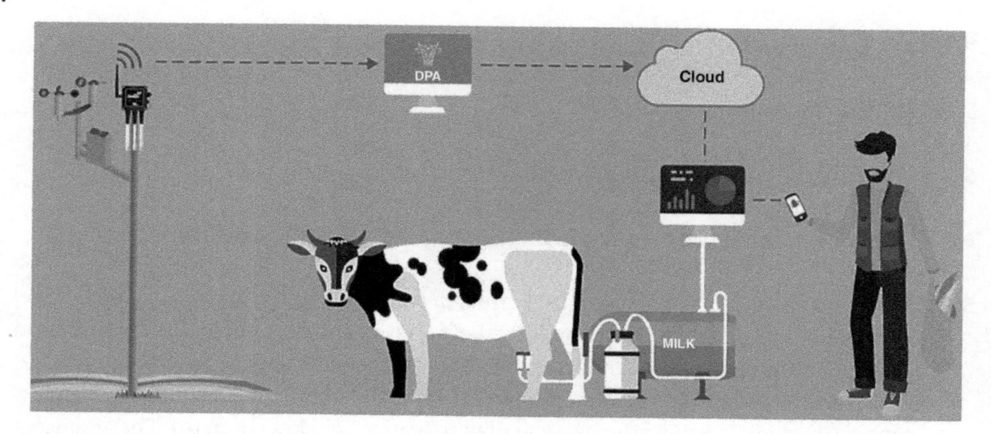


**Figure 5.31** Figure for Question 5.5. Source: Libelium [42].

and results of this analysis (in the form of graphs) are available on mobile and desktop devices of livestock staff (as shown in Figure 5.31), which ultimately help them in decision-making about cattle feed and proper time of milking. For developing such kind of system, which IoT application protocol is suitable and why? Explain the proper working of your suggested protocol in this scenario.

## References


1 Raj, P. and Raman, A.C. (2017). *The Internet of Things: Enabling Technologies, Platforms, and Use Cases*. CRC Press.

2 Al-Fuqaha, A., Guizani, M., Mohammadi, M. et al. (2015). Internet of things: a survey on enabling technologies, protocols, and applications. *IEEE Communications Surveys and Tutorials* 17 (4): 2347–2376.

3 Hersent, O., Boswarthick, D., and Elloumi, O. (2011). *The Internet of Things: Key Applications and Protocols*. Wiley.

4 Kumar, N.M. and Mallick, P.K. (2018). The internet of things: insights into the building blocks, component interactions, and architecture layers. *Procedia Computer Science* 132: 109–117.

5 Karimi, K. and Atkinson, G. (2013). What the internet of things (IoT) needs to become a reality. White Paper, FreeScale and ARM, p. 1–16.

6 Serbanati, A., Medaglia, C.M., and Ceipidor, U.B. (2011). Building blocks of the internet of things: state of the art and beyond. In: *Deploying RFID-Challenges, Solutions, and Open Issues*, 351–366. InTech.

7 Grasso, J. (2004). *The EPCglobal network: overview of design, benefits, & security*. EPCglobal Inc., Position Paper, 24.

8 Chen, C.-L., Lai, Y.-L., Chen, C.-C. et al. (2011). *RFID ownership transfer authorization systems conforming EPC global Class-1 Generation-2 standards. IJ Network Security* 13 (1): 41–48.

**9** Jones, E.C. and Chung, C.A. (2016). *RFID and Auto-ID in Planning and Logistics: A Practical Guide for Military UID Applications*. CRC Press.

**10** *Z-Wave Alliance*. https://z-wavealliance.org/.

**11** *Z-Wave for Smart Homes*. https://www.z-wave.com/.

**12** Sesia, S., Toufik, I., and Baker, M. (2011). *LTE-the UMTS Long Term Evolution: From Theory to Practice*. Wiley.

**13** Hasan, M., Hossain, E., and Niyato, D. (2013). Random access for machine-to-machine communication in LTE-advanced networks: issues and approaches. *IEEE Communications Magazine* 51 (6): 86–93.

**14** Mackensen, E., Lai, M., and Wendt, T.M. (2012). Bluetooth low energy (BLE) based wireless sensors. In: *IEEE Sensors*. IEEE.

**15** Siekkinen, M., Hiienkari, M., Nurminen, J.K. et al. (2012). How low energy is bluetooth low energy? Comparative measurements with zigbee/802.15. 4. In: *IEEE Wireless Communications and Networking Conference Workshops (WCNCW)*, 1481–1486. IEEE.

**16** Howitt, I. and Gutierrez, J.A. (2003). IEEE 802.15. 4 low rate-wireless personal area network coexistence issues. In: *2003 IEEE Wireless Communications and Networking (WCNC)*. IEEE.

**17** Zheng, J. and Lee, M.J. (2006). A comprehensive performance study of IEEE 802.15. 4. *Sensor Network Operations* 4: 218–237.

**18** Gutierrez, J.A., Naeve, M., Callaway, E. et al. (2001). IEEE 802.15. 4: a developing standard for low-power low-cost wireless personal area networks. *IEEE Network* 15 (5): 12–19.

**19** Olyaei, B.B., Pirskanen, J., Raeesi, O. et al. (2013). Performance comparison between slotted IEEE 802.15. 4 and IEEE 802.1 lah in IoT based applications. In: *IEEE 9th International Conference on Wireless and Mobile Computing, Networking and Communications (WiMob)*, 332–337. IEEE.

**20** Ahmed, N., Rahman, H., and Hussain, M.I. (2016). A comparison of 802.11 ah and 802.15. 4 for IoT. *ICT Express* 2 (3): 100–102.

**21** Alliance, Z. (2010). *Zigbee alliance*. WPAN industry group, http://www.zigbee.org. The industry group responsible for the ZigBee standard and certification.

**22** Kinney, P. (2003). Zigbee technology: wireless control that simply works. In: *Communications Design Conference*, vol. 2, 1–7.

**23** Shelby, Z. and Bormann, C. (2011). *6LoWPAN: The Wireless Embedded Internet*, vol. 43. Wiley.

**24** Winter, T. (2012), Thubert P., Brandt A., et al.. RPL: IPv6 Routing Protocol for Low-Power and Lossy Networks. *RFC 6550*, 1–157.

**25** Tripathi, J., de Oliveira, J.C., and Vasseur, J.-P. (2010). A performance evaluation study of rpl: Routing protocol for low power and lossy networks. In: *IEEE 44th Annual Conference on Information Sciences and Systems (CISS)*, 1–6. IEEE.

**26** Cheshire, S. and Krochmal, M. (2013). *Multicast DNS*. RFC 6762, February.

**27** Cheshire, S. and Krochmal, M. (2013). *DNS-based service discovery*. RFC 6763, February.

**28** Pardo-Castellote, G., Farabaugh, B., and Warren, R. (2005). *An Introduction to DDS and Data-Centric Communications*. Real-Time Innovations.

**29** Schlesselman, J.M., Pardo-Castellote, G., and Farabaugh, B. (2004). OMG data-distribution service (DDS): architectural update. In: *IEEE MILCOM 2004. Military Communications Conference*, vol. 2, 961–967. IEEE.

**30** MQTT (2018). MQTT – Message Queuing Telemetry Transport Protocol. http://docs.oasis-open.org/mqtt/mqtt/v3.1.1/cos02/mqtt-v3.1.1-cos02.html.

**31** Yassein, M.B., Shatnawi, M.Q., Aljwarneh, S. et al. (2017). Internet of Things: survey and open issues of MQTT protocol. In: *IEEE International Conference on Engineering & MIS (ICEMIS)*, 1–6. IEEE.

**32** Constrained Application Protocol (CoAP). https://tools.ietf.org/html/rfc7252.

**33** Shelby, Z., K. Hartke, and C. Bormann (2014), The constrained application protocol (CoAP). RFC7252. Available online: http://www.rfc-editor.org/info/rfc7252.

**34** Bormann, C., Castellani, A.P., and Shelby, Z. (2012). *Coap: an application protocol for billions of tiny internet nodes. IEEE Internet Computing* 16 (2): 62–67.

**35** Vinoski, S. (2006). Advanced message queuing protocol. *IEEE Internet Computing* 10 (6): 87–89.

**36** Cui, P. (2017). *Comparison of IoT Application Layer Protocols*, MS Thesis.

**37** Advanced Message Queuing Protocol (AMQP). http://docs.oasis-open.org/amqp/core/v1.0/os/amqp-core-transport-v1.0-os.html.

**38** Extensible Messaging and Presence Protocol (XMPP). https://xmpp.org/rfcs/rfc3921.html.

**39** Extensible Messaging and Presence Protocol (XMPP). https://tools.ietf.org/html/rfc6120.

**40** Extensible Messaging and Presence Protocol (XMPP) - Stanza. https://xmpp.org/rfcs/rfc3921.html#stanzas.

**41** Libelium (2017). Smart irrigation system to improve kiwi production in Italy.

**42** Libelium (2019). How a dairy farm increased their milk production 18% with IoT and Machine Learning.

# 6

# IoT Cloud and Fog Computing

**LEARNING OBJECTIVES**

After studying this chapter, students will be able to:

- describe the components of IoT Cloud architecture.
- define the usage of application domains of IoT Cloud platforms.
- explain the layered architecture of Fog computing.
- distinguish Fog computing from other related terms (edge and mobile edge computing).
- discuss Fog deployment and Fog service models.
- employ Fog computing solution in different IoT systems.

## 6.1 IoT Cloud

IoT applications offer services, which cover all aspects of human life, i.e. home and building automation, smart cities, city waste management, smart grid, smart health, intelligent traffic management, health monitoring, emergency and surveillance services, supply chain, retail, smart industry, etc. According to one of the CISCO reports, it is estimated that by 2030, 500 billion smart devices will be interconnected on planet Earth [1, 2]. A large proportion of these connected smart things (having multiple embedded sensors and actuators) will be part of IoT systems. These smart things with equipped sensors generate a huge amount of data aka BigData, which is not possible to store on locally available hardware resources. Furthermore, commonly used data management tools are unable to manage and process this immense amount of data. Therefore, smart things in IoT systems ultimately demand smart and efficient mechanisms for the collection, storage, and processing of BigData. However, this massive amount of IoT data is valuable if analysis can be performed on it. The emerging paradigm of Cloud computing enables IoT systems to use third-party software and hardware components for the efficient and reliable management of BigData, which ultimately pave the way for BigData analytics to extract valuable knowledge from stored data [3]. Nevertheless, it is not efficient to store all kinds of data on Cloud. For example, in the case of latency-critical applications, it is impractical to send data to the faraway Cloud for processing. For such types of applications, Fog computing has been designed as

*Enabling the Internet of Things: Fundamentals, Design, and Applications*, First Edition.
Muhammad Azhar Iqbal, Sajjad Hussain, Huanlai Xing, and Muhammad Ali Imran.

an extension of Cloud computing for the provisioning of additional storage and compute resources close to end-user devices to perform parallel IoT data analytics tasks. Figure 6.1 shows a typical integration of IoT applications with Fog and Cloud computing where:

- Cloud computing offers BigData analytics over the Internet with the provisioning of flexible and scalable hardware and software resources.
- Fog computing offers data analytics on edge devices to facilitate real-time processing.

This chapter provides details related to Cloud and Fog computing with reference to their implications in IoT systems.

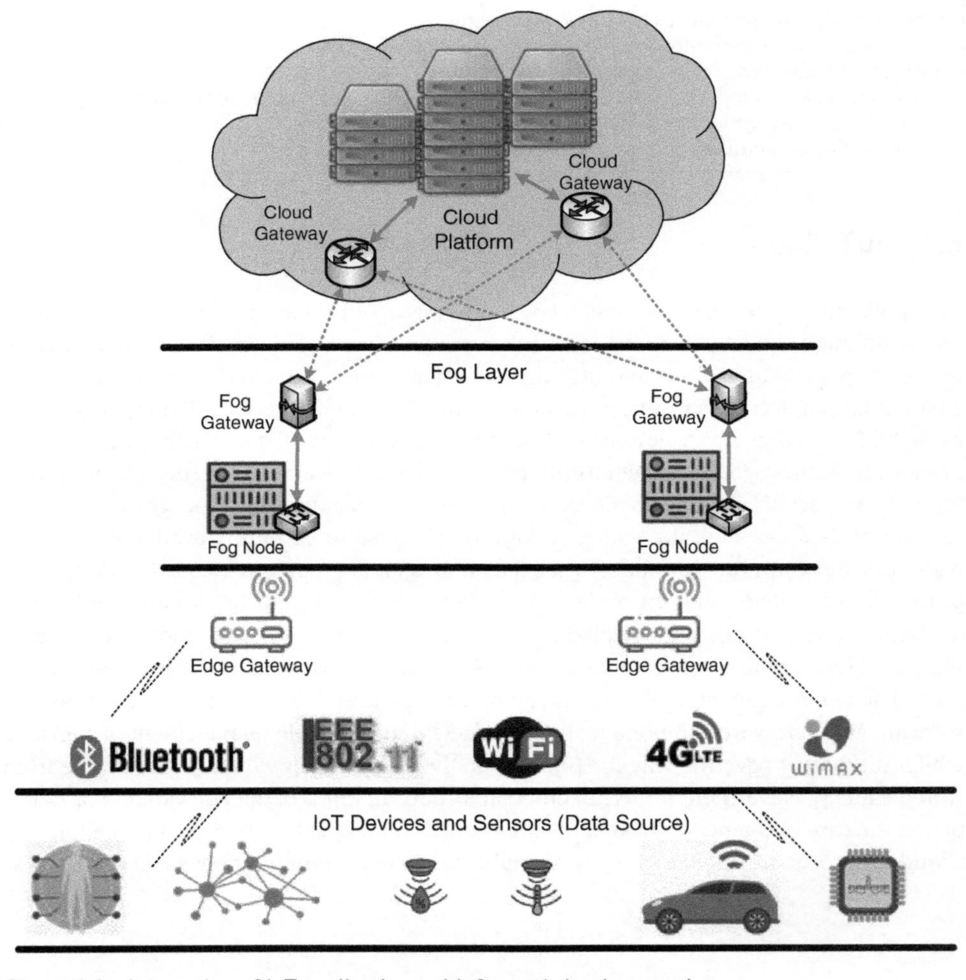


**Figure 6.1** Integration of IoT applications with fog and cloud computing.

### 6.1.1 Cloud Computing for IoT

Cloud computing and IoT have experienced an independent evolution [4]. IoT systems produce a massive amount of unstructured data that possesses all three characteristics, i.e. volume, velocity, and variety. Cloud, on the other hand, provides huge computing and storage capacity to deal with the IoT data and acts as an intermediate layer between IoT smart things and IoT applications. However, the complementary characteristics (shown in Table 6.1 [4, 5]) of both (IoT and Cloud) encourage their integration to realize the management of data produced by smart things in IoT systems.

Cloud computing provides efficient management schemes for BigData to extract valuable knowledge from it. However, the integration and utilization of Cloud in IoT systems are difficult because of the following challenges:

- *Standardization*: Standardizing of Clouds for a diverse range of IoT cloud-based services is challenging, which demands the involvement of various Cloud vendors.
- *Balancing*: IoT requirements are mostly different from what generally Cloud environments provide, and therefore balance is required to be achieved between these.
- *Management*: IoT systems and Cloud systems are different in resources and components, and therefore the parallel management of both is much challenging.
- *Security*: The security of IoT cloud-based services is much more challenging due to the differences in security mechanisms of IoT devices and Cloud platforms.

### 6.1.2 IoT Cloud Architecture

The IoT Cloud is aimed to be connected with millions of smart things or end-user devices, and therefore this massive-scale implementation involves the usage of appropriate computing technologies. At the architecture level, IoT Cloud is a network of high-performance virtual servers (i.e. application server, database server, and load balancers), which is established over a pool of virtual resources. These Virtual Machine (VM)-based configured servers work independently even running on the same physical machine in the available

**Table 6.1** Complementary characteristics of IoT and cloud.

| Characteristic | IoT | Cloud computing |
|---|---|---|
| Components | Things in Real world | Virtual resources of the digital world |
| Component Prevalence | Pervasive (things in real world available everywhere) | Ubiquitous (computing resources are usable from everywhere) |
| Computation Capability | Limited | Unlimited |
| Storage Capacity | Limited | Unlimited |
| BigData Relation | Source of BigData | Means to manage BigData |
| Role of the Internet | Point of convergence | Used to deliver services |

Sources: Pourqasem et al. [5]; Liu et al. [4].

Virtual Resource Pool as shown in Figure 6.2 [6]. Moreover, these virtual servers are able to retrieve and handle user requests and store, process, and manage massive amounts of data for analysis.

The implementation of IoT Cloud is shown in Figure 6.3, and the relevant details of IoT Cloud components, i.e. Virtual Resource Pool and VM-based configured servers, are given in the following.

#### 6.1.2.1 Virtual Resource Pool

A virtualization platform or hypervisor is required to establish a Virtual Resource Pool consisting of a number of VMs to directly handle CPU and memory resources of physical machines. Any kind of server (i.e. application server, database server, load balancer) can be implemented as a VM server. The Virtual Resource Pool consists of two components, i.e. [6]:

- Hardware resources are available in the form of several CPU, memory, and network connectivity on physical machines.
- Hypervisor software running on these physical machines, which provide operating system (OS) environment (known as VMs) and enable dynamic resource allocation. Regarding the implementation of the hypervisor, virtual OS, i.e. VMware vSphere, is preferred, which can access direct computing and memory resources of physical machines. IoT Cloud services are implemented on VMs to provide high performance at low cost.

#### 6.1.2.2 Application Server

Application Servers in IoT Cloud comprises of both Hypertext Transfer Protocol (HTTP) servers and Message Queue Telemetry Transport (MQTT) servers, which can be developed using different technologies (i.e. Node.js), which are able to provide high concurrency. Moreover, these technologies can perform I/O operations with a single asynchronous thread.

HTTP servers interact in a request–response manner and require a flexible web application framework (i.e. Express) to deploy web and mobile applications. Using HTTP, clients can make three types of requests, i.e.:

*GET*: To obtain server resources.

*POST*: To send information to the server.

*DELETE*: To delete certain resources on the server.

HTTP Servers receive client requests and send the response back to the clients.

However, HTTP is not suitable for resource-constrained IoT devices. On the other hand, for business services to the customers, application servers based on MQTT protocol [7, 8] are preferred in contrast to conventional application servers, which are based on HTTP protocol. MQTT is a lightweight publishing–subscription-based message transportation protocol and is specifically designed for IoT devices, which are constrained by computing, communication, and energy resources [9]. MQTT servers can be developed using MQTT-Connection (an open-source Node.js library). MQTT servers maintain long-lived Transmission Control Protocol (TCP) connections to improve real-time performance and use different levels of QoS.

The clustering of multiple application servers allows scaling of server programs across distributed parallel processors. For the clustering of HTTP servers, Parallel Multithreaded machines can be used. However, for a cluster of MQTT servers, an MQTT server operates as a master node to initiate other servers, which are considered as the slave nodes, and all run on the individual CPU cores.

#### 6.1.2.3 Database Servers

Depending on the nature of IoT application, the data is stored in relational (Structured Query Language [SQL]) or non-relational (NoSQL) databases. However, due to the non-suitability SQL databases for real-life IoT applications, NoSQL databases are highly recommended in IoT Cloud, which ultimately provide efficient real-time services for data storage.

Regarding implementation, Redis (a NoSQL database) is used in IoT Cloud, which in turn has several associated advantages, for example,

- Redis enables direct (key-value) data storage in a memory, which improves I/O speed.
- More than one Redis node can be configured to establish a Redis cluster. In Redis cluster, all nodes are connected to each other through TCP connections, which ultimately enhances the availability of database servers.
- The data retrieved in the Redis database is first hashed and then stored on the corresponding Redis node. Uniform distribution of hash slots of incoming data helps to balance the load on each Redis node in the Redis cluster. Hence, for end users, there exists no difference to access database from a single Redis node or Redis cluster.
- Redis is equally compatible with the HTTP and MQTT servers. Moreover, to achieve high concurrency, the MQTT load can be transferred to the Redis Cluster. For example, several clients can publish messages on certain topics to MQTT servers. MQTT servers transfer this payload of messages to a certain node of the Redis cluster, which shares this payload with other nodes in Redis Cluster, which are ultimately connected to correct subscribers for that topic.
- Redis cluster also supports transaction logging (recording of database history) and standby techniques (executing backup of certain Redis node in case of malfunctioning) to guarantee reliability.

#### 6.1.2.4 Load-balancing Servers

Load-balancing Servers are essential and beneficial in IoT Cloud as these:

- process requests in a scheduled way.
- distribute the workload on application/database servers.
- avoid congestion on application/database servers.
- achieve maximum utilization of available resources.

Regarding implementation, HAProxy (an open-source lightweight load balancer) is preferred, which offers HTTP- and TCP-based load balancing along with the support of high concurrency. For the distribution of requests over HTTP servers, the weighted round-robin strategy is configured in HTTP load balancers. To achieve efficient load balancing, HAProxy first binds requesters' URLs to available ports, and then following some predefined

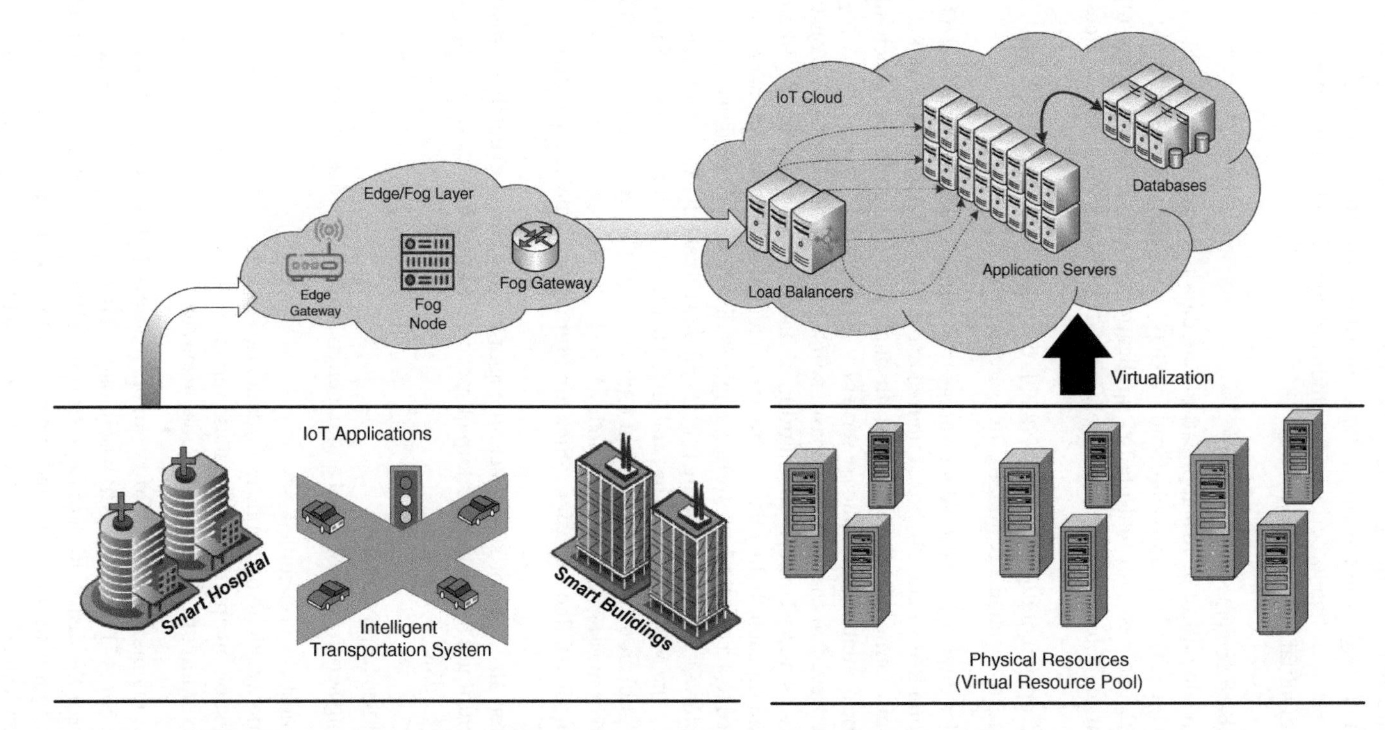


**Figure 6.2** VM-based configured servers on IoT cloud architecture. Source: Adapted from Hou et al. [6].

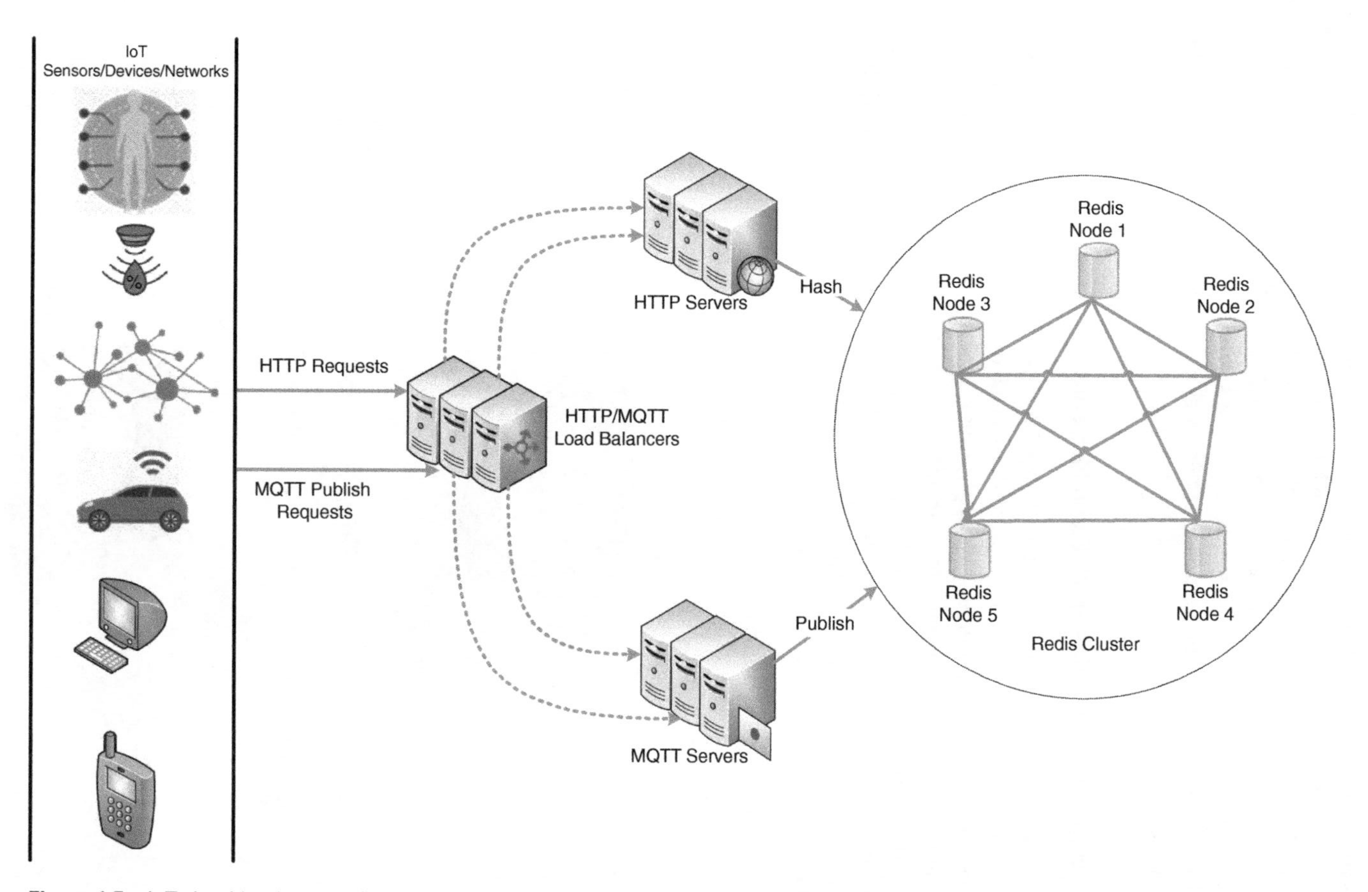


**Figure 6.3** IoT cloud implementation details. Source: Adapted from Hou et al. [6]

principle, it distributes these requests over HTTP servers. In MQTT-based load balancers, a TCP load-balanced mechanism is implemented to distribute the load over MQTT servers. HAProxy provides virtual Internet Protocol (IP) and port number to receive TCP packets. Upon receiving the TCP packet, HAProxy changes the destination IP and port number of TCP packets and forwards this to the target MQTT server (which keeps long-lived connections with clients). In the end, the Least Connections policy (server with the least number of active connections) is implemented in HAProxy as a load-balancing method to select an appropriate MQTT server.

### 6.1.3 Application Domains of IoT Cloud Platforms

The application domains of IoT Cloud platforms have been shown in Figure 6.4. Considering these application domains, several Cloud platforms with different levels of services have been developed to assist a diverse range of IoT applications. Table 6.2 provides a comparison of few IoT Cloud platforms with reference to their suitability (+) and appropriateness (√) in various application domains, i.e. Application Development (AD), Device Management (DM), System Management (SM), Heterogeneity Management (HM), Data Management (DaM), tools for Analysis (A), Deployment Management (DeM), Monitoring Management (MM), Visualization (V), and Research (R) [10].

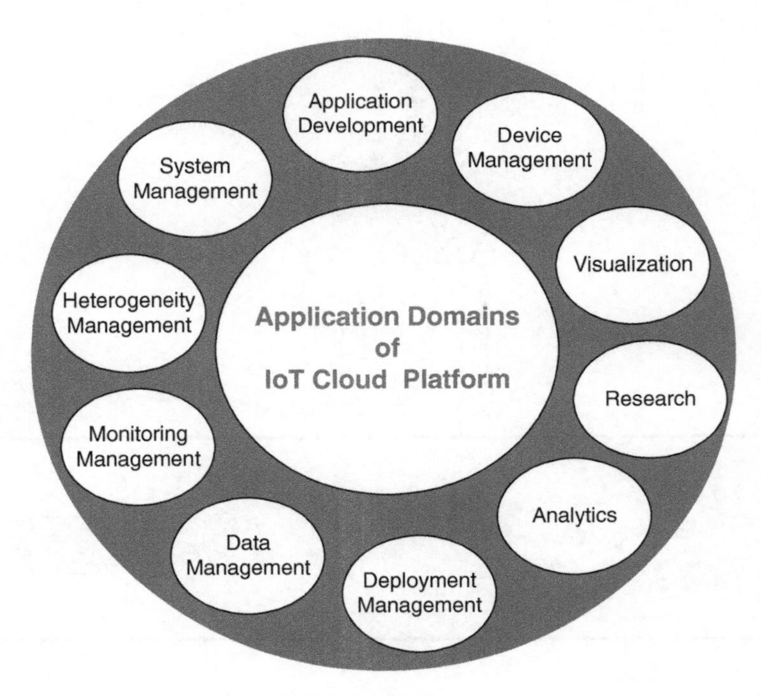


**Figure 6.4** Application domains of IoT cloud platform.

**Table 6.2** IoT application domains and comparison of cloud platforms.

| | AD | DM | SM | HM | DaM | A | DeM | MM | V | R |
|---|---|---|---|---|---|---|---|---|---|---|
| Arkessa | | √ | | | + | | | | | |
| Xively | √ | + | | | | | | √ | | |
| ThingWorx | √ | | | | + | | | √ | | |
| Nimbits | | | | | + | √ | | | | |
| SeeControl IoT | | + | | | | √ | | | √ | |
| SensorCloud | | + | | | | | | √ | √ | |
| Axeda | | √ | | | + | | | √ | | |
| IBM IoT | | √ | | | | | | | | + |

Source: Based on Ray [10].

## 6.2 Fog Computing for IoT

The concept of Fog computing has been introduced to extend the computing power and storage capacity of Cloud to the network edge [11]. Fog computing acts as a bridge between smart end-user devices and IoT Cloud to enhance the performance of IoT applications, which require low latency real-time response. Table 6.3 provides a comparison of Fog computing with Cloud computing [1].

From this comparison, it becomes clear that Fog computing improves various aspects of different IoT systems [12], i.e.:

*Low Latency:* Because of the proximity of computing and storage resources to the end users, fog computing offers services with better delay performance.

*Real-time Analytics Support:* Fog computing provides real-time interactions with end-user devices rather than the faraway processing in IoT Cloud. Therefore, Fog nodes play an important role in the real-time reception, processing, and forwarding of data (either to end-user devices or to IoT Cloud for BigData storage and analytics).

**Table 6.3** Comparison between fog and cloud characteristics.

| Cloud/Fog characteristics | Fog computing | Cloud computing |
|---|---|---|
| Scalability | Limited | Dynamic and High |
| Latency | Low | High |
| Location Awareness | Yes | No |
| Real-time Response | Yes | No |
| Mobility Support | Yes | No |
| Storage Capacity | Low | High |
| User-defined Security | Yes | No |

*Geographical Distribution:* Resources, services, and applications provided by Fog are distributed and can be deployed anywhere.

*Scalability:* The provisioning of computing and storage resources at different geographical locations supports the efficient deployment of large-scale sensor networks.

*Mobility Support:* Considering the direct connectivity with mobile devices, Fog computing enables mobility methods.

*Location Awareness:* The distributed deployment of Fog nodes is able to support location awareness.

*Heterogeneity:* Fog computing supports the processing of data received from end devices designed by different manufacturers.

*Interoperability:* Fog components can accomplish tasks utilizing resources available on different domains and service providers.

### 6.2.1 Difference from Related Computing Paradigms

Edge Computing, and Mobile Edge Computing (MEC) are computing paradigms which are related to Fog Computing [13]. It is important to know the uniqueness of Fog computing from these computing paradigms.

#### 6.2.1.1 Edge Computing

In Cloud computing, resources in Cloud datacenters are geographically centralized and are located far away from end-user devices. This paradigm is not suitable for latency-sensitive IoT applications as clients have to endure large round-trip delay and network congestion. Edge computing technology emerges to resolve the problems associated with centralized Cloud computing paradigms [14, 15]. Edge computing (compared to Cloud computing) provides local processing at the edge of the network. This local processing near the end users mitigates the computational stress of Cloud and reduces the latency of response time in IoT systems [16, 17]. End devices (i.e. smart things, smart/mobile phones, Personal Digital Assistant (PDA), data collector, etc.), edge devices (i.e. switches, bridges, routers, base stations, wireless access points, etc.), and edge servers equipped with specialized capabilities support edge computation. With reference to its ability of localized processing, Edge computing provides faster response to service requests and mostly resists to send raw data toward the core network. Nevertheless, Edge computing does not provide IaaS, PaaS, and SaaS and is more concerned with prompt response to end-user devices.

#### 6.2.1.2 Mobile Edge Computing (MEC)

MEC (one of the key enablers of the modern evolution of cellular base stations) offers edge servers and base stations to be operated together and can be considered as an extension of the Edge computing paradigm [18–20]. MEC with its optional connectivity to Cloud datacenters aims in the faster provisioning of cellular services and flexible access to network information about content distribution [21].

Although the terms Edge computing, mobile edge computing, and Fog computing have been used interchangeably in literature, there exists a subtle difference, which makes Fog computing different from these two.

Fog computing besides enabling edge computation at the network edge can be expanded to the core network as well [22]. In simple words, Fog computing's computational infrastructure consists of both edge and core networking components (i.e. switches, routers, regional servers, etc.). Like Edge computing, edge network components of Fog computing are placed closer to smart things/devices/sensors to enable IoT data to be stored and processed within local vicinity to improve service delivery latency for real-time IoT applications. However, unlike Edge computing, Fog computing provides Cloud-based services, i.e. IaaS, PaaS, and SaaS, to the network edge as shown in Figure 6.5. Due to this Fog computing has been regarded as a well-structured paradigm for IoT systems in comparison to Edge computing.

## 6.2.2 Architecture of Fog Computing

Fog computing approach offers Cloud operations to be performed at the edge devices (having limited computing and storage capacity) of the network in a distributed manner between end-user devices and IoT Cloud. The primary aim of Fog is to address the problem of high latency for time-sensitive IoT applications.

The fog architecture consists of the following six layers [12, 23–25] as shown in Figure 6.6.

### 6.2.2.1 Physical and Virtualization Layer

This layer consists of smart things, physical sensor nodes, virtual sensor nodes of wireless sensor networks, and virtual sensor networks. Geographically distributed sensors (attached to physical objects) sense environment, collect data, and send it to the monitoring layer via gateways.

### 6.2.2.2 Monitoring Layer

The Monitoring layer involves the monitoring of:

- the availability status of sensors, fog, and network nodes.
- the resource utilization status of nodes.
- the status of activity/task performance on each node as well as the time and requirement of each task.
- the energy consumption status of each (sensor and fog) node.
- status of IoT services deployed on the infrastructure of IoT systems.

### 6.2.2.3 Preprocessing Layer

The preprocessing layer is responsible for the analysis, filtering, and trimming of collected data to extract useful information from it.

### 6.2.2.4 Temporary Storage Layer

This layer deals with the temporary storage of data before forwarding it to the IoT Cloud. Data is removed from Fog nodes after being successfully transmitted to the IoT Cloud.

### 6.2.2.5 Security Layer

The layer applies encryption, decryption, and integrity measures on data to protect it from tampering.

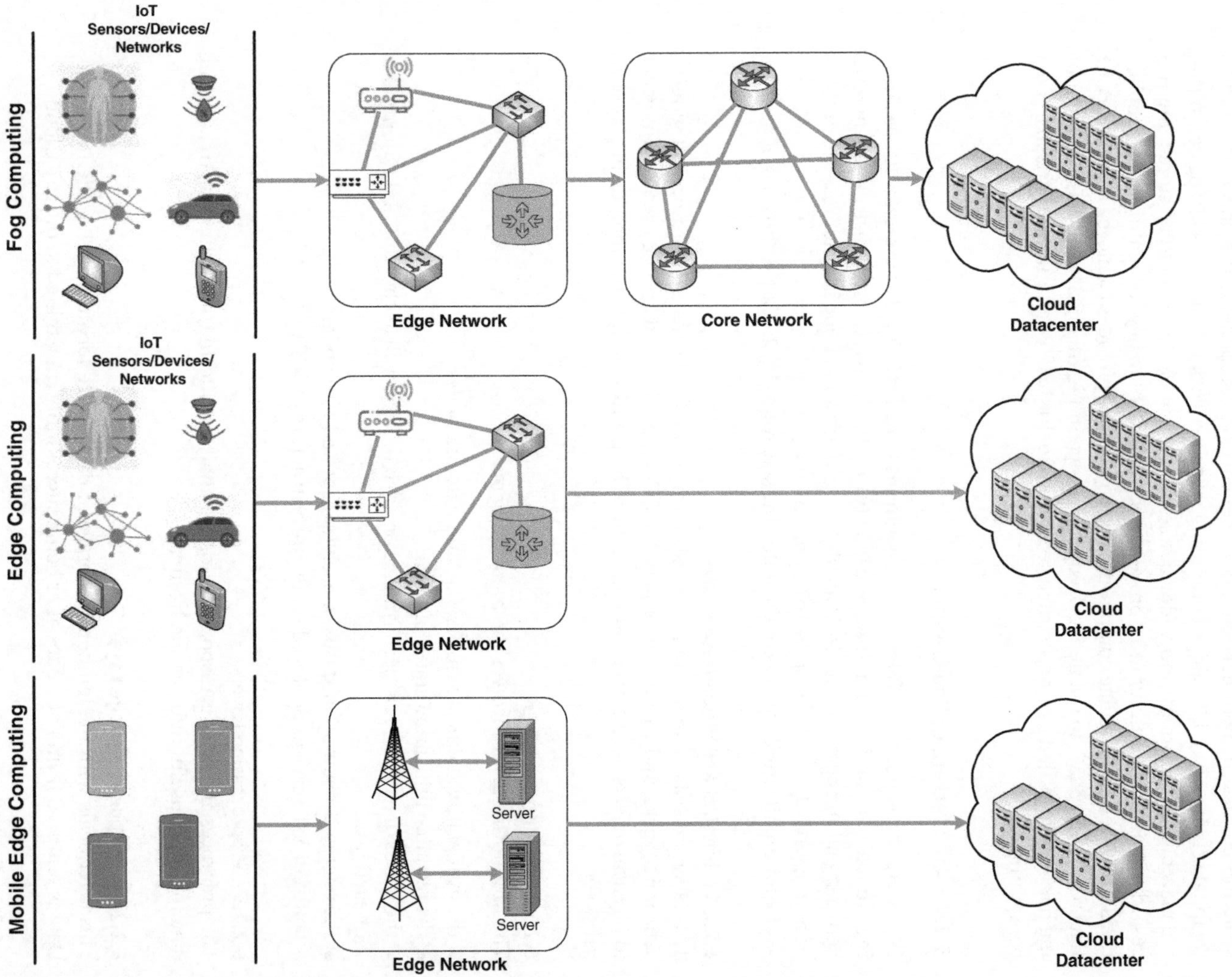

**Figure 6.5** Comparison of fog computing with edge/MEC computing.

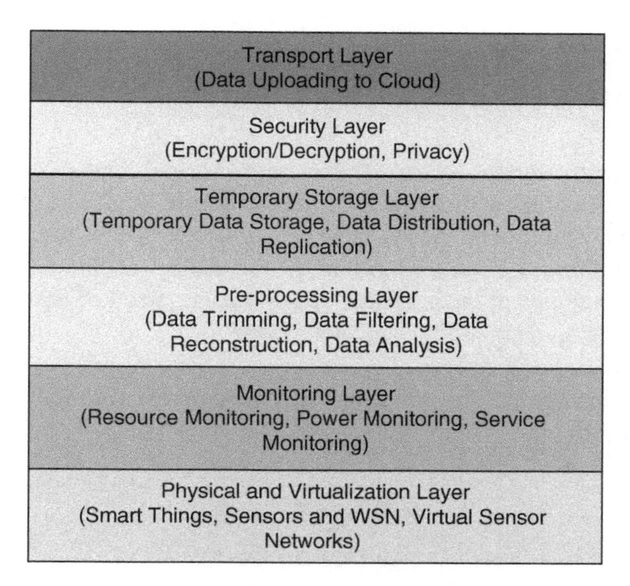


**Figure 6.6** Layered architecture of fog computing.

#### 6.2.2.6 Transport Layer

The transport layer is responsible for the uploading of preprocessed data on IoT Cloud for BigData analysis. Only a portion of collected data is transferred from sensing nodes to IoT Cloud via a gateway device.

### 6.2.3 Fog Deployment Models

Based on the ownership of Fog computing infrastructure and underlying resources, Fog models [26] can be described as:

*Private Fog:* Created, owned, managed, and operated by a private organization and/or third party. The Fog resources are exclusively deployed within the premises of that organization to be used by the different business units in that organization.

*Public Fog:* Created, owned, managed, and operated by a government organization and/or academic institutes to be used by the general public.

*Community Fog:* Created, managed, and operated by community organizations and/or third party and resources are exclusively deployed to be used by the members of specific communities having shared concerns.

*Hybrid Fog:* Combines the use of private/public/community fog with public/private Clouds. Hybrid Fog scales the resources (computing, storage) of Fog platforms and is suitable for real-life use cases, which demand resources other than those available in Fog infrastructure.

### 6.2.4 Fog Service Models

Similar to Cloud computing, Fog computing provides virtualized resources to its users in three ways [26], i.e.:

*FogIaaS (Fog-infrastructure-as-a-service)* offers hardware (i.e. computing, storage, communication, etc.) resources without revealing the details of physical hardware. End users are able to deploy utilities on the available pool of virtualized resources.

*FogPaaS (Fog-platform-as-a-service)* built on the top of FogIaaS and offers operating software and/or development environment. It provides a quick and cost-effective solution for software development, testing, and deployment.

*FogSaaS (Fog-software-as-a-service)* offers software applications without installing them on their personal computer. The services are accessed from the web browser remotely through a network.

## 6.3 Case Study – Vehicles with Fog Computing

Several IoT applications can be considered as Fog-based applications due to their substantial dependency on Fog infrastructure. An Intelligent Transportation System (ITS) is one of the examples of IoT systems, which exploit the advantages associated with Fog computing. Fog can be implemented in different scenarios and in different ways to support different fragments of ITS. ITS system consists of various components, i.e. Vehicular Ad hoc Networks (VANETs), Roadside Units (RSUs), Fog nodes, and Cloud datacenter. Vehicles in VANETs interact with RSUs to transfer and access safety/infotainment information and services from distributed Fog nodes and centralized Cloud, respectively. Few scenarios and applications have been discussed in this section.

### 6.3.1 VANETs and Fog Computing

Fog computing plays a significant role in several VANET applications, which have specific requirements of mobility, location awareness, and latency-sensitive response [12].

In the VANET scenario, Fog computing using Road-Side Infrastructure (RSI) creates a distributed computing platform to deal with data processing and storage services. RSI consists of RSUs, Base Station (BS), and Radio Tower. Fog layer acts as near-end computing proxies between the on-road vehicles and vehicular Cloud. Fog layer is involved in three types of communications (as shown in Figure 6.7), i.e.:

- *RSU to Fog Server:* Vehicles communicate with RSUs and RSUs communicate with each other as well as with the Fog server to propagate safety and infotainment messages.
- *Fog Server to Fog Server:* Peer Fog servers at different locations communicate with each other and are responsible to manage a pool of resources to a localized area for collaborative service provision and content delivery.
- *Fog Server to Cloud:* Fog servers transfer localized information to vehicular Cloud, which aggregates the received information and ultimately performs centralized computation.

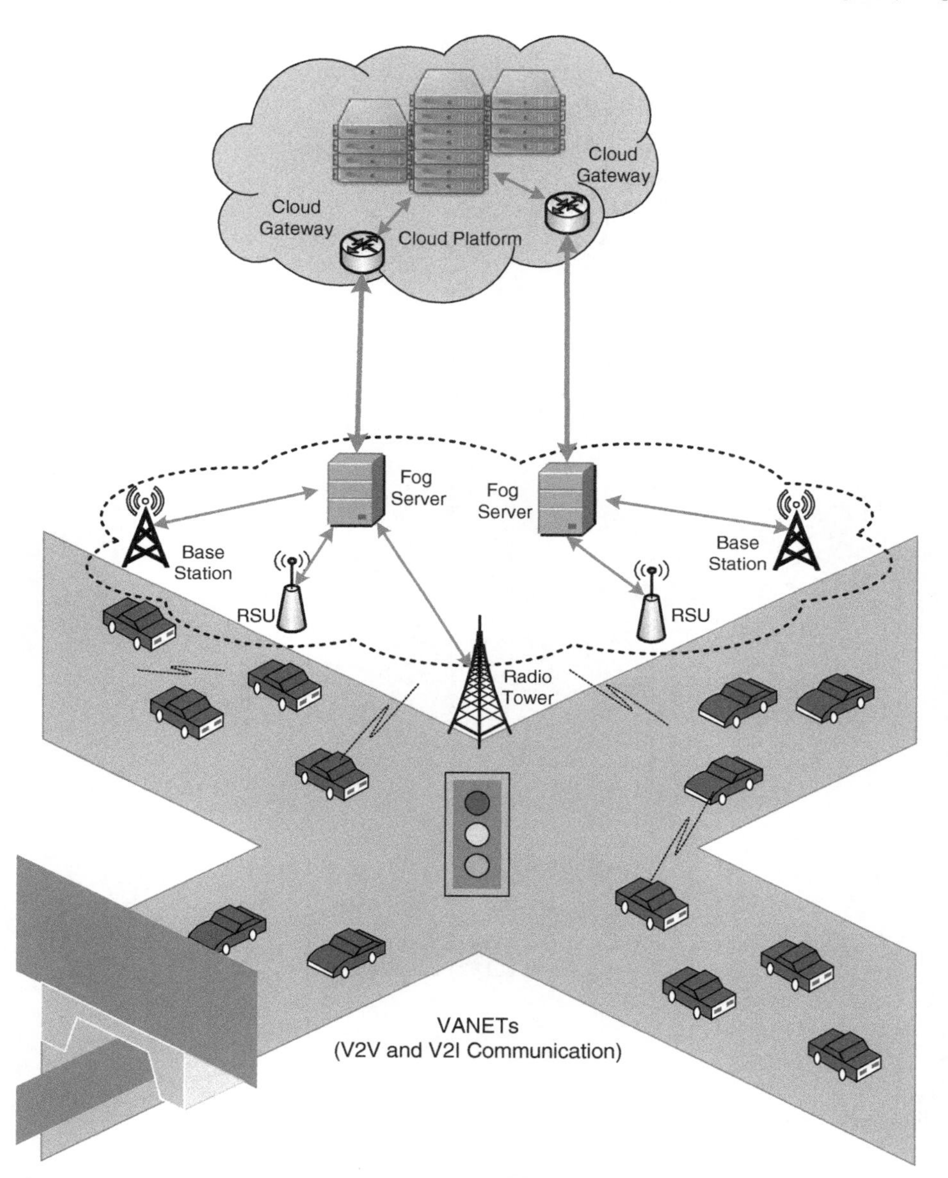


**Figure 6.7** Fog computing architecture for VANETs.

### 6.3.2 Dynamic Traffic Light Signal Management

One of the main applications of VANETs is the Dynamic Traffic Light System (DTLS), which optimally manages road traffic at traffic signals with the help of Fog computing. Traffic signals are distributed; therefore for this location-specific and time-sensitive application, Fog computing solution is more realistic than Vehicular Cloud Computing (VCC) solution. Fogs are implemented at traffic signal junctions (shown in Figure 6.8) to compute

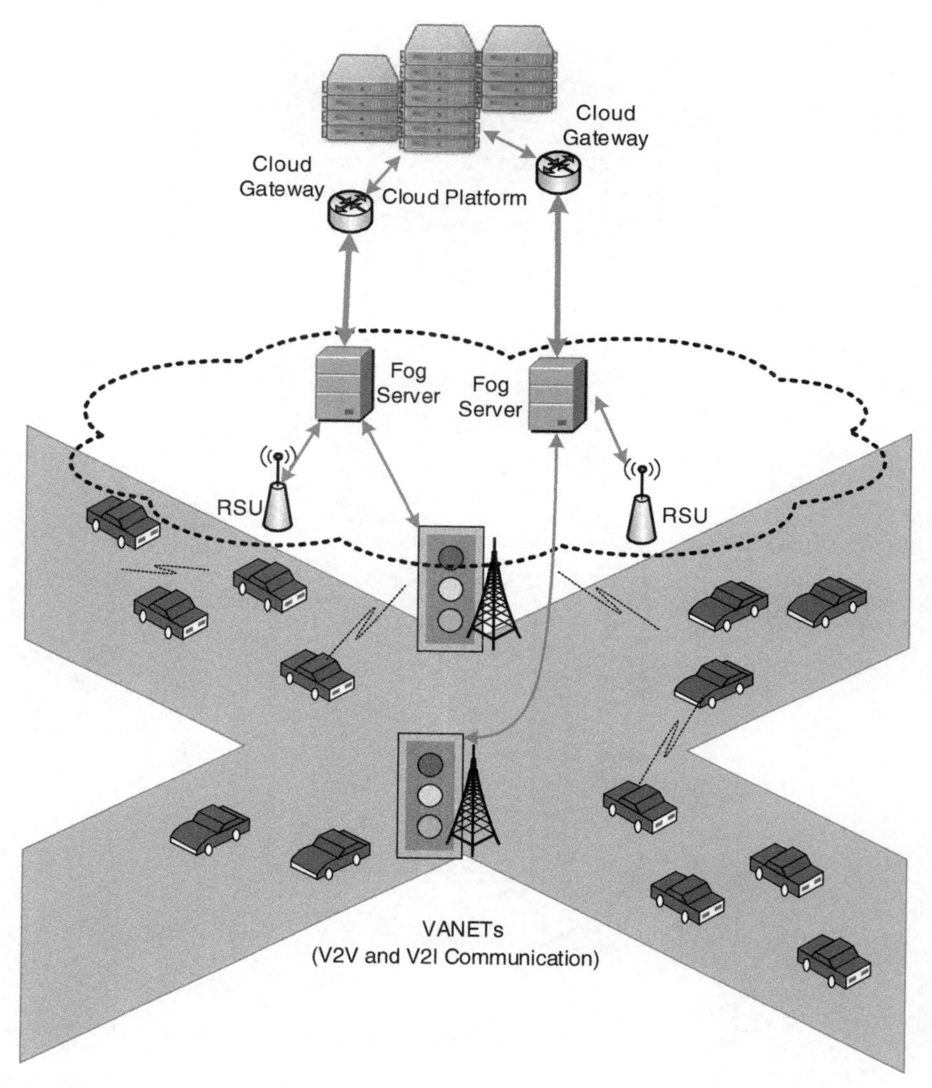


**Figure 6.8** Traffic light system with fog.

the duration of each light signal for traffic around that junction. DTLS provides a dynamic solution by allotting time to each traffic signal based on traffic situation around the junction at any instant of time. Moreover, traffic lights connected to Fog servers are also able to send warning messages to vehicles approaching traffic junctions.

### 6.3.3 Parking System

Fog computing helps to find empty parking spaces in rush hours to relieve traffic congestion. Fog nodes can be installed in local areas (i.e. shopping malls, restaurants, banks, parks, etc.), and these nodes collect and store information about the status of parking slots (reserved or free). This information is delivered to RSUs, which ultimately direct drivers to use the most suitable available parking slot.

### 6.3.4 Content Distribution

Fog computing helps to distribute safety and infotainment information in the intelligent traffic system. For example, vehicles can obtain information from a fog server about the blocking of a certain section of the road. In this way, Fog is preferred over Cloud to save lives by avoiding collisions and accidents as it does not allow high latency, which is one of the characteristics of a centralized Cloud.

### 6.3.5 Decision Support System

The intelligent decision support system can be implemented using Fog computing to improve drivers' safety. For example, Fog servers exchange localized information with each other and help the decision support system to monitor the safety of drivers.

## Review Questions

**6.1** Illustrate the architecture of IoT Cloud.

**6.2** What is the difference between Fog computing and Edge computing?

**6.3** Explain the architecture of Fog Computing.

**6.4** In how many ways Fog models can be implemented?

**6.5** Describe Fog service models.

**6.6** Considering your expertise in IoT systems, city administration contacts you to propose an IoT solution for the monitoring of environmental parameters in the city which is experiencing high pollution coming directly from road traffic. The fundamental aim of this project is to control and analyze the management of city traffic through the consideration of:

**A** air-quality sensors network.
**B** data collection at Fog nodes.
**C** data collection at Cloud for analysis.
**D** the communication network (consisting of Bluetooth/Wi-Fi devices).

Illustrate the working of your proposed IoT-based solution while considering all details related to Fog and Cloud architectures.

**6.7** Consider vehicle-to-roadside unit (V2R) communication use case in VANET where a driverless car is supported by Fog nodes.

**A** Briefly describe how Fog computing can support this use case.

**B** Identify two application scenarios in which a driverless car can get benefits from Fog computing.

**C** Determine which of the following motivating factors is more important in the context of VANET with Fog computing:

**(i)** Real-time Data Analysis

**(ii)** Data Storage

**D** One of the IoT applications running on a car has generated 32 MB (Mega-Bytes) data, which needs to be processed within one second. You have two options:

**(i)** Send the data to Fog node over a duplex link with 128 MB/s (Mega-Bytes/second) and 50 ms latency. The processing of the data on this Fog device takes 10 ms.

**(ii)** Send the data directly to Cloud over a duplex link of 64 MB/s and 25 ms latency. The processing of the data on Cloud takes 5 ms

Assume the same amount of processed data is sent back to the application after processing, i.e. 32 MB (Mega-Bytes). Which option would you choose to meet the timing constraint? Explain your reasoning.

## References

1 Dang, L.M., Piran, M., Han, D. et al. (2019). A survey on internet of things and cloud computing for healthcare. *Electronics* 8 (7): 768.

2 CISCO Solutions (2018). *Edge-to-Enterprise IoT Analytics for Electric Utilities*. Document ID: 1517290075438108. Cisco Systems Inc. Available at: https://www.cisco.com/c/en/us/solutions/collateral/data-center-virtualization/big-data/solution-overview-c22-740248.html.

3 Al-Fuqaha, A., Guizani, M., Mohammadi, M. et al. (2015). Internet of things: a survey on enabling technologies, protocols, and applications. *IEEE Communications Surveys and Tutorials* 17 (4): 2347–2376.

4 Liu, Y., Dong, B., Guo, B. et al. (2015). Combination of cloud computing and internet of things (IOT) in medical monitoring systems. *International Journal of Hybrid Information Technology* 8 (12): 367–376.

5 Pourqasem, J. (2018). Cloud-based IoT: integration cloud computing with internet of things. *International Journal of Research in Industrial Engineering* 7 (4): 482–494.

6 Hou, L., Zhao, S., Xiong, X. et al. (2016). Internet of things cloud: architecture and implementation. *IEEE Communications Magazine* 54 (12): 32–39.

7 Yassein, M.B., Shatnawi, M.Q., Aljwarneh, S. et al. (2017). Internet of Things: survey and open issues of MQTT protocol. In: *IEEE International Conference on Engineering & MIS (ICEMIS)*, 1–6. IEEE.

8 MQTT (2018). Message queuing telemetry transport protocol. http://docs.oasis-open.org/mqtt/mqtt/v3.1.1/cos02/mqtt-v3.1.1-cos02.html.

**9** Tang, K., Wang, Y., Liu, H. et al. (2013). Design and implementation of push notification system based on the MQTT protocol. In: *International Conference on Information Science and Computer Applications (ISCA 2013)*. Atlantis Press.

**10** Ray, P.P. (2016). A survey of IoT cloud platforms. *Future Computing and Informatics Journal* 1 (1–2): 35–46.

**11** Solutions, C.F.C. (2015). *Unleash the Power of the Internet of Things*. Cisco Systems Inc.

**12** Atlam, H.F., Walters, R.J., and Wills, G.B. (2018). Fog computing and the internet of things: a review. *Big Data and Cognitive Computing* 2 (2): 10.

**13** Mahmud, R., Kotagiri, R., and Buyya, R. (2018). Fog computing: a taxonomy, survey and future directions. *Internet of Everything*: 103–130.

**14** Garcia Lopez, P., Montresor, A., Epema, D. et al. (2015). *Edge-Centric Computing: Vision and Challenges*. NY, USA: ACM New York.

**15** Shi, W., Cao, J., Zhang, Q. et al. (2016). Edge computing: vision and challenges. *IEEE Internet of Things Journal* 3 (5): 637–646.

**16** Salman, O., Elhajj, I., Kayssi, A. et al. (2015). Edge computing enabling the Internet of Things. In: *2015 IEEE 2nd World Forum on Internet of Things (WF-IoT)*. IEEE.

**17** Varghese, B., Wang, N., Barbhuiya, S. et al. (2016). Challenges and opportunities in edge computing. In: *IEEE International Conference on Smart Cloud (SmartCloud)*. IEEE.

**18** Sun, X. and Ansari, N. (2016). EdgeIoT: mobile edge computing for the internet of things. *IEEE Communications Magazine* 54 (12): 22–29.

**19** Mao, Y., You, C., Zhang, J. et al. (2017). A survey on mobile edge computing: the communication perspective. *IEEE Communications Surveys and Tutorials* 19 (4): 2322–2358.

**20** Hu, Y.C., Patel M., Sabella D. et al. (2015). Mobile edge computing—a key technology towards 5G. ETSI white paper, 11(11), 1–16.

**21** Cau, E., Corici, M., Bellavista, P. et al. (2016). Efficient exploitation of mobile edge computing for virtualized 5G in EPC architectures. In: *4th IEEE International Conference on Mobile Cloud Computing, Services, and Engineering (MobileCloud)*. IEEE.

**22** Bonomi, F., Milito, R., Zhu, J. et al. (2012). Fog computing and its role in the internet of things. In: *Proceedings of the first edition of the MCC workshop on Mobile cloud computing*, 13–16. ACM.

**23** Mukherjee, M., Shu, L., and Wang, D. (2018). Survey of fog computing: fundamental, network applications, and research challenges. *IEEE Communications Surveys and Tutorials* 20 (3): 1826–1857.

**24** Muntjir, M., Rahul, M., and Alhumyani, H.A. (2017). An analysis of internet of things (IOT): novel architectures, modern applications, security aspects and future scope with latest case studies. *International Journal of Engineering Research and Technology* 6 (6): 422–447.

**25** Aazam, M. and Huh, E.-N. (2014). Fog computing and smart gateway based communication for cloud of things. In: *IEEE International Conference on Future Internet of Things and Cloud*. IEEE.

**26** Ahmed, A., Arkian H., Battulga D. et al. (2019). Fog computing applications: taxonomy and requirements. arXiv preprint arXiv:1907.11621.

# 7

# IoT Applications

| LEARNING OBJECTIVES |
| --- |
| After studying this chapter, students will be able to:<br>• describe the main applications of IoT.<br>• explain the implementation details of various IoT application domains.<br>• propose IoT system respecting practical constraints. |

## 7.1 Application Domains of IoT

Immense potentials associated with IoT have enabled the actualization of a huge number of applications in different domains to improve the quality of human lives. Domestic automation, smart transportation, smart agriculture and farming, smart manufacturing and industry automation, smart education, public safety and military, retail and hospitality, government and corporate sectors, and energy conservation are the main environments in which IoT plays a vital role [1–3]. In recent times, these environments equipped with intelligent devices of different capabilities are able to improve the standard of human life in different capacities. Table 7.1 illustrates the example of scenarios associated with the aforementioned IoT domains.

## 7.2 IoT and Smart Home

A smart home can be defined as digitally engineered domestic life with the use of IoT technologies to anticipate and respond to the needs of home residents while providing comfort, convenience, entertainment, security, and connection to the outside world [4, 5]. The fundamental purpose of smart home is to improve the routine life of home residents. In the near future, modern homes would be equipped with smart electronic home appliances, smart furniture, smart power outlets, and smart meters to control lightning, air quality, heating, ventilation, air conditioning (HVAC), security, and energy through the use of smart grids [6].

*Enabling the Internet of Things: Fundamentals, Design, and Applications*, First Edition.
Muhammad Azhar Iqbal, Sajjad Hussain, Huanlai Xing, and Muhammad Ali Imran.

**Table 7.1** IoT application domain and example scenarios.

| IoT application domain | Example |
|---|---|
| Domestic Automation | Comfortable living, home automation, home security and monitoring, smart appliances, children protection, video surveillance, infotainment, etc. |
| Healthcare and Well-being | Smart hospital services, remote patient monitoring, elderly assistance, disabled assistance, medical equipment/pharmaceuticals, ambulance tracking and management, remote diagnostics and examinations, medical records management, assisted living, remote caregiver assistance, mobile assistance, etc. |
| Smart Transportation | Connected automobiles, smart mobility, road monitoring, vehicle sharing, automated cars, automatic payment systems, parking system, proactive schedule maintenance of automobiles, safety application, infotainment applications, traffic management applications, traffic signals management, smart trains, smart planes, smart boats, Intelligent Transportation System (ITS), etc. |
| Smart Agriculture and Farming | Field/farm/pasture monitoring and management, monitoring and management of agricultural production and feed, livestock management and tracking, field/farm irrigation, monitoring food processing, management, expiry, and automation of ordering, delivery, and billing of agricultural products, etc. |
| Industry Automation, Smart Manufacturing, and Logistics | Smart manufacturing including material identification and product deterioration, warehouse management, monitoring of industrial plants, luggage management, shipping tracking, and boarding operations, process/equipment monitoring, monitoring employees and suppliers' premises, etc. |
| Smart Education | Monitoring and security of educational institutes, remote education, and e-learning, automation of student/faculty identity services, etc. |
| Public Safety and Military | Weapon identification and tracking, sniper detection, territory surveillance through the video camera, radar and satellite, emergency plan and rescue, etc. |
| Retail and Hospitality | Inventory management, shopping operations, fast payment, anti-theft and fraud, support for multiple languages, etc. |
| Energy Conservation | Smart grid for power generation, distribution, and management, smart metering, smart water management, smart load-balancing with batteries and storage systems' coordination, etc. |
| Government and Private Sector Applications | Smart cities, smart buildings, smart communities, real-time environmental monitoring, municipal water, and sewer monitoring, remote service delivery and compliance monitoring, maintenance, and management of historical places and parks, etc. |

### 7.2.1 IoT-based Smart Home Framework

The generic IoT-based smart home framework is hierarchical and consists of different levels, i.e. household things networks, Cloud, utility, third party, and user interfaces as shown in Figure 7.1 [7]. At the first level, smart household goods equipped with sensors and wireless communication interfaces are able to sense and transmit data to home/residential hub

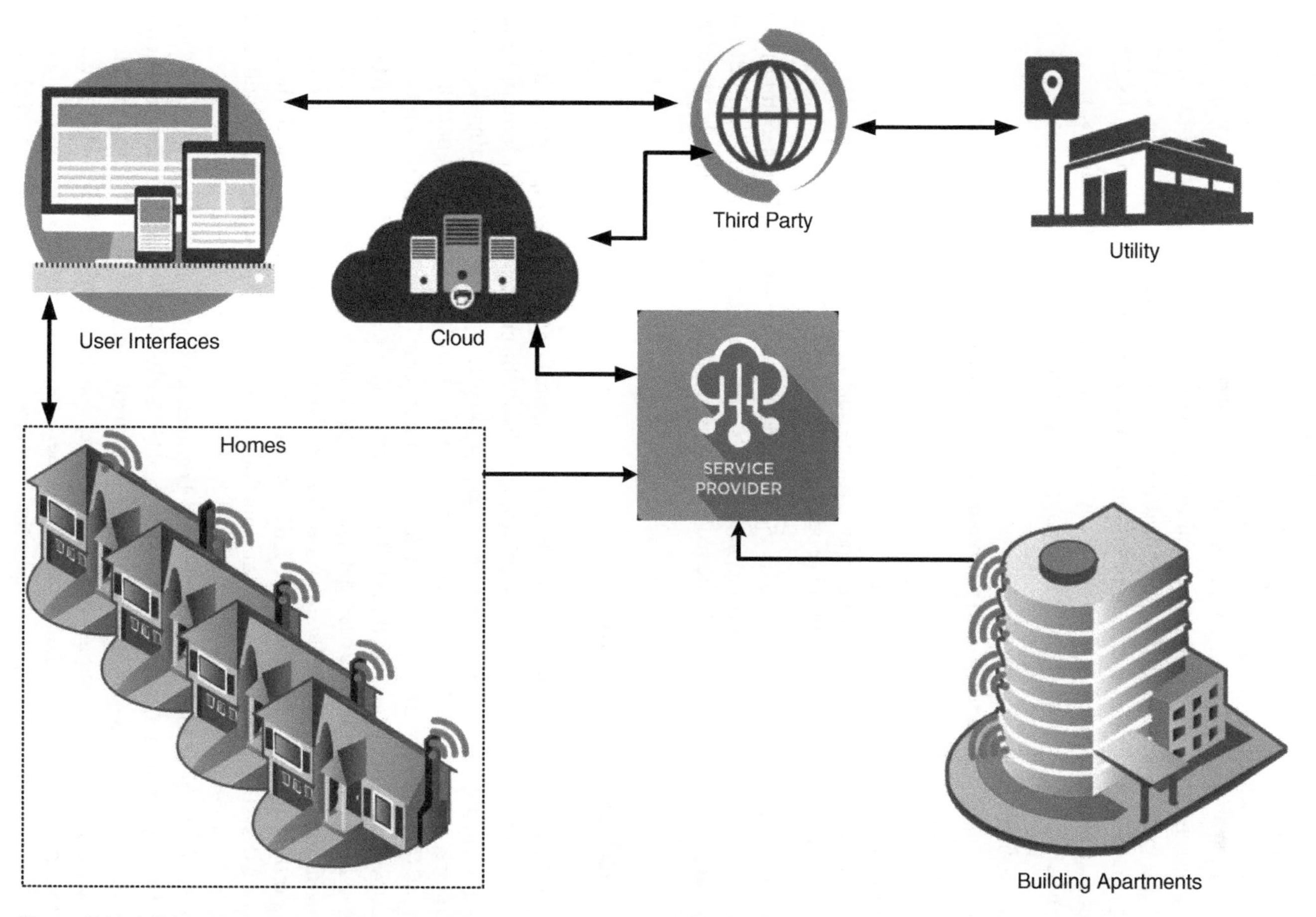


**Figure 7.1** IoT-based smart home framework. Source: Based on Stojkoska and Trivodaliev [7].

(i.e. desktop, tablet, smart meter, or smartphone). Home/Residential hub has storage capacity, processing, and communication interface for the routing, and transmission of received data packets to the required outside destination through the Internet. At the second level, the Cloud provides gigantic storage and processing infrastructure which are used by third-party application for data accumulation, analysis, and offering of utilities' (electricity/gas) production, distribution, and billing to the end users. User interfaces at third level through the use of efficient visualization techniques represents household consumption, and usage results in the form of graphs, notifications, and recommendations. This is ultimately beneficial for consumers to control the usage of devices and utilities.

The benefits of IoT-based smart home applications include [8]:

- Home energy conservation
- Cost reduction for basic needs (remote health monitoring and treatment alerts)
- Comfort and entertainment
- Remote security
- Automatic payment of utility bills
- In-time recommendations about goods damage and utility usage
- Increased reliability of household goods

However, technology acceptance, device heterogeneity, hardware failures, communication failure, large data flow, information leakage, security attacks, and battery life cycles are major challenges associated with the realization of IoT-based smart homes at large scale.

## 7.3 IoT and Healthcare

Current healthcare systems face new challenges associated with the rapid growth of the aging population and overall intensification of fatal chronic diseases. All human beings belonging to any social status identify health as a top priority issue for a sustainable society. Therefore, health issue has been centrally positioned within the 2030 agenda of United Nations General Assembly [9]. The 2030 agenda puts considerable emphasis on the usage of innovative approaches based on ICT to accommodate a much broader range of health-related issues. Proactive wellness, detection and prevention of diseases at early stages, and realization of a Ubiquitous Health Care System (UHCS) are some of the major concerns behind the emphasized use of the ICT. The main components of UHCS (one of the examples of IoT systems) are biomedical sensors, local server (coordinator machine), and medical server as shown in Figure 7.2.

Referring to biomedical sensors, WBAN becomes an important research topic for IoT-based health-related applications [10–13]. In UHCS, WBAN is based on heterogeneous, miniaturized, and low-power sensor nodes, which are designed for collecting the physiological parameters (vital signs) of the human body, i.e. blood pressure, heartbeat rate, respiratory rate, temperature, glucose level, peripheral capillary oxygen saturation ($SpO_2$), etc. WBAN sensors can be placed in (implanted), out (wearable), and around (in-house) human body depending on requirements. WBAN, using an available network of wireless Access Points (AP) or Base Stations (BS), promotes assisted living with the primary objectives of

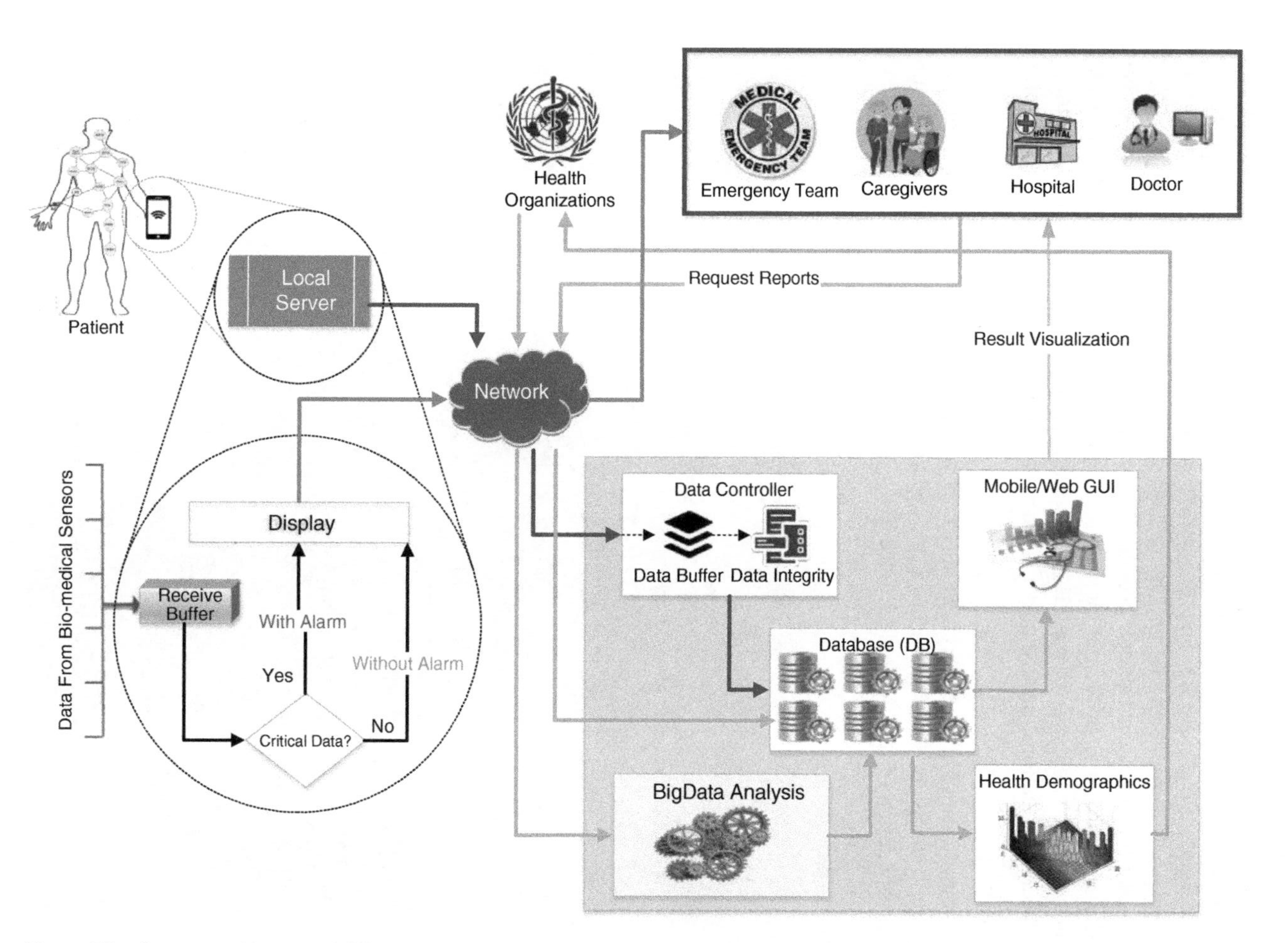


**Figure 7.2** System architecture IoT-based healthcare system.

collection and transmission of vital signs information of the human body to designated medical servers through a coordinator node also called local server. In addition, the mobile phone acting as coordinator-node/local-server/IoT gateway provides health assistance through alarm generation to doctors, carers, and emergency teams available in critical situations as shown in Figure 7.2. The medical servers and databases are responsible for data archiving and data processing, which are used by hospitals, clinics, doctors, carers, and government health organizations to construct and synthesize health demographics based on a large number of data streams for thousands of remote patients using BigData analytics.

The complete realization of the UHCS framework, while fulfilling the demands of major healthcare applications, necessitates addressing the following challenges [14] (also shown in Figure 7.3)

- Communications and networking issues such as antenna design, short-range communication, interference with other colocated networks, energy-efficient MAC, and routing protocol design of the biomedical sensors.

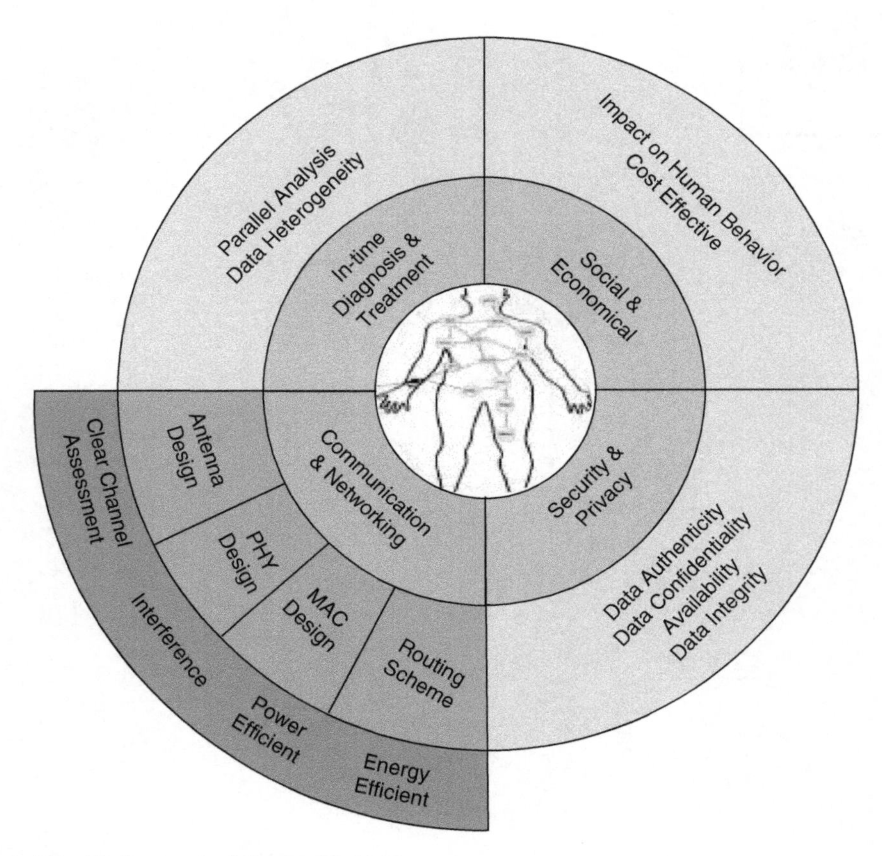


**Figure 7.3** Challenges for IoT-based healthcare system.

- Social and economic concerns such as the impact on human behavior toward social acceptance or rejection of mobile, web, and WBAN technology on the basis of ease and cost-effectiveness.
- Real-time vital signs anomaly detection demands processing of the data streams, of heterogeneous nature, to extract critical patterns of physiological parameters.
- Security and privacy issues, (i.e. secure group management, confidentiality, privacy, integrity, authorization, and authentication).

In a smart hospital, in-door patients are equipped with concerned body sensors to collect vital signs' information with related context information, i.e. location, date, and time to make accurate inferences about unusual patterns. The smart hospital architecture consists of three components i.e., medical sensor/actuator network, distributed e-Health gateways, and back-end Cloud computing platform [15] as shown in Figure 7.4:

- Medical sensor/actuator with communication interfaces transmits information (both from body and ward/room).
- E-health gateways at different geographical sites in hospitals support edge computing, heterogeneous communication, protocol conversion, data aggregation, filtering, dimensionality reduction, etc.
- Back-end cloud computing platform implements data warehousing, analytics, and visualization techniques.

## 7.4 IoT and Smart Mobility

In the near future, IoT has certain goals about the realization of the smart cities through the use of different technologies, i.e. smart things sensors, Cloud, smart grid, and mobile networks. Considering mobility, IoT is very suitable for the collection of data from sensors attached to mobile devices, i.e. sensors available in vehicles. This type of data collection in IoT Cloud and at the edge devices for local processing promotes smart transportation or the implementation of an ITS. Therefore, smart transportation or ITS has been considered as an example of an IoT-based application, which supports inter-vehicle communication (IVC) among smart vehicles to promote telematics for the provisioning of a range of applications and/or services. IVC is possible due to the technological progress in VANETs. In VANETs, vehicles equipped with wireless interfaces and homo-/heterogeneous radio technologies, are connected and are able to communicate with each other, known as vehicle-to-vehicle (V2V) or car-to-car (C2C) communication, as well as with fixed roadside equipment known as vehicle-to-infrastructure (V2I), car-to-infrastructure (C2I) or vehicle-to-road (V2R) communications on highways and city roads. These communications (V2V, C2C, V2I, C2I, or V2R) also transport collected data (e.g. parking information, traffic density, driver behavior, etc.) from vehicles to vehicular Cloud through gateways as shown in Figure 7.5. Further, this massive stored information at vehicular Cloud with the help of IoT edge computing infrastructure can be used by city ITS control center to:

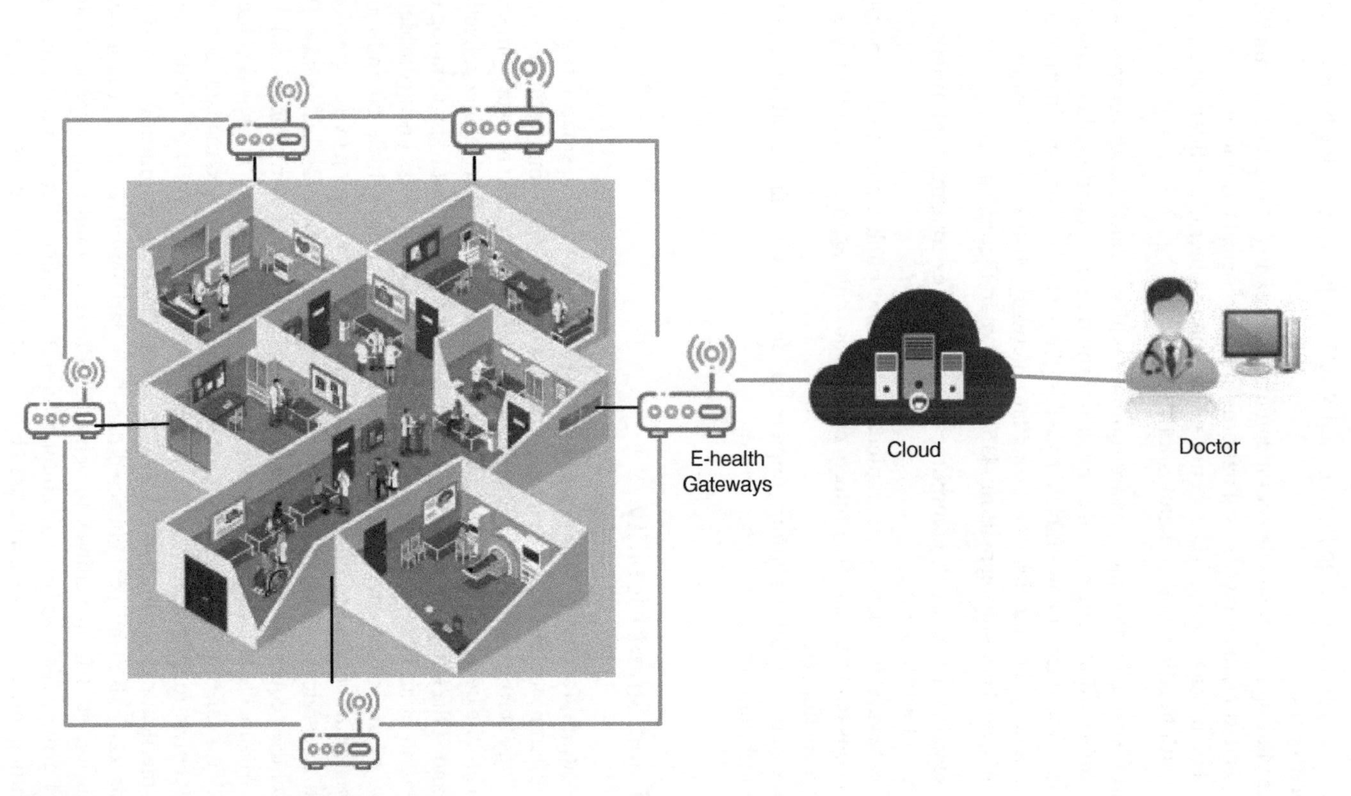


**Figure 7.4** Smart hospital healthcare system.

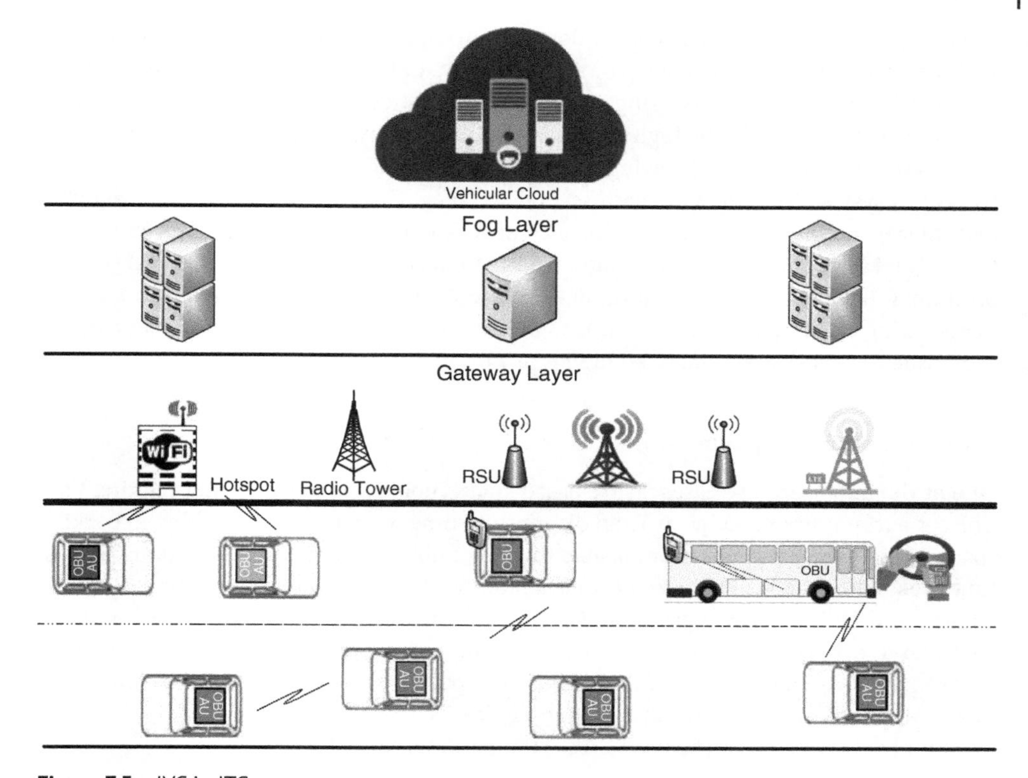


**Figure 7.5** IVC in ITS.

- Avoid traffic congestion on roads.
- Give safety notifications to drivers.
- Reduce carbon footprints in city.
- Detect illegal parking.
- Check toll plaza payments (which are collected with the help of RFID-based systems).
- Observe a number of road accident and their causes.
- Provide relevant notifications about required infotainment data by passengers.
- Provide relevant information to tourists.

IoT-based ITS applications can be classified into three different categories, i.e. safety applications, traffic monitoring/management applications, and infotainment applications [16, 17]. Safety applications are important to reduce the number of accidents on highways and city roads. Examples of VANET safety applications include all driver assistance applications, i.e. overtaking support, emergency braking notification, optimal route selection, accident information, security distance, and traffic jam warnings, to avoid accidents and ultimately promote road safety. Infotainment applications include applications, which provide comfort, Internet-connectivity, and entertainment to on-road drivers and passengers. Traffic monitoring and management applications are useful in a sense that it provides information to drivers about particular road traffic conditions with an ultimate goal to avoid road congestion for the smooth flow of traffic. A summarized view of applications within various domains (i.e. safety, infotainment, and traffic management) of ITS is shown in Figure 7.6a. However, possessing unique characteristics (i.e. predictable mobility, location information availability, high computational capability, frequent fragmentation, and low energy constraints) and having diverse requirements for various applications, IoT-based ITS systems have to cope with several challenges concerning technical (scalability and interoperability of required protocols, network-partitioning/link-disconnectivity due to high mobility, IP address configuration, and mobility management, etc.), social (impact on human behavior), economical (quantification of the cost–benefit relationship, equipment selling or equipment marketing), application (data heterogeneity), and security/privacy issues, [18, 19] as summarized in Figure 7.6b.

### 7.4.1 Car Parking System

One of the important use cases of IoT-based smart mobility is related to car parking [20]. The car parking system helps to avoid dispute and time wastage for car parking in parking spaces. Smart car parking system is able to store, publicize, and update its status (in real time) related to the availability and unavailability of parking space. Moreover, it allows on-spot and online parking payment. On the other hand, using smartphone, a client can view and reserve suitable place for car parking. The parking system architecture consists of many hardware and software components, i.e. parking sensors, local processing unit, mobile application, and Cloud as shown in Figure 7.7.

A number of sensors can be used to assist in parking, e.g. [21]:

- *Camera:* Available in parking slots and are able to detect, process, and transfer information of free parking slots.
- *Ultrasonic Sensors:* Used in the parking area and are able to detect the distance between objects and have been widely used in many parking solutions because of their precision.
- *Infrared Sensor:* Can detect the distance between objects through time calculation of emitted and reflected signal and can be used with ZigBee technology.
- *Magnetometer Sensors:* Through their capability of measuring the magnetic field, these are used to perceive the existence of vehicles in parking slots.
- *Mobile Phones:* Due to the availability of accelerometer, gyroscope, and GPS sensors, mobile phones are capable to acquire information about available parking slots in the parking area.

Other than sensors, RFID tag or ID card is used in car for the authentication of vehicle information.

A Car Parking Processing Unit (CPPU) is a microcontroller (i.e. Arduino, Raspberry Pi, AdaFruit Flora, etc.) contains:

- RFID reader to authenticate information about the vehicle
- Memory to store obtained sensor information from the parking area

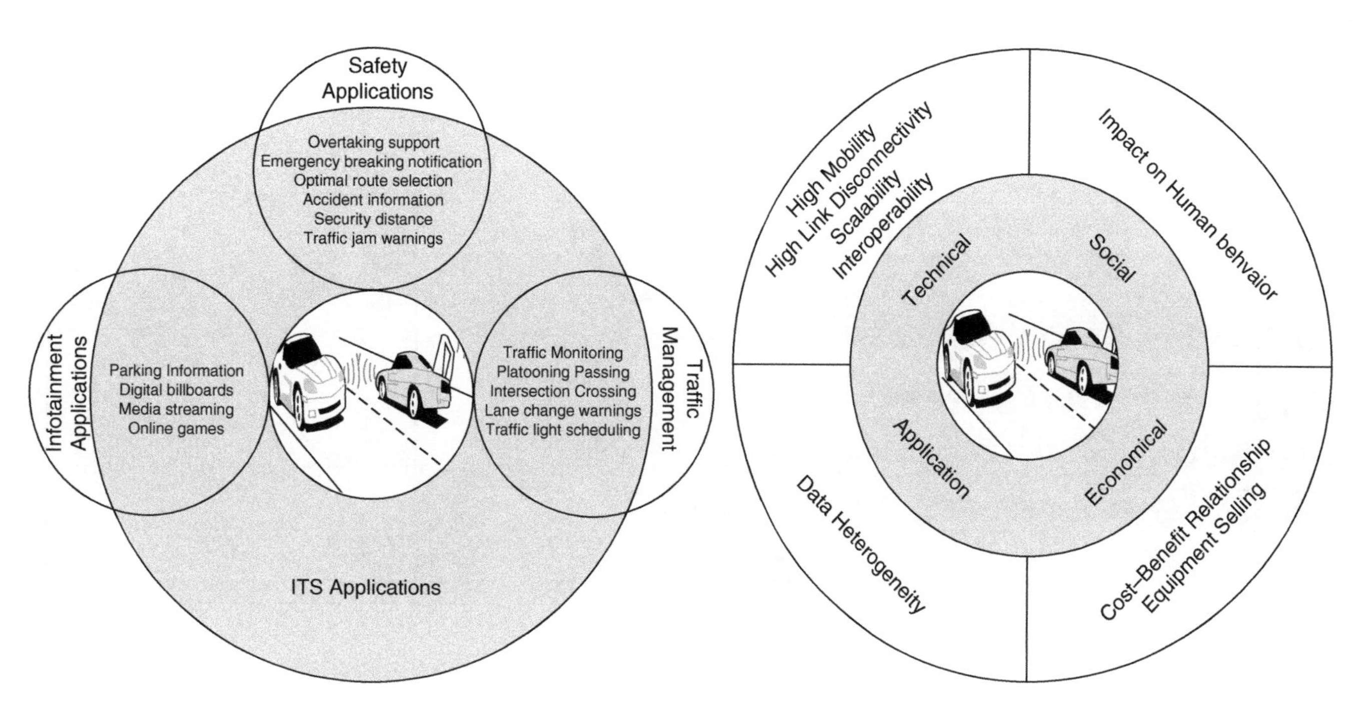


**Figure 7.6** (a) ITS application domains and (b) challenges.

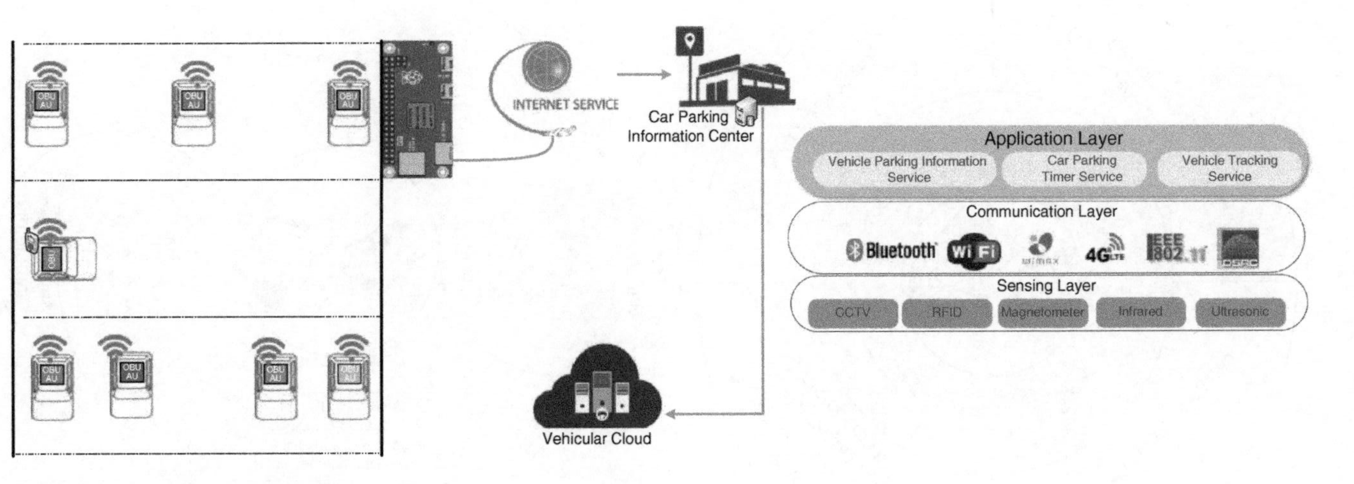


Figure 7.7 Car parking system.

- Display screen to show the parking status of the parking area

*Car Park Database Server* is used to store acquired information and the status of parked vehicles.

Car Parking Network infrastructure consists of both wired and wireless technologies. Wireless technology supports direct communication or transmission of information from sensors or CPPU to Car Park Database Server. However, sometimes CCPU is connected to core network devices (i.e. gateways) through wired links to relay received information.

*Client and Manager Software* are applications running on the client's mobile phone and parking manager's desktop to reserve and visualize parking space, respectively.

*Cloud-based Server* provides a massive database to store, maintain, and backup for quick recovery of all parking records.

The car parking system with related layered architecture has been shown in Figure 7.7.

## 7.5 IoT and Agriculture

Smart agriculture is an automated ICT realization for plant and animal farming with the use of IoT [22, 23]. Smart agriculture is essential to improve crop yield and animal farming for meeting the demands of rapid human population growth. IoT-based smart agriculture practices through precise information and accurate decision support have enabled farmers to improve prevailing solutions of conventional farming, i.e. land-crop suitability, irrigation, and insect/pest control. Through seamless integration of advanced computing technologies (i.e. RFID, sensor, fog/edge, Cloud, visualization, etc.), IoT assists in agriculture, and the key driving factors related to IoT adoption in agriculture are [24, 25] the following:

- Enhancement of crop quality and growth capacity through farmhouse management
- Resource optimization automation in terms of irrigation and the use of fertilizers, pesticides, and herbicides
- High crop yield
- Easy tracking, monitoring, and management of livestock

IoT-based agricultural systems ensure the health and high production of crops and animals, which have been used in human consumption [26]. The major instances [27] in which the IoT technology is helping farmers at different stages of crop and animal growth have been discussed in the following.

### 7.5.1 Major Instances of Crop Growth and IoT

Before crop sowing, comprehensive soil testing is essential to acquire field-specific information, i.e. soil type, topography, nutrient level, moisture quantity, fertilizer amount, etc. Soil testing provides a clear insight into the physical, biological, and chemical status of soil to recognize deficiency factors, which can affect crop production. IoT can play a significant role in the automation of soil monitoring through the development of sensors.

For efficient and high production, accurate estimation about crop water demands are required to be known, which ultimately depends on different factors, i.e. soil type, crop type, irrigation method, etc. Traditional methods of irrigation (i.e. flood and furrow irrigation) and irregular irrigation adversely affect the quality and quantity of field crops. IoT systems can be deployed for optimal use of irrigation water while sensing precise air and moisture level in the soil.

Efficient crop production is also dependent on three fundamental nutrients known as Nitrogen, Phosphorus, and Potassium (abbreviated as NPK). Any nutrient imbalance affects plant health, i.e. Nitrogen deficiency affects leaf growth, Phosphorus for fruit development, and Potassium for stem growth. IoT technology with the provisioning of sensors for the accurate estimation of soil nutrient levels can save time and provide high crop yield.

Crop diseases significantly affect crop yield. With the implication of image processing, IoT-based disease monitoring systems can save time and crop through quick sensing, evaluation, and treatment.

Imbalanced use of pesticides affects the growth of crops and are also harmful to humans. IoT provides real-time monitoring and controlling of pests through the use of sensors, which precisely identify the areas of insect/pest attack in the crop field. Drone technology further assists in precise pesticide spray to effected crop areas. In addition, automated insect/pest traps have been used for trapping and transferring of insect/pest captured data to Cloud for further evaluation.

Proper prediction and an alarming crop harvesting schedule is possible through an IoT-based crop monitoring system and significantly increases crop production. Automatic yield monitoring about various aspects, i.e. grain size, grain quantity, grain mass flow, grain moisture, etc. is helpful to record the performance of crop and associated practices.

Similar to plant/crop farming, IoT applications assist livestock farming including:

- Management of animal shed environment (e.g. remote or automatic opening/closing of farms)
- Management of animal information (e.g. animal age, sex, size, behaviour, demographic information, etc.)
- Monitoring of animals' physiological information (i.e. body temperature, heartbeat, respiratory rate, weight, etc.)
- Monitoring of animals' nutritional status (i.e. food and water intake)
- Monitoring of animals' health (e.g. vaccination status, disease prevention and control through tagging of infected animals [using RFID], and isolation from healthy animals)
- Monitoring and management of shed environment (i.e. detection of dung and cleanliness)
- Monitoring of animals' passage (i.e. Tag/GPS tracking in the large meadow)
- Management of intelligent animal farm processes (i.e. feed scheduling, automatic feeding, control of feed time, etc.)
- Water quality for aquatic life (i.e. temperature, dissolved oxygen, water pH value, etc.)

### 7.5.2 IoT Architecture of Smart Agriculture

The five-layer IoT-based architecture of agriculture is shown in Figure 7.8.

*Perception Layer:* This layer consists of RFID tags/readers and wireless networks of various sensors and advanced agricultural equipment, i.e.:

- Environment Sensors (temperature, humidity, and wind speed sensors)

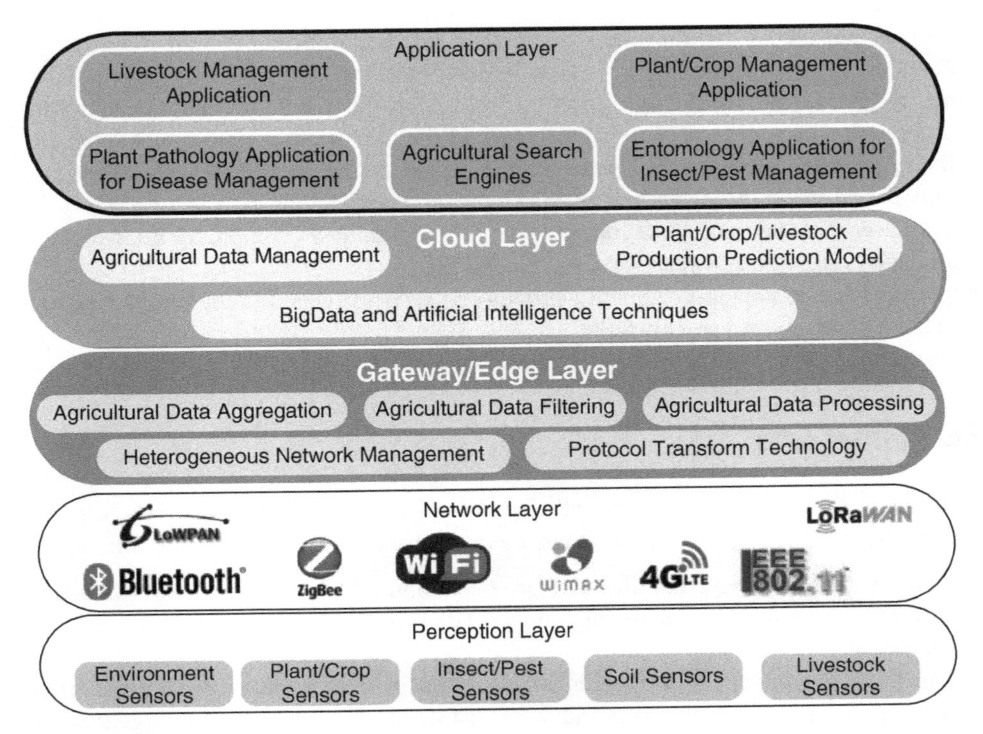


**Figure 7.8** IoT-based agriculture architecture.

- Plant Sensors (leaf sensor, stem sensor, root sensor, plant temperature sensor, twig sensors, fruit growth sensors, sap flow sensors, etc.)
  - *Leaf sensors:* Phytometric devices to measure water deficiency in plants through monitoring moisture levels in leaves
  - *Stem sensors:* Measure the changes in the stem diameter of the plant
  - *Temperature sensors:* Devices to percept plant temperature through the detection of photoreceptor Phytochrome B [28, 29]
- Humidity Sensors to measure the moisture level in the air and in the soil
- Soil Sensors (fertilizer, humidity, and soil temperature sensors)
- Weed detection and identification
- Water level sensor for irrigation
- Insect/pest sensors
- Livestock sensors (RFID tag detector and vital sign sensors)

The sensed information obtained from these sensor nodes is sent to the network layer.

*Network Layer:* This layer mostly consists of various wireless communication devices to form a network, which is responsible to transmit information to the Internet. Different types of wireless technologies (Bluetooth, ZigBee, Wi-Fi, and 2G/3G/4G/5G) can be used to transfer field information to the higher layers of Cloud and Applications. This layer is also responsible for the transmission of control instruction from the higher layer to the actuators at the perception layer to take necessary actions.

*Gateway/Edge Layer or Middleware layer:* IoT solution involves heterogeneous devices with different configurations and specifications, and the role of the middleware is related to the aggregation, filtering, and processing of received data. In addition, this layer also provides edge computing with local processing on received data.

*Cloud Layer:* This layer supports massive storage and intelligent computation for BigData and decision-making, respectively.

*Application Layer:* This layer provides agricultural services related to the monitoring and management of crops and animals.

## 7.6 Smart Grid

Addressing the problems of energy efficiency services to consumers of the traditional power grid, IoT technology promotes the growth of smart grids. The smart grid introduces:

- Smart meters to record/monitor energy consumption and improve the interaction between consumers and energy suppliers related to the monitoring and billing consumed energy
- Distributed energy generators for efficient utilization of several available energy resources
- Renewable energy resources (i.e. wind, waves, sunlight, etc.) for the electricity generation
- Electric vehicles to enhance energy storage and reduction of $CO_2$ emission

With the implication of these approaches, the smart grid is capable to provide interactivity, efficiency, and reliability [30, 31]. Using IoT technology, smart meters are implemented in houses and apartments, which are ultimately connected to a smart grid. This network of smart meters across a geographical area interacts with energy suppliers to provide real-time energy consumption and electricity prices to concerned consumers [30, 32]. Through this interaction, energy suppliers and customers are able to optimize electricity generation and energy consumption, respectively. In addition, IoT Cloud and Edge computing infrastructure support the large volume of collected data storage and local processing (to execute effective operations) near the edge devices, respectively. The core functionalities of the smart grid are optimal energy routing, price adjustment, and security provisioning to preserve the integrity and privacy of collected data [33].

Both wired and wireless network technologies have been used to connect several geographically distributed smart grid appliances, i.e. camera, scanner, smart meters, switches, transformers, reclosers, actuators, field testing devices, etc. [34]. Using different IoT communication protocols, these smart grid appliances (having different bandwidth and latency requirements) are able to exchange information. Figure 7.9 shows the use of different communication technologies from renewable energy resources to data centers.

One of the examples of IoT applications in the smart grid is the monitoring of power transmission lines [35]. Different sensors to observe conductor temperature, wind vibration, and conductor galloping have been deployed on high voltage transmission lines. The

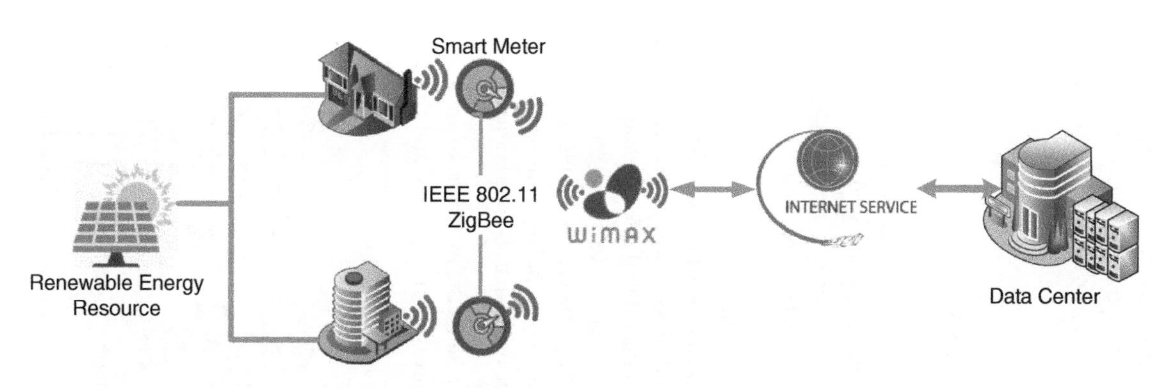


**Figure 7.9** Communication technologies from renewable energy resource to data center.

communication between these sensors and other IoT devices takes place through wired and wireless networks as shown in Figure 7.10.

Electric vehicles in Vehicle-to-Grid (V2G) technology can be considered as the mobile power sources for distributed grids to provide electricity to the grid. In V2G, electric vehicles can be considered as gridable vehicles and are capable to store and discharge electricity from renewable energy sources as shown in Figure 7.11. V2G technology is environment friendly and is able to manage and provide services to the electricity network in terms of price arbitrage.

## 7.7 IoT-based Smart Cities

The smart city is an ultimate integration of various IoT applications (i.e. smart health, smart transportation, smart buildings, smart grid, smart waste management, environmental monitoring, etc.) as shown in Figure 7.12. A smart city paradigm facilitates city municipality to manage public resources and services and improves citizens' quality of life through the provisioning of quality services at a reduced cost [36, 37]. A number of smart city definitions have been reported in literature, and one of the definition states that smart city use information and computing technologies to establish infrastructure components and services of a city including city administration, public safety, health, transportation, education, and utilities in a more intelligent and effective way [38, 39]. The term "smart city" has many conceptual relatives, i.e. digital city, intelligent city, ubiquitous city, information/knowledge city, creative city, humane city, learning city, etc. However, these relatives are different from each other in terms of creative, technology, and social perspectives [38]. The multilayer generic architecture of IoT-based smart cities is shown in Figure 7.12.

*Layer 1:* The bottom layer represents the existence of smart systems covering the domains of ITS, ubiquitous healthcare system, security and fire protection systems, smart grid

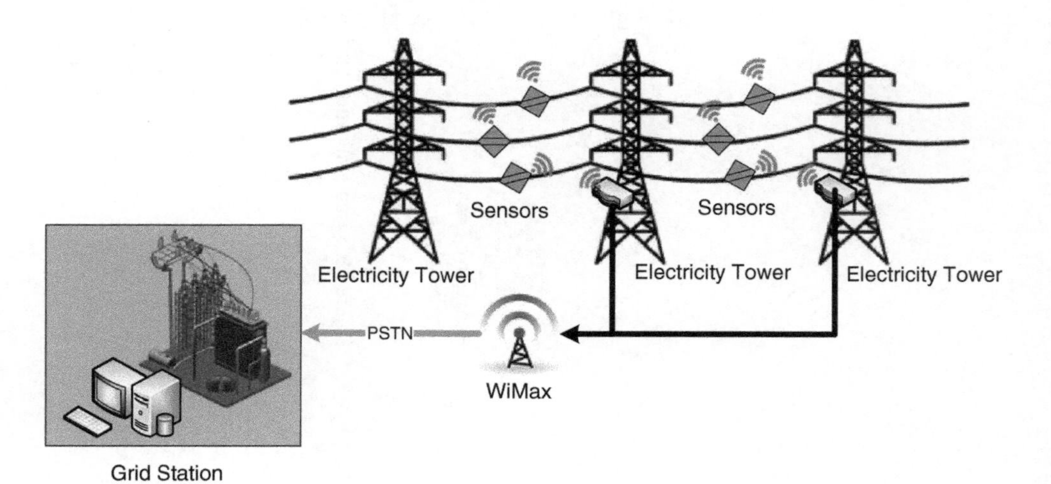


**Figure 7.10** Wireless sensor technology to monitor powerline.

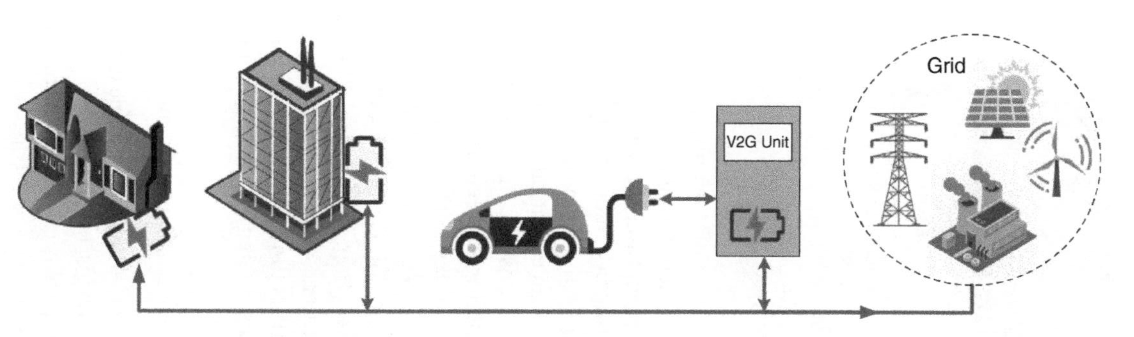


**Figure 7.11** V2G technology concept home and office/industry tags can be put on home and office/industry.

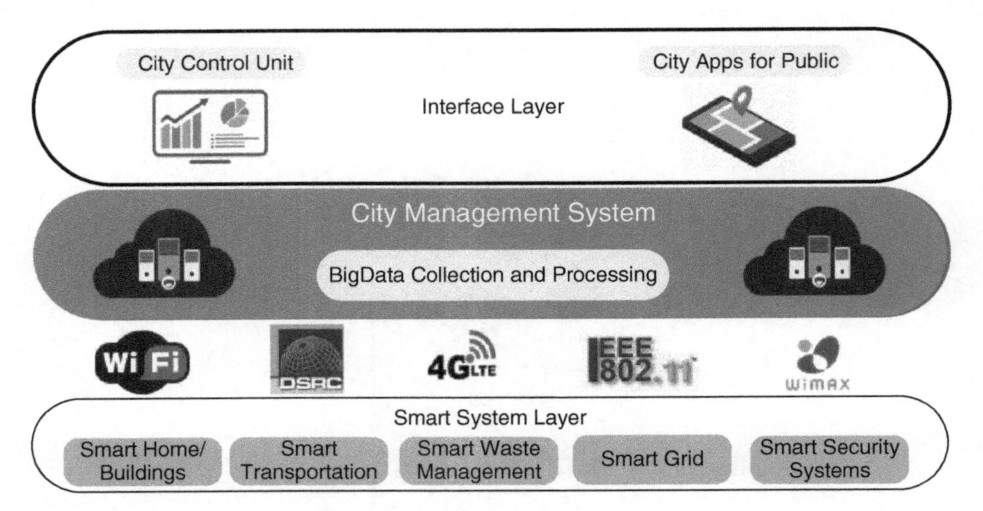


**Figure 7.12** Generic architecture of smart city.

system (related to electricity, gas, central heating, and water services), tax payment system, etc. These smart-systems with different types of sensor networks are able to receive information from city environments and transfer it to the integrated information center for further processing. For example, static traffic light sensors and mobile sensors in vehicles in ITS are able to share gathered information with municipality administration, which ultimately monitor city road conditions, air pollution, and garbage collection service. Similarly, the smart grid ensures monitoring and optimal energy production through the use of smart metering and sensors in electric vehicles available in smart buildings and on roads of the city, respectively.

*Layer 2:* The gathered sensor BigData is large in volume, heterogeneous in nature, and collected at the Integrated Information Center or City Operating platform. This data is transferred from the Integrated Information Center to supplementary platforms (i.e. Cloud center, Management center, Control center, etc.).

*Layer 3:* Citizens and city authorities through the Internet and mobile service platforms are capable to access the services offered by these platforms through web and mobile applications.

The typical challenges for IoT-based smart city applications are:

- Reliability (data collection are sometimes not reliable especially in case of data collection from sensors on highly mobile nodes, i.e. vehicles)
- Highly scalable (large storage space is required to store large-scale information)
- Heterogeneity (specific hardware and software is required to integrate heterogeneous data)
- Social acceptance (user participation demands incentives for data sharing)
- BigData analysis (huge volume of data management)
- Security/Privacy (data collection and analysis at a common place is subjected to attacks)

## 7.8 IoT and Smart Education

The use of IoT in education is highly encouraged to propagate knowledge with the provisioning of smart learning environments at all levels of academia, i.e. school, college, and university. The pervasive use of portable and mobile devices, the development of digital boards, and the easy availability of electronic books are the main driving factors of the use of IoT in education [40, 41]. The IoT in education assists academia through:

- The designing of digital campus (with the provisioning of improved physical/virtual educational infrastructure)
- The modification of classical to smart teacher–student interaction, which enhances the learning quality (with the provisioning of advanced technological tools for continuous monitoring and analysis)

The IoT-based smart campus framework consists of several smart systems [42, 43], for example:

- Smart ID cards (for automatic identification of students/teachers/staff members)
- Ubiquitous availability of wireless network (i.e. Wi-Fi)
- Automatic student attendance system
- Smart Classroom:
  - Automatic cleaning by robots
  - Smart HVAC system
  - Existence of Digital whiteboards
- Personalized learning via smart devices
- Wi-Fi-enabled door/window locks to support entrance and exit
- Smart parking
- Smart bus tracking system
- Smart payments
- Smart surveillance security system
- Automatic audio/video lecture recording and storage on Campus Cloud
- Free availability of educational apps and programs on campus Cloud
- Online testing
- Existence of smart trash receptacles and the use of robots for hygiene
- Smart tracking of campus inventory stock
- Ubiquitous existence of international collaboration for all students
- Smart e-health system or healthcare center
- Beacons availability for emergency situations
- Existence of network analytics for the monitoring of network behavior

IoT-based smart campus helps teachers, students, and staff members in many ways [42, 44], for example,

- Improvement in quality of learning environments
- Time saving (with the replacement of conventional teaching methods to advanced methods)
- Connection among international teachers and students' communities

- Enhanced management and better security of educational institutes
- IoT practice saves energy and promotes reinvestment of saving in the education sector

## 7.9 Industrial IoT

The terms Industrial Internet on Things (IIoT) typically pertains to the interconnectivity of sensors and instruments with the computing applications of manufacturing industries. In simple words, IIoT refers to the connectivity of machines and power utilities available in factories. Therefore, IIoT through the use of IoT technologies enables the monitoring of industrial plants to mitigate the risk of accidental failure of machinery used in industrial plants. For example:

- Sensors have been attached to the conveyer belt of transportation containers (especially dealing with hazardous material) to alarm the control center in case of emergency situations.
- Smart factory warehousing applications.
- Smart logistics, manufacturing process, and freight/asset tracking monitoring applications.
- Industrial HVAC and energy optimizing applications.
- Security alarm applications in the industrial area.
- Automation of production line that is able to track the production, maintenance, and shipping information of all manufacturing goods in an industry as shown in Figure 7.13.

IIoT is different from typical IoT in scale as it deals with several endpoints in manufacturing plants. Integration with legacy technology, lack of standardization, and high cost of implementation are the challenges that are unique to IIoT.

## Review Questions

**7.1** Explain the role of IoT in different domains of real life.

**7.2** Briefly describe the working of an IoT-based smart home.

**7.3** What are the challenges associated with the implementation of an IoT-based healthcare system?

**7.4** Explain the role of IVC in ITS.

**7.5** Explain the architecture of IoT-based agricultural system.

**7.6** What are the advantages associated with the vehicle to grid technology?

**7.7** What are the typical challenges for IoT-based smart city applications?

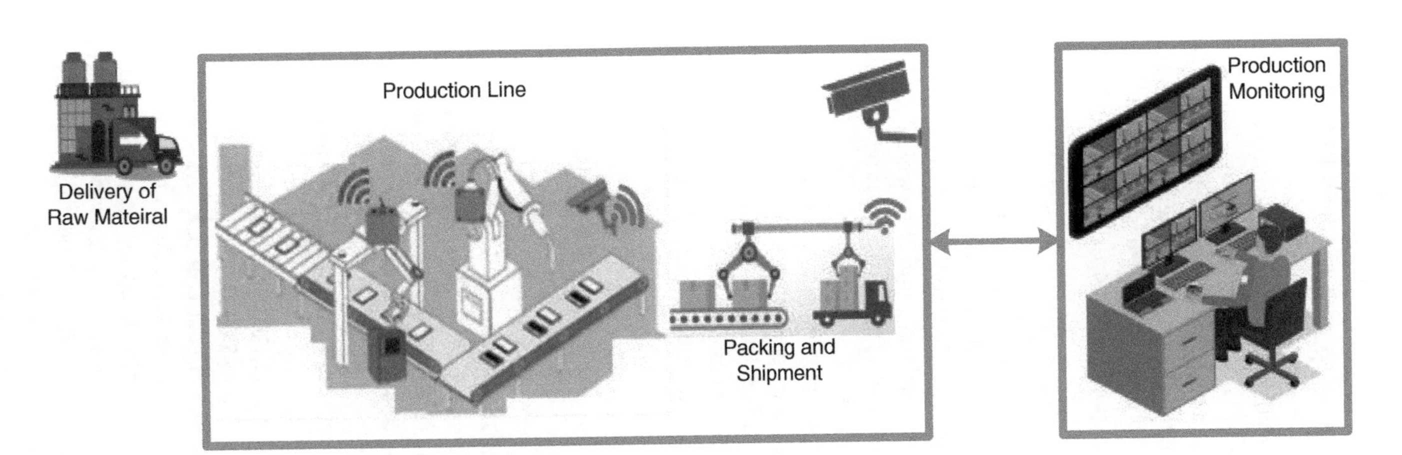


**Figure 7.13** Tracking and monitoring of production line.

**7.8** With the help of an architecture diagram, explain the role of IoT in education?

**7.9** How the use of IoT technology mitigates the risk of industrial plant failure?

**7.10** Plant diseases and insect/pest attacks significantly decrease the quality and quantity of Cotton crops. Propose an IoT-based system that is used by farmers and researchers to monitor and control diseases and insect/pest attacks on cotton crops. Consider the advancements in different technologies, i.e. image processing for detection of diseases, sensing of insects/pests, drone-based spray of pesticide/insecticides, automated traps, etc., as part of proposed IoT system, and explain the working of your solution.

**7.11** The detection, collection, and management of garbage in a metropolitan are the primary concerns, which impact the health of individuals in that environment. The traditional way of garbage detection, collection, and management is very difficult and requires more human effort, time, and cost, which can be replaced with IoT-based technologies. City municipality is interested in an IoT-based solution, which provides a web-based graphical view of:

**A** Collected garbage level in garbage bins in different locations of metropolitan
**B** Status of toxic gases in garbage bins
**C** Mobility status of garbage trucks
**D** Status of city recycling plants

Propose an IoT-based solution and illustrate the placement and working of all components of the designed system.

## References

1 Sethi, P. and Sarangi, S.R. (2017). Internet of things: architectures, protocols, and applications. *Journal of Electrical and Computer Engineering* 2017: 1–25.
2 Al-Fuqaha, A., Guizani, M., Mohammadi, M. et al. (2015). Internet of things: a survey on enabling technologies, protocols, and applications. *IEEE Communication Surveys and Tutorials* 17 (4): 2347–2376.
3 Borgia, E. (2014). The internet of things vision: key features, applications and open issues. *Computer Communications* 54: 1–31.
4 Park, S.H., Won, S.H., Lee, J.B. et al. (2003). Smart home–digitally engineered domestic life. *Personal and Ubiquitous Computing* 7 (3–4): 189–196.
5 Alaa, M., Zaidan, A.A., Zaidan, B.B. et al. (2017). A review of smart home applications based on internet of things. *Journal of Network and Computer Applications* 97: 48–65.
6 Harper, R. (2006). *Inside the Smart Home*. Springer Science & Business Media.
7 Stojkoska, B.L.R. and Trivodaliev, K.V. (2017). A review of internet of things for smart home: challenges and solutions. *Journal of Cleaner Production* 140: 1454–1464.
8 Wilson, C., Hargreaves, T., and Hauxwell-Baldwin, R. (2017). Benefits and risks of smart home technologies. *Energy Policy* 103: 72–83.

9 Nations, U. (2015). Transforming our world: the 2030 agenda for sustainable development. General Assembly 70 session.

10 Wu, T., Wu, F., Redoute, J.-M. et al. (2017). An autonomous wireless body area network implementation towards IoT connected healthcare applications. *IEEE Access* 5: 11413–11422.

11 Movassaghi, S., Abolhasan, M., Lipman, J. et al. (2014). Wireless body area networks: a survey. *IEEE Communication Surveys and Tutorials* 16 (3): 1658–1686.

12 Khan, J.Y. and Yuce, M.R. (2010). Wireless body area network (WBAN) for medical applications. In: *New Developments in Biomedical Engineering*, vol. 31, 591–627. InTechOpen.

13 Ullah, S., Higgins, H., Braem, B. et al. (2012). A comprehensive survey of wireless body area networks. *Journal of Medical Systems* 36 (3): 1065–1094.

14 Hanson, M.A., Powell, H.C. Jr., Barth, A.T. et al. (2009). Body area sensor networks: challenges and opportunities. *Computer* 42 (1): 58–65.

15 Rahmani, A.M., Gia, T.N., Negash, B. et al. (2018). Exploiting smart e-health gateways at the edge of healthcare internet-of-things: a fog computing approach. *Future Generation Computer Systems* 78: 641–658.

16 Karagiannis, G., Altintas, O., Ekici, E. et al. (2011). Vehicular networking: a survey and tutorial on requirements, architectures, challenges, standards and solutions. *IEEE Communication Surveys and Tutorials* 99: 1–33.

17 Ting Cheng, H., Shan, H., and Zhuang, W. (2010). Infotainment and road safety service support in vehicular networking: from a communication perspective. *Mechanical Systems and Signal Processing*: 2020–2038.

18 Hartenstein, H. and Laberteaux, K. (2010). *VANET Vehicular Applications and Inter-Networking Technologies*. West Sussex, UK: Wiley Online Library.

19 Blum, J.J., Eskandarian, A., and Hoffman, L.J. (2004). Challenges of intervehicle ad hoc networks. *IEEE Transactions on Intelligent Transportation Systems* 5 (4): 347–351.

20 Khanna, A. and Anand, R. (2016). IoT based smart parking system. In: *IEEE International Conference on Internet of Things and Applications (IOTA)*. IEEE.

21 Barriga, J.J., Sulca, J., León, J.L. et al. (2019). Smart parking: a literature review from the technological perspective. *Applied Sciences* 9 (21): 4569.

22 Patil, K. and Kale, N. (2016). A model for smart agriculture using IoT. In: *IEEE International Conference on Global Trends in Signal Processing, Information Computing and Communication (ICGTSPICC)*. IEEE.

23 Verdouw, C., Wolfert, S., and Tekinerdogan, B. (2016). Internet of things in agriculture. *CAB Reviews: Perspectives in Agriculture, Veterinary Science, Nutrition and Natural Resources* 11 (35): 1–12.

24 Shi, X., An, X., Zhao, Q. et al. (2019). State-of-the-art internet of things in protected agriculture. *Sensors* 19 (8): 1833.

25 Marković, D., Koprivica, R., Pešović, U. et al. (2015). Application of IoT in monitoring and controlling agricultural production. *Acta Agriculturae Serbica* 20 (40): 145–153.

26 Yan-e, D. (2011). Design of intelligent agriculture management information system based on IoT. In: *IEEE Fourth International Conference on Intelligent Computation Technology and Automation*. IEEE.

**27** Ayaz, M., Ammad-Uddin, M., Sharif, Z. et al. (2019). Internet-of-things (IoT)-based smart agriculture: toward making the fields talk. *IEEE Access* 7: 129551–129583.

**28** Sakamoto, T. and Kimura, S. (2018). Plant temperature sensors. *Sensors* 18 (12): 4365.

**29** Legris, M., Klose, C., Burgie, E.S. et al. (2016). Phytochrome B integrates light and temperature signals in *Arabidopsis*. *Science* 354 (6314): 897–900.

**30** Lin, J., Yu, W., and Yang, X. (2015). Towards multistep electricity prices in smart grid electricity markets. *IEEE Transactions on Parallel and Distributed Systems* 27 (1): 286–302.

**31** Lin J., Yu W., Yang X., et al. (2012) On false data injection attacks against distributed energy routing in smart grid. IEEE/ACM Third International Conference on Cyber-Physical Systems.

**32** Zhang, X., Yang, X., Lin, J. et al. (2015). Towards efficient and secured real-time pricing in the smart grid. In: *2015 IEEE Global Communications Conference (GLOBECOM)*. IEEE.

**33** Yu, W., Griffith, D., Ge, L. et al. (2015). An integrated detection system against false data injection attacks in the smart grid. *Security and Communication Networks* 8 (2): 91–109.

**34** Al-Ali, A. (2015). Role of internet of things in the smart grid technology. *Journal of Computer and Communications* 3 (05): 229–233.

**35** Liu, J., Li, X., Chen, X. et al. (2011). Applications of Internet of Things on smart grid in China. In: *IEEE 13th International Conference on Advanced Communication Technology (ICACT2011)*. IEEE.

**36** Lin, J., Yu, W., Zhang, N. et al. (2017). A survey on internet of things: architecture, enabling technologies, security and privacy, and applications. *IEEE Internet of Things Journal* 4 (5): 1125–1142.

**37** Arasteh, H., Hosseinnezhad, V., Loia, V. et al. (2016). Iot-based smart cities: a survey. In: *IEEE 16th International Conference on Environment and Electrical Engineering (EEEIC)*. IEEE.

**38** Nam, T. and Pardo, T.A. (2011). Conceptualizing smart city with dimensions of technology, people, and institutions. In: *Proceedings of the 12th Annual International Digital Government Research Conference: Digital Government Innovation in Challenging Times*.

**39** Washburn, D., Sindhu, U., Balaouras, S. et al. (2009). Helping CIOs understand "smart city" initiatives. *Growth* 17 (2): 1–17.

**40** Majeed, A. and Ali, M. (2018). How Internet-of-Things (IoT) making the university campuses smart? QA higher education (QAHE) perspective. In: *IEEE 8th Annual Computing and Communication Workshop and Conference (CCWC)*. IEEE.

**41** Zhu, Z.-T., Yu, M.-H., and Riezebos, P. (2016). *A research framework of smart education. Smart Learning Environments* 3 (1): 4.

**42** Abdel-Basset, M., Manogaran, G., Mohamed, M. et al. (2019). *Internet of things in smart education environment: supportive framework in the decision-making process. Concurrency and Computation: Practice and Experience* 31 (10): e4515.

**43** Sari, M.W., Ciptadi, P.W., and Hardyanto, R.H. (2017). Study of smart campus development using internet of things technology. In: *IOP Conference Series: Materials Science and Engineering*. IOP Publishing.

**44** Gul, S., Asif, M., Ahmad, S. et al. (2017). A survey on role of internet of things in education. *IJCSNS* 17 (5): 159.

# 8

# IoT Security

**LEARNING OBJECTIVES**

After studying this chapter, students will be able to:

- describe security constraints in IoT systems
- elaborate security requirements of IoT systems
- classify the nature of IoT attacks
- analyze the security threats at each layer of IoT architecture
- design secure IoT system for specific application

## 8.1 IoT Systems and Security Constraints

With the proliferation and increased utilization of low cost IoT devices, societies will become more connected and hence will be more susceptible to cyber-attacks. The implementation of security mechanism in IoT systems is different and more challenging than application of security in conventional wired and wireless networks. For better comprehension about the issues of IoT security, as a prerequisite it is essential to overview the working and connectivity of fundamental components of IoT ecosystems (i.e. smart things, coordinator, IoT gateway, network [internet/Internet], IoT Cloud, IoT applications, and end-user devices) as shown in Figure 8.1.

*IoT Smart Thing*: Smart things (consisting of sensors, communication interfaces, operating systems, and actuators) are mainly responsible for the collection of sensor data and its transmission to the connected coordinator or gateway.

*Coordinator*: Coordinator is responsible for the management of associated multiple smart devices (or sometimes sensors).

*Networking Infrastructure*: Different types of wired and wireless networks including Wi-Fi, Bluetooth, ZigBee, LoRAWAN, etc. enable the connectivity and communication of smart things to the Internet Protocol (IP)-based network infrastructure through IoT Gateways.

*IoT Gateway*: IoT gateway can be a dedicated physical device or software that assists connectivity between devices and Cloud. At one end, IoT gateway is connected to sensor devices or coordinator and on the other side with the IoT Cloud. Sensor-acquired data

*Enabling the Internet of Things: Fundamentals, Design, and Applications*, First Edition.
Muhammad Azhar Iqbal, Sajjad Hussain, Huanlai Xing, and Muhammad Ali Imran.

moving toward the IoT Cloud passes through the gateway that preprocesses the sensor data at the edge. Preprocessing on large volumes of sensor data involves the compression of aggregated data to reduce transmission costs. IoT gateway performs the translation of different network protocols to support interoperability of smart things and connected devices.

In literature, sometimes the terms coordinator and gateway have been used interchangeably, and sometimes coordinator also performs the job of gateway, i.e. preprocessing and transmission of data to the Cloud.

*IoT Cloud*: IoT Cloud is a network of high-performance servers that stores, processes, and manages massive amounts of data for analysis.

*End-User Devices*: IoT applications are accessible to the user on mobile and other computing devices. Modern technology related to mobile and web interfaces has high significance as it offers a considerable interactive and user-friendly style to improve customer experience.

*IoT Applications*: IoT applications available on IoT Cloud are globally accessible to users. Covering all domains of IoT applications (e.g. from simple home/building automation to complex industrial automation), IoT security includes the physical device security, data security, network security, and Cloud security to protect IoT systems from a broad spectrum of IoT specific attacks. However, the actual practice of security approaches on diverse range of IoT devices is challenging and demands the acquaintance of associated security constrains.

Concerning the implication of security in IoT systems, besides the identification of IoT system components, it is also important to consider the end-to-end view of IoT systems to

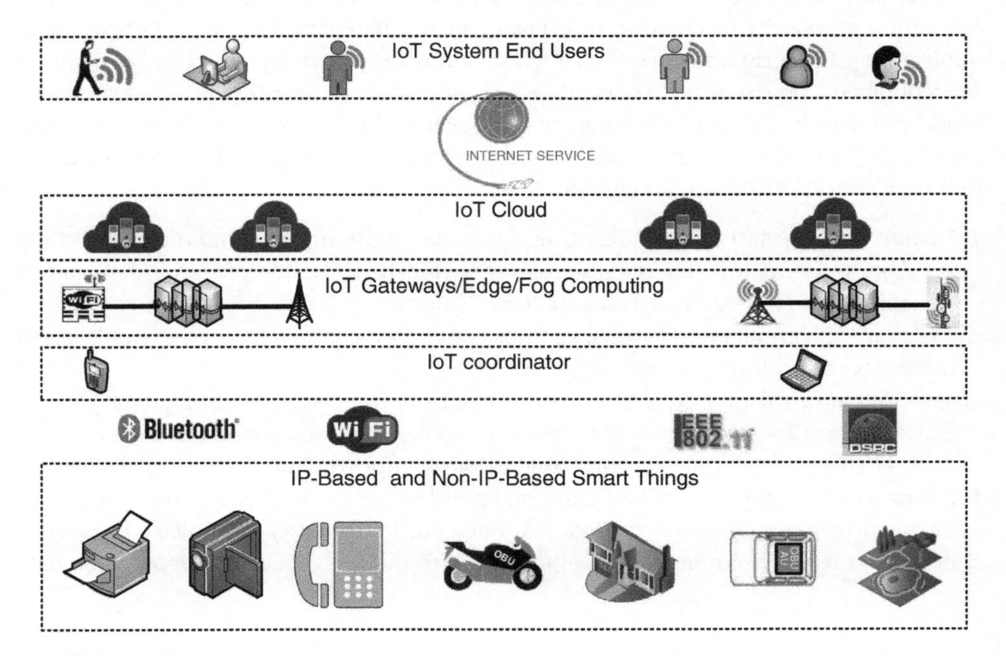


**Figure 8.1** IoT ecosystem.

identify the correspondence of basic functionalities to various components of IoT Systems [1], for example:

- Upgrading of smart thing firmware (e.g. any piece of code for sensing)
- Pairing of smart things with concerned controller device (e.g. IoT Coordinator)
- Binding involves the configuration of smart things through controller (binding of thing to the Internet through controller)
- Local Authentication of controller to a port that is open on a smart thing
- Local Control of smart things through commands from controller (e.g. IoT Coordinator and local wireless connection)
- Remote Authentication and Remote Control are required in case the controller is connected to smart things through the Internet
- Sensing and Notification of smart things to controller
- Data Analytics performed on Edge devices or Fog nodes
- BigData Analytics performed on Cloud

Considering functionalities of IoT components, it becomes evident that IoT systems are exposed to two types of threats, i.e. *threats against IoT* and *threats from IoT*, and are especially very attractive for hackers [1].

Examples of threats against IoT includes:

- IP camera hacking through buffer overflow attacks [2]
- A Distributed Denial-of-Service (DDoS) attack on Dyn Servers [3]
- Botnet attack to hack IoT devices [4]
- SQL injection attacks

Threats from IoT include cross-site scripting attack that are launched to access private data/resources in IoT systems and privacy risk of people from Unmanned Aerial Vehicle (UAV) [5].

In addition to these two types of threats, IoT devices at sensing level are fundamentally resource-constrained in terms of hardware, software, and communication. These limitations of IoT devices must be taken into account before developing security mechanisms in IoT system. Moreover, these limitations hinder the employment of traditional cryptographic algorithms in IoT systems. The main security constraints are based on limitation of associated IoT hardware, software, and network/communication equipment [6].

### 8.1.1 IoT Security Constraints Based on Hardware Limitations

- IoT sensing devices have energy and computing limitation in terms of small battery and low-power CPU with low clock rate, respectively. Therefore, computationally expensive security algorithms cannot be implemented on these devices.
- Sensing devices in IoT systems have limited storage capacity and hence demand memory efficient security approaches.
- Tamper-resistant security algorithms are preferred for remotely deployed unattended IoT devices.

### 8.1.2 IoT Security Constraints Based on Software Limitations

- Robust communication protocol stack and dynamic security patches are difficult to be implemented on thin embedded operating systems installed on IoT sensing devices.

### 8.1.3 IoT Security Constraints Based on Communication Limitations

- Device heterogeneity, scalability, presence of multiple communication interfaces/protocols, and portability/mobility characteristics limit the employment of conventional security protocols in IoT systems.

These IoT security constraints of exclusive nature are required to be addressed to prevent the risks and threats of personal information misuse. In addition to the consideration of these limitations of IoT devices, numerous other IoT security requirements are essential to be addressed [7–9].

## 8.2 IoT Security Requirements

IoT security requirements can be of different types, i.e. [6]:

- Information-level security requirements
- Access-level security requirements
- Functional-level security requirements

### 8.2.1 Information-level Security Requirements

*Confidentiality*: Data accessibility to only authorized individuals and assurance about the protection of data privacy and proprietary information (anonymity or hiding data source) is known as confidentiality. It hinders the eavesdropping and interference of unauthorized users. In IoT systems, data confidentiality is essential as the sources of large volumes of data are RFID devices that are exposed to neighboring devices. Several approaches of secure key management have been proposed in literature to achieve data confidentiality in IoT systems [10, 11].

*Integrity*: Integrity prevents the tampering of sensor data and modifications of IoT devices by unauthorized individuals and other smart objects. Due to the sensitivity of sensed data, integrity is crucial in IoT applications because the forged data or incorrect feedback can be hazardous for the normal operation of IoT systems. To implement integrity, a number of secure data schemes (i.e. false data filtering techniques [12] and blockchain-based integrity) have been proposed in literature [10]. Due to high overhead, conventional integrity approaches are not suitable for energy-constrained IoT sensing devices. Therefore, low-power data integrity techniques are also proposed [13].

*Non-repudiation*: Non-repudiation ensures the certainty against the denial of sent messages/data claims.

*Freshness*: It confirms the recency of sent or received messages/data in IoT systems.

### 8.2.2 Access-level Security Requirements

*Authorization*: Authorization guarantees the access of authorized users/devices to access resources.

*Identification and Authenticity*: Identification ensures the authorization of devices in an IoT system, and authentication ensures the credibility of information/transaction and legitimacy of requester's IoT applications/devices. Due to the involvement of various factors (i.e. device heterogeneity, scalability, and complexity), conventional authentication approaches (i.e. password-protection, preshared secrets, and public-key cryptosystems) are not feasible to implement in IoT systems. Therefore, for the identification and authentication of IoT devices/information, different mechanisms have been developed [14–16].

*Access control*: It confirms the authenticated IoT devices' ability of accessing authorized resources only.

### 8.2.3 Functional Security Requirements

*Availability*: Availability refers to the assurance of IoT information and computing resources' (i.e. sensors, computing system, network, and storage) availability at the time of demand or in the case of power loss and failures. In IoT systems, availability is critical as IoT services are required to provide and receive data in real time. Secured and efficient routing techniques have been proposed for guaranteed delivery of messages in IoT systems [17]. Few routing schemes have been proposed to avoid Denial-of-Service (DoS) attacks [18].

*Exception Handling and Resiliency*: Exception handling and resiliency ensures liveliness and normal working of IoT devices in case of hardware malfunctioning and software glitches.

*Self-organization*: This confirms the adequate level of security in IoT systems even in cases of devices' failure or energy drainage.

## 8.3 Security Challenges

Considering the security constraints and requirements of machine-to-machine and human-to-machine interactions in IoT systems, several challenges are required to be addressed:

- *Efficient Cryptography Techniques*: Cryptographic algorithms/techniques must be efficient enough to execute on resource-constrained IoT devices.
- *Interoperability*: Security procedure must not limit the functional capability of IoT device.
- *Scalable Solution*: Security mechanism must be able to cope with the scalability of IoT systems.
- *Privacy Protection*: Personal data must not be identifiable to attackers.

- *Resilience to Physical Attacks*: Protection is required from hardware theft/damage and natural disaster.
- *Autonomous Control*: Autonomous configuration settings mechanism of IoT devices is required to be developed.
- *Cloud Security*: Large volumes of personal information stored on IoT devices demand high security

## 8.4 Taxonomy of IoT Security Threats/Attacks

IoT security attacks and threats have been classified in a number of ways. Figure 8.2 provides a comprehensive classification of IoT security attacks [6, 19, 20].

### 8.4.1 IoT Security Attacks Based on Device Category

IoT devices can be categorized as low-power devices and high-power devices. IoT attack launched with low-power devices can be disastrous enough to change the normal behavior of devices in IoT system. For example, home appliances can be controlled remotely through short-range wireless capabilities of low-power smart watches. On the other hand, high-power devices (workstations, laptops, desktops, etc.) are also capable to launch attacks from anywhere through the use of the Internet which result in fatal errors in IoT systems.

### 8.4.2 Attacks Based on Access Level

Passive and active security attacks affect the confidentiality and integrity of IoT systems, respectively. Passive attacks threaten the confidentiality of IoT system through the monitoring of ongoing transmission and eavesdropping without an interruption. Active attacks are launched to disrupt the network communication and alter the information of ongoing transmission. In other words, active attacks threaten both the confidentiality and integrity of IoT systems [6, 21, 22].

### 8.4.3 Attacks Based on Attacker's Location

Internal attacks are launched by attackers that are residing inside the network of IoT system and execute malicious programs of different nature. On the other hand, external attacks are launched outside of native network (through public network) with unauthorized remote access.

### 8.4.4 Attacks Based on Attack Strategy

Physical attacks include the physical impairment or damage to devices' configuration and properties in IoT system. Contrary, logical attacks affect the functionality of IoT system without physical involvement.

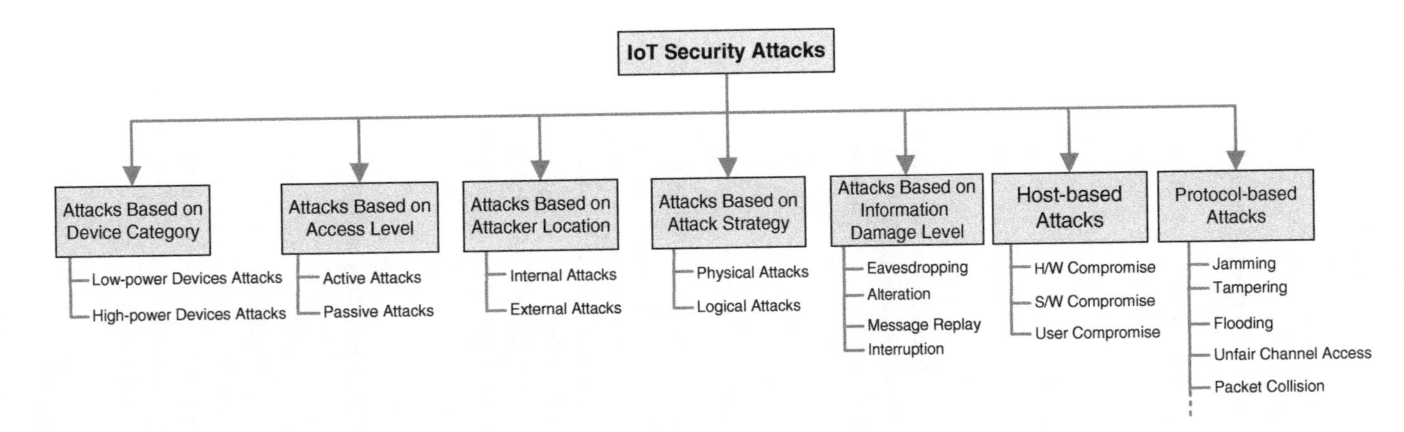


**Figure 8.2** Classification of IoT security attacks.

### 8.4.5 Attacks Based on Information Damage Level

Information damage at different levels is possible through man-in-the-middle attack, i.e.:

*Eavesdropping*: Passive listening of ongoing transmission.
*Alteration*: Unauthorized access and tampering of information.
*Fabrication*: Deliberate way of introducing false information to create misperception in ongoing communication.
*Message Replay*: Affects the communication in terms of message freshness.
*Interruption*: Unavailability of services through disruption or disaster.

### 8.4.6 Host-based IoT Attacks

These attacks target private sensitive information that is part of hardware and embedded software on IoT devices including:

*User Compromise*: Entrapping of user to disclose security credentials.
*Software Compromise*: Exploiting the vulnerabilities of system's software on IoT device.
*Hardware Compromise*: Tampering of IoT device hardware to extract embedded information.

### 8.4.7 Protocol-based Attacks

These types of attacks imperil service availability by deviating either from the normal working of standard communication protocols, i.e. at different layers of Transmission Control Protocol (TCP)/IP or from communication stack (including Physical, Data Link, Network, Transport, and Application). Jamming, tampering, packet collision, unfairness of channel access, wormhole, and flooding are examples of these kinds of attacks [7, 23]. In addition, attacks related to the disruption of standard protocol (i.e. malicious attack on data aggregation protocol, key management protocol, information extraction protocol, etc.) are also part of this category.

## 8.5 IoT Architecture and IoT Security

Concerning the requisites of specific IoT application, a robust security mechanism is required to be identified and developed. However, IoT security problems and possible solutions can be best described at different layers of IoT architecture. Therefore, the three-layer IoT architecture (consisting of Perception, Network, and Application layers) mentioned in Chapter 1 of this book has been taken to explain IoT security problems and possible countermeasures as shown in Figure 8.3.

### 8.5.1 Perception Layer Security

Perception layer at the bottom of IoT architecture is responsible for the collection of various types of information through physical sensors or components of smart things (i.e.

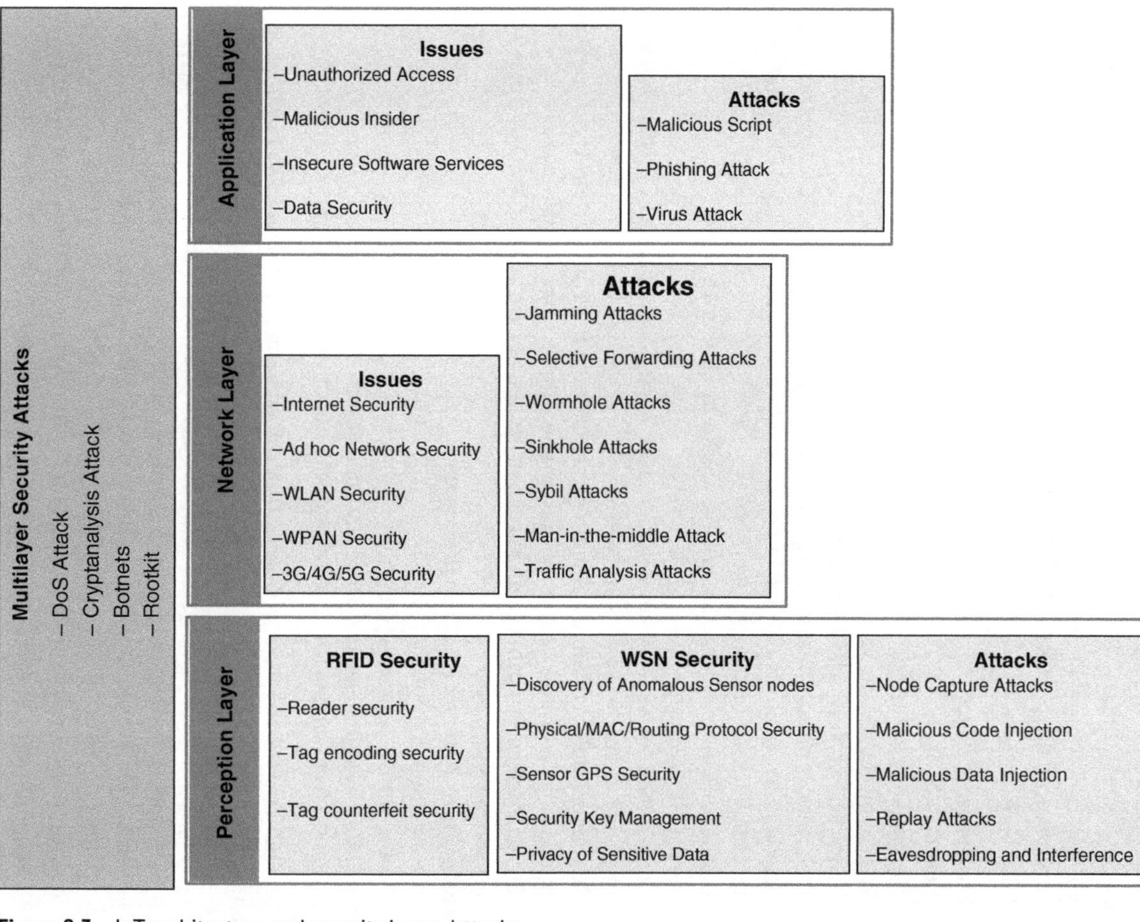


**Figure 8.3** IoT architecture and security issues/attacks.

RFID, sensors, Implantable Medical Devices [IMDs], Global Positioning System [GPS], etc.). In other words, the main functionality at perception layer is related to the recognition and perception of environmental factors (i.e. temperature, humidity, pressure, light, sound, etc.) through the use of low-power and nanoscale technology in smart things. Moreover, this layer controls the transmission or exchange of processed information to upper network layer via service interfaces. Perception node and perception network are two parts of perception layer that is used for data acquisition and communication, respectively.

Concerning security, following challenges are required to be addressed at perception layer: [7]

- RFID Security Issues:
  - Reader security
  - Tag encoding security
  - Tag counterfeit security
- Wireless Sensor Network (WSN) security issues
  - Discovery of physically attacked anomalous (or faulty) sensor nodes to restrict further degradation of sensor network by using faulty node detection algorithm and decentralized intrusion detection algorithm
  - Physical/Medium Access Control (MAC)/Routing Protocol Security
  - Sensor GPS security
  - Management of security key (including key generation, storage, distribution, updating, and destruction of security key)
  - Selection/Implementations of appropriate cryptographic algorithms (i.e. Low-power public key algorithms), e.g. NtruEncrypt and Elliptic Curve Cryptography
  - Privacy of sensitive data (i.e. medical data from body sensors and IMD devices) while preserving anonymity
- RFID-Sensor Network (RSN) security issues:
  - Sensor + tag security
  - Sensor + RFID tag reader security
- Prevention from damage in case of natural disasters
- Prevention from damage/misuse of physical infrastructure from human activity

The following are the types of attacks that can be launched at perception layer [10], for example,

- *Node Capture Attacks*: Include the controlling of IoT device(s) via hardware tampering or physical node replacement to capture sensitive and important information of IoT system.
- *Malicious Code Injection Attacks*: Attacker injects malicious code that changes the execution course to perform illegal functions in IoT system, i.e. controlling of system and transmission of sensitive information to the outside network. These attacks are possible through improper handling and validation of inputs and outputs of a system.
- *Malicious Data Injection Attacks*: Injection of false data in captured node to affect the normal working of IoT system.
- *Replay Attacks*: Adversary exploits malicious node in IoT system to steal and transmit legitimate identification information to the destination host for further correspondence.

- *Eavesdropping and Interference*: Wireless communication among various IoT devices is serious vulnerability in terms of eavesdropping and interference. Through eavesdropping, unauthorized users or malicious nodes are able to retrieve private and sensitive information. Encryption, key management, and noise filtering techniques can protect system from these types of attacks.
- *Sleep Deprivation Attacks*: Sensing devices in IoT system are energy-constrained and are automated to follow alternate sleeping and awaking period to save energy. Sleep deprivation attacks disrupt the sleeping schedule and keep the node awake till the drainage of whole node energy. Energy-harvesting techniques can be implemented to utilize energy from environmental sources.

Concerning security implementation at perception layer, human-based physical attacks and natural disaster threats are required to be addressed and managed in the following ways:

- Secure sensor design
- Secure sensor deployment
- Secure infrastructure
- Efficient user authentication approach (biometric or smart card) to implement for legitimate access to physical devices and confidential information
- Implement efficient accessibility control mechanisms
- Efficient implementation of trust management
- Efficient hardware failure recovery schemes

### 8.5.2 Network Layer Security

The middle layer in a three-layer IoT architecture is Network (also known as Communication) layer. Network layer accepts processed information from perception layer and reroutes the received data to distant application interface(s) by using integrated networks, the Internet and other communication technologies. A number of communication technologies (i.e. WLAN, Wi-Fi, Long-Term Evolution [LTE], Bluetooth Low Energy [BLE], Bluetooth, 3G/4G/5G, PSTN, etc.) is integrated in IoT gateways that handle heterogenous types of data to or from different things to applications and vice versa. In addition to network operations, the Network layer in some cases enhances to perform information operations within the Cloud. Regardless of the presence of appropriate encryption techniques at IoT Perception layer, it is indispensable to deal network attacks that are responsible for the exchange of messages between different components of IoT systems.

The well-known IoT network layer attacks are:

- *Jamming Attacks*: Any kind of these attacks (constant, reactive, random, and deceptive) hampers node communication through the utilization of communication channel.
- *Selective Forwarding Attacks*: Malicious devices are used to demolish the established routing paths in network through planned prohibition of the transmission of few packets, for example:
  - Blackhole attack (malicious nodes do not allow passing of any packet through them)
  - Neglect and Greed attack (malicious nodes restrict the forwarding of few packets)

- *Sinkhole Attacks*: Endanger the availability of services/system by advertising a certain path in network to directing all communication traffic on that path.
- *Wormhole Attacks*: Like Sinkhole attacks, these target the availability of the system where two malicious nodes start communication directly (ignoring intermediate relay nodes) with private links to accomplish deceitful one-hop transmission.
- *Sybil Attacks*: Malicious devices create multiple legitimate identities and through their false impersonation control the network and able to transmit wrong information in IoT system.
- *Hello Flood Attacks*: Malicious nodes exploit specific network joining "hello" packet formats to claim itself a legal neighbor in IoT system.
- *Traffic Analysis Attacks*: Various software analysis packages (i.e. Wireshark, Cloudshark, Omnipeek, Kismet, etc.) consist of two components, i.e. sniffer and protocol analyzer that are used to capture network traffic and to perform decoding of packets for analysis purposes. These tools can be used by adversaries to get confidential information of IoT system. Strong encryption schemes can prevent the leakage of confidential information in an IoT system.
- *Man-in-The-Middle (*MiTM*) Attacks*: Several attacks of these types (i.e. session hijacking, Address Resolution Protocol [ARP] poisoning, Domain Name Server [DNS] spoofing, Secure Socket Layer [SSL] spoofing, etc.) can be launched by an intruder for the purpose of illegal monitoring of ongoing transmission between two nodes in IoT system. Through the implication of encryption mechanisms and Intrusion Detection and Prevention System (IDPS), these attacks can be prevented.
- *Spoofing Attacks*: In IP spoofing and RFID spoofing attacks, the attacker spoofs valid network IP address and valid RFID tag information, respectively, to gain full control of IoT system to interrupt its normal working. Through secure identification, authentication, and trust management, spoofing problems can be handled.
- *Routing Protocol Attacks*: The adversary manipulates the available routing information at IoT devices to create routing loops that eventually causes either increase network latency or failure of packet delivery at destination in an IoT system. Secure routing and securing of node identification or IP addresses information leakage can protect IoT system from such attacks.

IoT protocols are implemented by different technologies or at different layers of TCP/IP protocol stack, i.e. IEEE 802.15.4, ZigBee, RPL, LoRaWAN, etc. to ensure confidentiality and integrity of devices' communication in IoT system.

- IEEE 802.15.4 provides security mechanism at MAC layer by utilizing Advanced Encryption Standard (AES) to ensure ultimate security of higher layers of TCP/IP protocol stack. In addition, related to access control, 802.15.4 has implemented access control lists.
- ZigBee along with the utilization of 802.15.4 at MAC layer has also implemented security mechanism at network and application layers.
- In addition, ZigBee offers two security modes, i.e. standard security mode (without network key encryption) and high security mode (with network key encryption).
- RPL has provided routing security through the utilization of secure RPL packets and ensures the authenticity of information.

- LoRaWAN has provided security through the provisioning of two security layers which deal with the authenticity and privacy issues of end nodes.

### 8.5.3 IoT Application Layer Security

The application layer at the top of the three-layer IoT architecture is responsible for the provisioning of services requested by the users of any IoT-based smart system (smart health system, intelligent transportation system, smart building, smart industry, and smart city, etc.). In addition to the user-requested services, application layer provides data services (i.e. BigData storage, data mining, data management, predictive modeling, etc.) to perform semantic data analysis. Therefore, IoT application layer can further be divided into two sublayers, i.e. support layer and service layer.

#### 8.5.3.1 Security Threats at Support Layer of IoT Applications

The main technologies at IoT Application Support sublayer are Fog and Cloud computing and main security issues are discussed as follows [7]:

*Unauthorized Access*: Concerns with the theft of authorized users' credential information to access and utilize IoT system resources illegally. Secure authentication and efficient access control techniques can be used to inhibit these attacks.

*Malicious Insider*: This kind of security issue arises due to granting of unusual level of trust on Edge resources and Cloud provider.

*Insecure Software Services*: Provisioning of services (VMs, APIs, web applications, etc.) can be affected by malwares. However, these services are required to be protected by Cloud providers.

*Data Storage Risk*: Data stored at edge and Cloud is prone to high security and privacy threats.

#### 8.5.3.2 Security Threats at Service Layer of IoT Applications

The security of IoT applications is highly dependent on all security mechanisms existing at the lower layers of IoT security architecture. However, the well-known IoT Application layer attacks are [10] given in the following:

*Malicious Scripts*: Several online malicious scripts (in JavaScript, Active-X) are commonly used to affect the normal functioning of IoT systems. Static/dynamic malicious code script detection schemes (i.e. honeypot) are used to protect IoT system against these kinds of attacks.

*Phishing Attacks*: These include the spoofing of user credentials (i.e. used identification, password, credit card information, etc.) through the use of phishing websites and infected e-mails in IoT systems. Secure authorization and intelligent vigilant surfing techniques can be applied to mitigate the effect of these kind of attacks.

*Virus Attacks*: Malicious virus attacks (i.e. Trojan Horse) can be controlled through antivirus programs and firewall protection mechanisms in IoT systems.

The support sublayer security mechanism controls the legitimacy of authorized users (through authentication and access control systems), which are interested to utilize available services. The service layer security mechanism includes the protection of application software, OS, and end-user interfaces through the utilization of high-level programming languages, i.e. Java, JavaScript, C++, and Python, etc. which assists to avoid insecure programming.

## 8.6 Multilayer Security Attacks

- *DoS Attacks*: By immense consumption of available resources, a perpetrator can render IoT services unavailable to genuine users of the IoT system.
- *Cryptanalysis Attacks*: Adversary obtained plain and ciphertext to infer the process and use of encryption keys.
- Spyware is a kind of malware attack that can be performed at all layers of IoT architecture stack and used to gather sensitive information (i.e. network traffic, confidential user credentials, and internal usage habits) of IoT systems.
- *Botnets*: A bot (robot) can be used as malware to perform DoS and payload attacks at layers of IoT architecture stack.
- *Rootkit*: This is a kind of malware that can be installed on an IoT device and grant administrator access privileges to an unauthorized user of IoT system. It may consist of antivirus disablers, password stealers, keystroke loggers, etc. and can be executed in all layers of IoT stack.

## 8.7 IoT Application Scenarios and IoT Security

IoT application scenarios encompass different domain of real life, i.e. smart home, smart healthcare, smart vehicle, and smart city. Depending upon the nature of hardware, software, communication techniques, and integration of different technologies, security attacks can be launched in a variety of ways on these application scenarios. Moreover, in each IoT scenario, the attacker/hacker's motivation is unique, and its severity depends on the nature of motivation that may range from simple data access to safety of human life in healthcare and automotive application domains [24]. Discussion about IoT-specific attacks is complementary to IoT security attacks that have been mentioned in previous sections.

### 8.7.1 Smart Home Security

Home automation is possible through connection of several digital appliances through the use of IoT. IoT technology (consisting of sensors and communication) related to home automation may involve (local or remote) management of home devices to provide comfort and safety to home residents. Mostly, end users communicate through smart phones to control home appliances. Exploiting security breaches related to involved communication technologies, attacker can easily compromise the users' security and privacy of home devices and related data, respectively [25]. Table 8.1 illustrates security issues of smart home in terms of the vulnerabilities associated with few smart home devices.

**Table 8.1** Smart home devices, functionality, and associated security threats.

| Smart device | Functionality | Potential security threat |
|---|---|---|
| Smart Lock | • Lock/unlock without physical key<br>• Lock/unlock through mobile device or web interface<br>• Automatically lock after a specified period of time<br>• Alarms ringing on forced entry or break-ins | • Lock/unlock by attackers to enter/exit from home<br>• Changing of lock/unlock password remotely<br>• Turn off the alarm in case of break-in |
| Smart Bulb | • Light bulb controllable remotely through mobile application<br>• Scheduling of turning on/off and coloring of light bulbs | • Control the turning on and off behavior of lights<br>• Overload power system by turning on unnecessary lights |
| Voice Automated Device | • Turn devices on or off based on voice Commands | • Steal private credentials from voice data<br>• Issue voice commands to order unwanted stocks by voice commands<br>• Steal voice data as credentials for use in other voice command systems |
| Smart Vacuum Cleaner | • Automatically map home layout and conduct automatic and scheduled cleaning in dry or wet mopping modes | • Monitor room activities and stealing of home layout |
| Smart Refrigerator | • Create grocery list and send order to shops through the Internet<br>• Set expiration data and send related alerts to residents<br>• Suggest recipes based on available ingredients | • Send order with modified grocery list<br>• Modify expiration date of food items in refrigerator or ruin food items by changing temperature |
| Smart Toilet | • Allow users to remotely set water temperature and pressure<br>• Sense and adjust right water amount to clean itself or for flushing wastes<br>• Notify residents about needed supplies (e.g. toilet paper, soap, and air freshener) | • Turn water tap on and leave water flowing without any need<br>• Remotely control smart toilet's lid and flush nozzles |

### 8.7.2 Smart Healthcare Security

The implication of IoT in healthcare provides several opportunities to help patients, doctors, caregivers, pharmacies, research organizations, and government healthcare authorities by offering different services; for example:

- patient monitoring
- real-time data sharing in emergency situations
- sharing of medical/device/drug/patient data by hospitals
- sharing of medical supplies record in case of epidemic by pharmacies

- collection, analysis, management, and sharing of healthcare data by hospitals and government authorities

Concerning smart healthcare realization, RFID, body sensors, and various communication technologies have been used for [24]:

- the identification of patients and doctors
- locating patients and doctors
- collection, processing, transmission, and storage of patients' vital sign data in real time by means of implanted/wearable devices

Nevertheless, these IoT-based healthcare systems are vulnerable to attacks that can even endanger patient's life if security mechanism is not effectively implemented. The following are the main factors that make these system more appealing targets for attackers [26, 27]:

- Existence of wireless communications between implanted and wearable devices
- Pervasive usage of RFID technology (tags and readers)
- Lacking support of complex security algorithms due to the scarcity of energy and computing resources of IoT devices

To empower the usage of IoT devices in smart health, it is important to understand the types of attacks and respective countermeasures from healthcare perspective. Concerning IoT healthcare system, attacks can be launched either by internal or external attackers which are mostly interested to get private information of patients or to make system unavailable through DOS attacks [28]. Table 8.2 presents the types of different security attacks on healthcare system with associated consequences.

**Table 8.2** Security attacks and potential threats to healthcare systems.

| Attack | Approach | Potential security threat |
|---|---|---|
| DoS Attack | • Overburdens the healthcare system with unknown traffic that consumes available resources | • IoT services become unavailable to genuine users (patients, caregivers, doctors, etc.) of healthcare system |
| Sensor Attack | • Control sensor activity<br>• Replaces real sensor nodes with fake nodes | • Engage sensors in frequently joining and leaving of networks to drain out the energy of network nodes<br>• The tampering of patients' data |
| Routing Attack | Modify routing table information | • Delivery of patient information to false destination (i.e. fake doctor or fake hospital)<br>• Drop packets to not reach to intended destination |
| Replay Attack | Malicious node evaluates the system to steal legitimate identification information | • Build trust and get unauthorized access to healthcare system to access private data |
| Select Forwarding Attack | Malicious devices pretend as destination nodes | • Drops data packets or do not allow passing of any packet through them |

### 8.7.3 Smart Vehicle Security

Smart cars, besides ensuring comfort and safety of drivers and passengers, improve road safety. Smart cars contain several electronic control units (ECU), i.e. engine control unit, on-board diagnostics port, telematics unit and wireless communication unit to support heterogenous connectivity with other vehicles, RSUs, and vehicular Cloud via gateways. Concerning the aspects of vehicle security threats, these elements can be grouped into three categories [29], i.e.:

- Under the hood elements including ECU, communication unit, and gateway. The elements include in this category can be used at the preparation phase of the attack for reverse engineering purposes.
- Devices connected to car including mobile phone belonging to car owner/service user and third-party user devices (in case of car sharing). Especially the mobile devices connected to vehicles are potential threat points to car security. For example, rooting or jailbreaking of mobile devices lead to the option of ongoing communication eavesdropping that ultimately enable direct-access attacks.
- Communication technologies (i.e. WiFi, GSM, etc.) have vulnerabilities that can be exploited to perform spoofing and/or MiTM attacks to extort personal information from vehicular Cloud.

Table 8.3 provides summary of the functionalities of few security attacks that can be launched in vehicular environment [30].

**Table 8.3** Security attack functionality in smart vehicle environment.

| Attack type | Functionality |
|---|---|
| MiTM Attack | Session hijacking (obtain legitimate key to steal smart car, network data collection, etc.) |
| DoS, DDoS Attack | Leads to Network failure or take over the control of smart car |
| Relay Attack | Launch Passive Keyless Entry and Start (PKES) systems for theft. Messages are relayed between smart key and the smart car |
| Jamming Attack | Communication channel of smart vehicle is disturbed through heavily powered signal of equivalent frequency |
| Blackhole Attack | Suspected smart car receives the packets but declines to forward to other vehicles |
| Eavesdropping Attack | Attacker obtains the confidential details of smart vehicles (vehicle identity, owner identity, data location, etc.) and may disclose to nonregistered users |
| GPS Spoofing | Attacker creates false GPS location information to hide its position and to dodge other vehicles about its availability at a particular location |
| Sybil Attack | A suspected node produces many fake identities to manipulate other vehicles behavior |

Source: Based on Sheikh et al. [30].

### 8.7.4 Smart City Security/Privacy Concerns

The conception of smart cities encompasses the monitoring, administration, management, and governance of whole city infrastructures related to health, education, traffic, energy resources, and natural environment through integrated IoT systems. These IoT systems ultimately consist of highly advanced integrated technologies (i.e. sensor, wireless networks, embedded systems, Fog and Cloud computing, etc.) which are linked with end-user devices. Therefore, the security and privacy of information flow within smart cities is important and difficult to implement. Influencing factors that mainly identify issues of information security and privacy in smart cities include Technological, Governance, and Socioeconomic factors [31]. Security and privacy concerns related to these factors have been summarized in Table 8.4.

## Review Questions

**8.1** Explain the categories of security constraints of IoT systems.

**8.2** How are the information-level security requirements different from functional-level security requirements?

**8.3** Describe the classification of security attacks in IoT systems.

**8.4** Explain the types and functionality of security attacks at Perception, Network, and Application layers of IoT architecture.

**Table 8.4** Influencing factors and security/privacy concerns in smart cities.

| | | |
|---|---|---|
| IoT Technological Factors | RFID Technology | Eavesdropping, Spoofing, Jamming, DoS |
| | WSN Technology | DoS, MiTM, Relay Attacks |
| | Communication Technology | Authentication Attacks in Wi-Fi, Bluetooth, ZigBee, etc., Protocol Attacks, DoS, MiTM |
| | Mobile Phones Technology | Threats from GPS, Bluetooth, Wi-Fi, etc., Threats from Social Networking, Botnets, Malwares |
| | Smart Grid | DoS, Message Replay Attacks, Malicious Data Injection Attacks |
| Governance Factors | Critical Infrastructure | Threats to Health Sector, Threats to Energy and Power Supply, Disaster Management Issues |
| | Mobility | Location privacy |
| | Utility | Misuse of data, exploitation of resources |
| Socioeconomic Factors | Individual Privacy | Social Networking Issues, Smart Phone Usage, Location Privacy |
| | E-Commerce | Cyber Attacks, Spoofing, Frauds, Data Integrity |
| | Banking | Cyber Crimes, Phishing, Frauds, Data Integrity |

**8.5** Briefly describe the functionality of three multilayer security attacks.

**8.6** Illustrate security issues of smart home devices (shown in Figure 8.4) in terms of associated vulnerabilities.

**8.7** Consider that attackers exploit the vulnerability in keyless entry/start systems using a digital theft technique called the relay attack (as shown in Figure 8.5).

**A** Explain the step-by-step working of a relay attack in this scenario.

**B** How you can secure smart vehicles from a relay attack?

**C** How will security be enabled when the key is lost temporarily or permanently?

**Figure 8.4** Figure for review question 8.6. Source: Sicari et al. [32].

**Figure 8.5** Figure for review question 8.7. Source: Evans [33].

**8.8** Considering IoT-complaint keyless entry systems for smart home:

Explain how it works with WiFi and Bluetooth enabled communication devices.

Explain how security would be enabled if the key is damaged or lost temporarily/ permanently.

Describe how this scenario is different from smart key fob for vehicles.

## References


1 Ling, Z., Liu, K., Xu, Y. et al. (2017). An end-to-end view of IoT security and privacy. In: *IEEE Global Communications (GLOBECOM) Conference*, 1–7. IEEE.

2 Chirgwin, R.(2016). Get pwned: Web CCTV cams can be hijacked by single HTTP request-server buffer overflow equals remote control. www.theregister.co.uk/2016/11/30/iot_cameras_compromised_by_long_url.

3 Hilton, S. (2016). Dyn analysis summary of Friday October 21 attack. https://dyn.com/blog/dyn-analysis-summary-of-friday-october-21-attack.

4 Antonakakis, M., April, T., and Bailey, M. (2017). Understanding the mirai botnet. In: *26th USENIX Security Symposium*, 1093–1110. USENIX Association.

5 Ling, Z., Liu K., Xu Y. et al. (2018). IoT security: an end-to-end view and case study. arXiv preprint arXiv:1805.05853, 2018.

6 Hossain, M.M., Fotouhi, M., and Hasan, R. (2015). Towards an analysis of security issues, challenges, and open problems in the internet of things. In: *IEEE World Congress on Services*, 21–28. IEEE.

7 Grammatikis, P.I.R., Sarigiannidis, P.G., and Moscholios, I.D. (2019). Securing the Internet of Things: challenges, threats and solutions. *Internet of Things* 5: 41–70.

8 Kharchenko, V., Kolisnyk, M., Piskachova, I. et al. (2016). Reliability and security issues for IoT-based smart business center: architecture and Markov model. In: *2016 Third International Conference on Mathematics and Computers in Sciences and in Industry (MCSI)*, 313–318. IEEE.

9 Hassan, W.H. (2019). Current research on internet of things (IoT) security: a survey. *Computer Networks* 148: 283–294.

10 Lin, J., Yu, W., Zhang, N. et al. (2017). A survey on internet of things: architecture, enabling technologies, security and privacy, and applications. *IEEE Internet of Things Journal* 4 (5): 1125–1142.

11 Capkun, S., Buttyán, L., and Hubaux, J.-P. (2003). Self-organized public-key management for mobile ad hoc networks. *IEEE Transactions on Mobile Computing* 2 (1): 52–64.

12 Yang, X., Lin, J., Yu, W. et al. (2013). A novel en-route filtering scheme against false data injection attacks in cyber-physical networked systems. *IEEE Transactions on Computers* 64 (1): 4–18.

13 Aman, M.N., Sikdar, B., Chua, K.C. et al. (2018). Low power data integrity in IoT systems. *IEEE Internet of Things Journal* 5 (4): 3102–3113.

14 Wachter, S. (2018). Normative challenges of identification in the Internet of Things: privacy, profiling, discrimination, and the GDPR. *Computer Law and Security Review* 34 (3): 436–449.

**15** Liu, J., Xiao, Y., and Chen, C.P. (2012). Authentication and access control in the internet of things. In: *IEEE 32nd International Conference on Distributed Computing Systems Workshops*, 58–592. IEEE.

**16** Chuang, M.-C. and Lee, J.-F. (2013). TEAM: trust-extended authentication mechanism for vehicular ad hoc networks. *IEEE Systems Journal* 8 (3): 749–758.

**17** Airehrour, D., Gutierrez, J., and Ray, S.K. (2016). Secure routing for internet of things: a survey. *Journal of Network and Computer Applications* 66: 198–213.

**18** Maheswari, S.U., Usha, N., Anita, E.M. et al. (2016). A novel robust routing protocol RAEED to avoid DoS attacks in WSN. In: *IEEE International Conference on Information Communication and Embedded Systems (ICICES)*, 1–5. IEEE.

**19** Rizvi, S., Kurtz, A., Pfeffer, J. et al. (2018). Securing the internet of things (IoT): a security taxonomy for IoT. In: *IEEE 17th International Conference on Trust, Security and Privacy in Computing and Communications/12th IEEE International Conference on Big Data Science and Engineering (TrustCom/BigDataSE)*, 163–168. IEEE.

**20** Yang, Y., Wu, L., Yin, G. et al. (2017). A survey on security and privacy issues in internet-of-things. *IEEE Internet of Things Journal* 4 (5): 1250–1258.

**21** Alam, S. and De, D. (2014), Analysis of security threats in wireless sensor network. arXiv preprint arXiv:1406.0298.

**22** Mayzaud, A., Badonnel, R., and Chrisment, I. (2016, 2016). A taxonomy of attacks in RPL-based Internet of Things. *International Journal of Network Security, ACEEE a Division of Engineers Network* 18 (3): 459–473.

**23** Mahmood, Z. (2019). *Security, Privacy and Trust in the IoT Environment*. Springer.

**24** Chen, K., Zhang, S., Li, Z. et al. (2018). Internet-of-things security and vulnerabilities: taxonomy, challenges, and practice. *Journal of Hardware and Systems Security* 2 (2): 97–110.

**25** Chang, Z. (2019). IoT device security locking out risks and threats to smart homes. In: *Trend Micro Research*. https://documents.trendmicro.com/assets/white_papers/IoT-Device-Security.pdf.

**26** Poslad, S. (2011). *Ubiquitous Computing: Smart Devices, Environments and Interactions*. Wiley.

**27** Abie, H. and Balasingham, I. (2012). Risk-based adaptive security for smart IoT in eHealth. In: *Proceedings of the 7th International Conference on Body Area Networks*, 269–275.

**28** Butt, S.A., Diaz-Martinez, J.L., Jamal, T. et al. (2019). IoT smart health security threats. In: *IEEE 19th International Conference on Computational Science and Its Applications (ICCSA)*, 26–31. IEEE.

**29** Bécsi, T., Aradi, S., and Gáspár, P. (2015). Security issues and vulnerabilities in connected car systems. In: *IEEE International Conference on Models and Technologies for Intelligent Transportation Systems (MT-ITS)*, 477–482. IEEE.

**30** Sheikh, M.S., Liang, J., and Wang, W. (2019). A survey of security services, attacks, and applications for vehicular ad hoc networks (VANETs). *Sensors* 19 (16): 3589.

**31** Ijaz, S., Shah, M.A., Khan, A. et al. (2016). Smart cities: a survey on security concerns. *International Journal of Advanced Computer Science and Applications* 7 (2): 612–625.

**32** Sicari, S., Rizzardi, A., Miorandi, D. et al. (2018). Securing the smart home: a real case study. *Internet Technology Letters* 1 (3): e22.

**33** Evans, C., Keyless car theft: what is a relay attack, how can you prevent it, and will your car insurance cover it? (2020). https://leasing.com/car-leasing-news/relay-car-theft-what-is-it-and-how-can-you-avoid-it.

# 9

# Social IoT

**Learning Objectives**

After studying this chapter, students will be able to:

- describe the nature of social relationships among IoT Devices.
- elaborate the functionality of different components of social IoT architecture.
- understand the applicability of social aspects of smart devices in IoT applications.

## 9.1 Smart Things to Social Things

Interactions among individuals of the same community or species have been defined as social behavior. In the real world, a number of animal species have been reported to cope with harsh environments by establishing social ties while living together. For example, honey bees in social relations behave defensively when intruder attacks their well-organized colonies in wax combs. Studies have shown that many animals are more successful in finding food if they search for food in groups [1]. In humans, social behavior is desirable as it not only helps to promote our physical/emotional well-being but also is essential to run any structured systems effectively. Considering the examples of animal species, human instincts, and advancements in the Internet of Things (IoT) technology, three categories of smart things (i.e. Res Sapiens, Res Agens, Res Socialis) [2] have been identified representing the following associated features:

- *Res Sapiens:* Smart things with increased interoperability and capability to work in IoT systems based on social networks.
- *Res Agens:* Smart things having interactions with the environment and represent pseudo-social behavior.
- *Res Socialis:* Smart things that can build own social network and able to collaborate with each other.

*Enabling the Internet of Things: Fundamentals, Design, and Applications*, First Edition.
Muhammad Azhar Iqbal, Sajjad Hussain, Huan Lai Xing, and Muhammad Ali Imran.

The advanced Res Socialis category of smart things with social relationship provides a new vision of IoT, where they are able to:

- Interact with other smart things with respect to ownership, co-work/location environment.
- Crawl and discover required services in IoT world consisting of billions of smart things.
- Advertise themselves for the provisioning of associated services.

Based on these facts, the enabling of social interaction of smart things in the IoT world gives rise to an emerging paradigm of promising Social Internet of Things (SIoT) [3]. SIoT ultimately encompasses the aspects of social networks and IoT. In other words, it can be seen that in the future, services will be associated with smart things and it would be easier to discover published services through navigating a social network of things in comparison to typical Internet discovery style that is not suitable to large-scale IoT [4, 5].

## 9.2 The Epitome of SIoT

In recent times, there have been many advances in the field of IoT technologies. These technological progressions in IoT have connected humans, physical things, and digital devices with each other in a consolidated way, as shown in Figure 9.1. The physical things with digital identification tags and sensors are basic components of IoT systems that are available to provide services to humans. However, the nature of human behavior is inclined toward the social usage of physical things to form communities where human–thing and thing-thing relationship become possible in terms of friendship and common goals/interests. The SIoT, an evolution of IoT, eases the process of service discovery and provides the

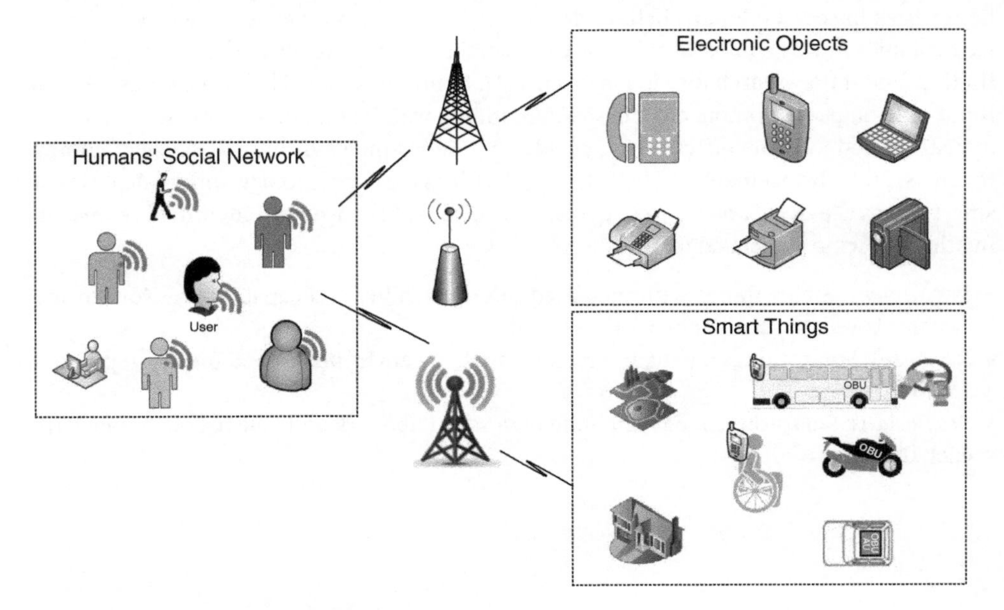


**Figure 9.1** Interaction of SIoT components.

foundations of autonomous interaction among smart things and also with humans in the IoT ecosystem [6].

The SIoT introduces the concept of smart thing relationships to obtain the following advantages:

- SIoT shapes the structure of IoT to assure efficient network navigability for effective service discovery.
- SIoT with effective discovery service guarantees the scalability of IoT ecosystems.
- SIoT improves the level of interaction among things in relationships with the provisioning of trustworthiness.

## 9.3 Smart Thing Relationships in SIoT

Depending on the nature of IoT application, the social structure of SIoT platform supports various types of relationships [5, 7], for example:

- *User–Thing Relationships (UTR)*: Where user and things association exists
  - *Things–Ownership Relationship (TOR)*: Smart things/objects (cell phones, smartwatch, tablets, iPad, etc.) belong to the same user.
  - *Social–Things Relationship (STR)*: Smart things in contact because of social relationships exist between owners (e.g. automatic exchange of contacts, images, audio/videos, etc. between cell phones of friends when they meet each other).
  - *Sibling–Things Relationship (SIBTR)*: Smart things of family members are in relationships for the exchange of daily activities.
  - *Guest–Things Relationships (GTR)*: Relationships between smart things of those individuals that spend time together at friends' place and declared as guests.
- *Thing–Thing Relationships (TTR)*: Where only things are in connection with each other
  - *Parental–Things Relationships (PTR)*: This type of relationship exists among smart things belonging to the same manufacturer and originated in the same batch. This strong relationship is not changeable over time except the obsolescence of smart object(s).
  - *Colocation–Things Relationships (CoLTR)*: This kind of relationship exists among smart things that are located at the same place to cooperate with each other for the sake of some common objectives in an IoT system. For example, Radio Frequency Identification (*RFID*) tags, sensors, actuators, etc. within a smart home/shop/office.
  - *Cowork–Things Relationships (CoWTR)*: Smart things that cooperate with each other within an IoT system. For example, body sensors, smartphones, home Internet, and caregiver alarms in IoT-based healthcare system.
  - *Guardian–Things Relationships (GRTR)*: A typical vehicle-to-infrastructure relationship existing between vehicle's On-Board Unit (OBU) and Road-Side Unit (RSU) of Vehicular Ad hoc Network (VANET) infrastructure for provisioning of safety and infotainment services.
  - *Stranger–Things Relationships (StTR)*: Things in relationships in public gatherings.

**Table 9.1** Smart thing relationship types and SIoT applications.

| Relationships type | SIoT application |
|---|---|
| TOR | Used in transportation and logistics by using the vehicles of the same owner |
| STR | Lane-change and collision warnings |
| SIBTR | Exchange of sports statistics in a stadium |
| GTR | Used in smart malls and smart restaurants for advertisements and food quality information, respectively |
| PTR | Smart connectivity among vehicles of the same manufacturers |
| CoLTR | Smart office, hospital, and parking |
| CoWTR | Telemedicine systems and Smart firefighter systems |
| GRTR | Early warning systems in vehicles for road accidents, road condition, and road blockage |
| StTR | Used in smart marketing |

*Source:* Based on Roopa et al. [7].

These types of smart thing relationships offer different types of SIoT applications [7] as shown in Table 9.1.

## 9.4 SIoT Architecture

In view of different IoT architectures, SIoT architecture has been presented accordingly, for example,

- Unified Architectural Model [8] has been used to explore the social features of smart objects.
- Based on integrated ubiquitous computing environments, SIoT architecture has been proposed while taking into account the interactions between humans and devices in the form of services [9].
- SIoT architecture has also been explained using the concept of the grouping of Virtual Entities [10].
- SIoT architecture has also been proposed while taking into account the principles of relational models [11].

However, here in this chapter, generic SIoT architecture proposed by Atzori et al. [4, 5] has been described that is consistent with the three-layer model of IoT architecture (consisting of perception, network, and application layers), described in Chapter 1. The generic SIoT architecture consists of the following three components i.e. (SIoT Server, SIoT gateway, and SocialThing).

### 9.4.1 SIoT Server

The SIoT server does not contain the Perception layer of IoT architecture and consists of only two architectural layers, i.e. Network layer and Application layer as shown in Figure 9.2.

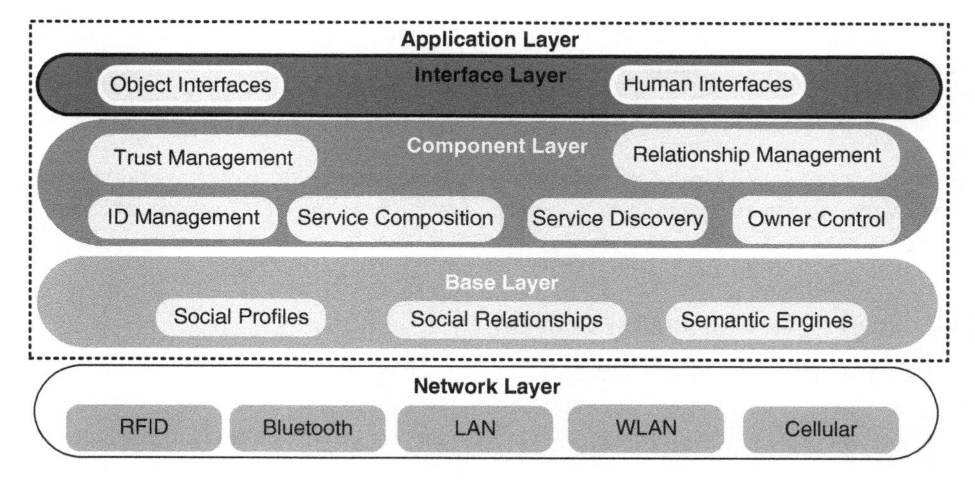


**Figure 9.2** Architecture layers of SIoT server.

#### 9.4.1.1 The Network Layer of SIoT Server

The Network layer of SIoT server is responsible for data transmission across the different networks through heterogeneous interfaces, i.e. Bluetooth, RFID, WLAN, LAN, etc.

#### 9.4.1.2 The Application Layer of SIoT Server

The Application layer of SIoT server is responsible for various functions and has been further divided into three sublayers, i.e. Base, Component, and Interface sublayers.

The *Base sublayer* is responsible for:

- storage and management of social data (member profiles, relationships, and relevant activities) of involved entities.
- storage and management of related social ontologies.
- extraction of semantic view of social activities from available ontologies.
- provisioning of machine-interpretable representation of functional and nonfunctional operations of smart things.

The *Component sublayer* consists of various modules that provide the implementation of basic functionalities of SIoT system:

- *ID Management Module*: For assignment of IDs to smart things.
- *Profiling Module*: For manual/automatic configuration of static/dynamic information of smart things.
- *Owner Control Module*: To allow information sharing and setting up of relationship types among smart things.
- *Relationship Management Module*: To incorporate intelligence about the starting, updating, and termination of smart thing relationships with each other.
- *Service Discovery Module*: To find things (with required services), which are associated with each other in social networks.
- *Service Composition*: To support the interaction between smart things for the provisioning or retrieval of information about specific services.

- *Trust Management Module*: To estimate the way of processing information through the use of social centrality measures.

#### 9.4.1.3 The Interface Sublayer

This layer is responsible for the provisioning of interfaces and APIs to objects and humans for accessing resources available in Cloud or deployed on multiples servers at various locations.

### 9.4.2 The SIoT Gateway and Social Things

Depending on the IoT application scenario, SIoT gateway and social thing consist of three layers, i.e. Perception (Sensing), Network, and Application Layers as shown in Figure 9.3.

*Perception Layer at SIoT Gateway*: This layer is optional and provides short-range communication services for integrated sensors that are optionally available on IoT gateways.

*Network layer at SIoT Gateway*: This layer is responsible for the provisioning of heterogenous communication interfaces to harmonize communication between smart things and SIoT server having different communication interfaces.

*Application layer at SIoT Gateway*: This layer consists of two modules, i.e.:

*Social Agent Module*: For object communications for the update of user profile management and discovery of required services

*Service Management Module*: Provides interfaces for humans to control the object behavior in social networks within SIoT systems

## 9.5 Features of SIoT System

The fundamental advantage of SIoT applications is the provisioning of discovering service in IoT systems. Therefore, the following features of applied discovery systems are the sine qua non for the typical working of an SIoT system:

- *Scalability and Mobility*: Discovery system must be able to address the problems of scalability, heterogeneity, interoperability, mobility, and dynamic connectivity of smart devices to whom services are associated.

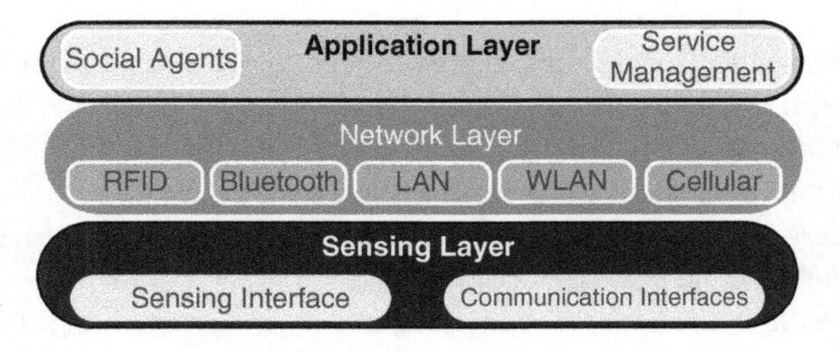


**Figure 9.3** Architecture layers of SIoT gateway/object.

- *Personalization*: Discovery system must be able to attune the results for user preferences.
- *Recommendation*: Discovery system must provide related service recommendation in case of unavailability of required service.
- *Response Time*: Discovery of required service must be accomplished within the minimum time delay.
- *Status Update*: Discovery system must be updated with current service status.
- *Results Accuracy*: Results of required queries must be accurate.
- *Security and Privacy*: Discovery system must provide user security and able to conceal sensitive user information.
- *Trustworthy*: Discovery system must show trustworthy results to its users.
- *Energy Management*: Discovery schemes must be energy efficient
- *Standardization*: Discovery system must be standardized enough to work with standard Transmission Control Protocol/Internet Protocol (TCP/IP) communication stack.

## 9.6 Social Internet of Vehicles (SIoV) – An Example Use Case of SIoT

Similar to SIoT that is basically an incarnation of IoT, Social Internet of Vehicles (SIoV) is an incarnation of Internet of Vehicles (IoV). SIoV represents the social network of vehicles and can be better described as a vehicular instance of SIoT [12]. Considering SIoT perspective, the SIoV architecture has been described in this section. However, for a detailed description and better understanding of SIoV architecture, it is required to have familiarity with the concepts of VANETs, IoV, and related terminologies. Therefore, next, we have briefly described the reference architecture of VANET and IoV along with the elaboration of differences between them.

### 9.6.1 Reference Architecture of VANETs

The current era of IoT realization encourages the evolution of vehicular communication systems from VANETs to the IoVs. In VANETs, vehicles equipped with wireless interfaces (of homo/heterogeneous radio technologies) are considered to transmit messages in Vehicle-to-Vehicle (V2V) or Vehicle-to-Infrastructure (V2I) modes on highways and city roads. In IoV, contrary to VANETs, a vehicle is considered as the full-fledged smart vehicle (consisting of powerful computing units, sophisticated sensors, advanced IP-based connectivity [13]) that is able to collect and share data from vehicles, roads, and surrounding environment. Therefore, IoV, with the provisioning of collection, computing, and sharing of data, offers services to assist other vehicles on the road in an efficient way [14]. Moreover, the communication architecture of VANETs is significantly different from IoVs. A reference architecture of VANETs deployment can include three domains, i.e. in-vehicle, ad hoc (V2V), and infrastructure (V2I) domain [15], as shown in Figure 9.4. Logically composed of two components (i.e. OBU and one or more Application Units [AUs]), the in-vehicle domain of VANETs represents a local network inside a vehicle. In vehicles, an OBU device is responsible for the (wired/wireless) communication, but on the other hand, AU devices (integrated or portable in vehicles) are responsible for the execution of applications while

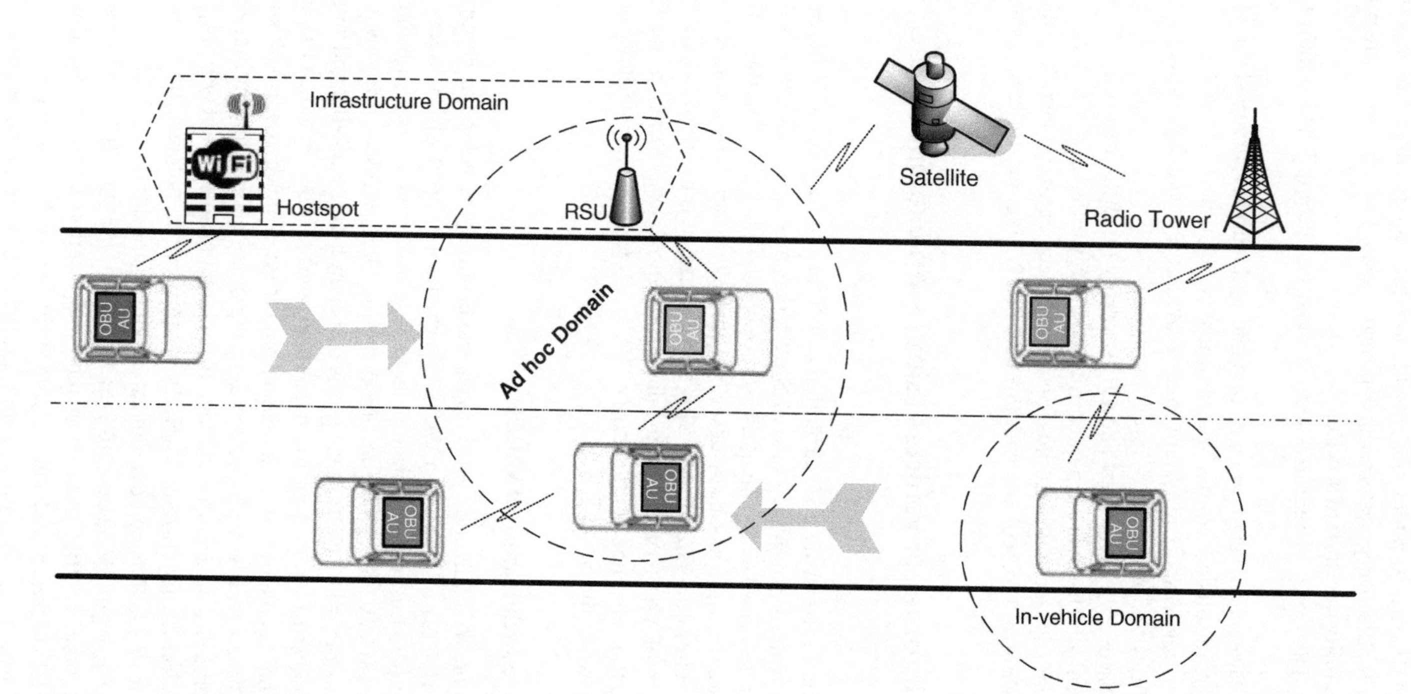


**Figure 9.4** VANETs reference architecture.

using the communication capabilities of the OBU. The ad hoc domain of VANETs refers to the communication among mobile vehicles (equipped with OBUs) as well as static roadside units (RSUs) deployed along the roads. The Infrastructure domain of VANETs comprises of RSUs and wireless hotspots. RSUs are able to communicate with each other and support the road safety while doing the job of application execution as well as sending/receiving/forwarding of data through multi-hops in ad hoc domain. Vehicles in VANETs are able to access the Internet via the components of infrastructure domain (i.e. RSUs and hotspots) as well as through cellular radio networks (in the absence of RSUs while interfaced within OBUs). A well-organized integration of vehicular and cellular network interfaces provides better data access services [16].

### 9.6.2 Reference Architecture of IoV

The IoV reference architecture is shown in Figure 9.5. In addition to V2V and V2I communications in VANETs, IoV vehicles with advanced sensing, computing, and communication technologies are able to support other types of communications, i.e. [14]:

- *Vehicle to Human (V2H)*: Interaction between vehicle and smart devices (smartphones, tablets, smartwatches, etc.), which appertain to humans (i.e. pedestrians, cyclists, divers, passengers, etc.).
- *Vehicle to Sensors (V2S)*: Communication of vehicle with attached sensors to monitor vehicle speed, tire pressure, tire alignment, engine oil status, etc.
- *Vehicle to Everything (V2X)*: Communication of vehicle with everything (e.g. road sensors, RSUs, digital billboards, pedestrians' devices, etc.) in surroundings to share concerned and useful information.

#### 9.6.2.1 Differences in Communication Standards

VANETs and IoVs are also different from each other in communication standards. VANETs support two communication standards known as WAVE and CALM:

- *Wireless Access in Vehicular Environment (WAVE) Standard* is based on Dedicated Short-Range Communications (DSRC) standard for accessing dynamic spectrum, Physical (PHY)/Medium Access Control (MAC) layer standards of IEEE 802.11p, and upper layer (Network and Transport) standards of IEEE1609.
- *Communication Access for Land Mobiles (CALM) Standard* is a combination of different wireless technologies, i.e. Infrared, GSM, GPRS, UMTS, etc. to support multiple ways of transmission.

On the other hand, in addition to WAVE and CALM, IoV vehicles support:

- *Bluetooth*: Short-range communication technology for personal devices and based on the IEEE802.15.1 standard. Bluetooth is used to connect mobile devices in vehicles, e.g. Bluetooth headset and rearview mirror.
- *ZigBee*: Short-range communication technology that is based on 802.15.4 and is used to connect internal sensors of a vehicle
- *4G/5G, Long-term Evolution (LTE)* offers low latency and high throughput to support safety and infotainment applications

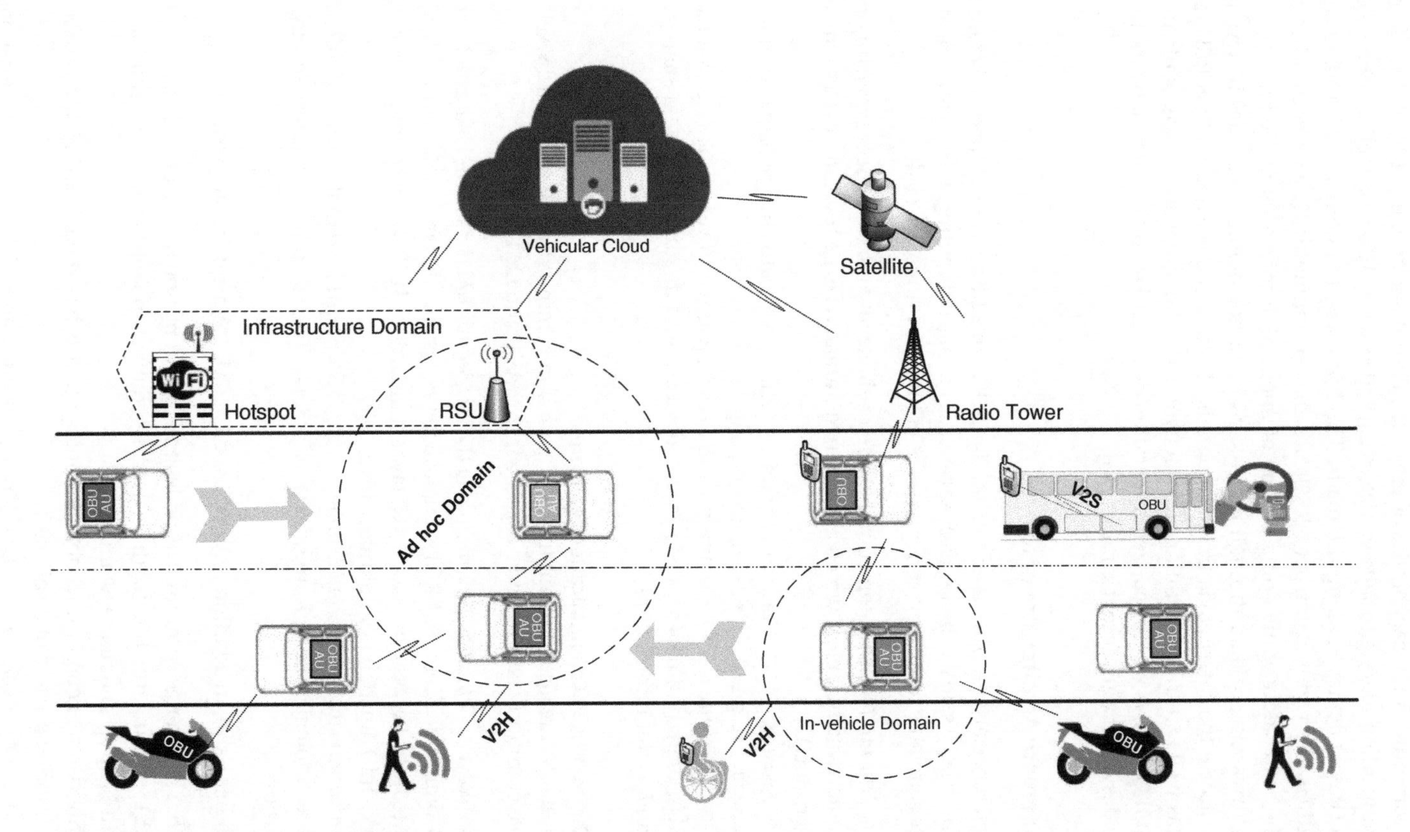


**Figure 9.5** IoV reference architecture.

- *Worldwide Interoperability for Microware Access (WiMax)* is based on IEEE802.16, having long transmission range and is able to provide QoS communication suitable for infotainment applications especially multimedia applications (as shown in Figure 9.6).

### 9.6.3 Reference Architecture of SIoV

SIoV layered architecture consists of four layers, i.e. Vehicle-Object Perception Layer, Gateway layer, Fog layer, and Cloud layer [12] as shown in Figure 9.7.

#### 9.6.3.1 Vehicle-Object Perception Layer (VOPL)

The Vehicle-object Perception Layers (VOPLs) consist of four sublayers, i.e. Physical, Intra-vehicle, Processing, and Application Layers.

The physical sublayer comprises powerful in-vehicle sensors, on-road sensors, and sensing devices that are directly or indirectly connected to vehicles (i.e. mobile devices carried by drivers, passengers, and pedestrian, etc.).

**In-vehicle Sensors**

A vehicle with all its sensing capabilities is the central part of SIoV architecture. An SIoV vehicle consists of various sensors (as shown in Figure 9.8), which are able to communicate with OBU. Normally, 60–100 in-vehicle sensors have been used in advanced vehicle and categories of these sensors have been shown in Table 9.1 [17, 18]. OBU in turn is responsible for intra-vehicle and inter-vehicle communication. Intra-vehicle communication includes communication with handheld devices of driver and passengers, and inter-vehicle communication involves communication with other vehicles directly or through RSUs.

**On-road Sensors**

In a transportation system, two categories of environmental sensors, i.e. intrusive and non-intrusive, have been deployed to collect environmental data and transfer it to SIoV vehicle.

Intrusive sensors are deployed on pavement surfaces, and non-intrusive sensors are deployed near or around road but not over it as shown in Figure 9.9.

Different types of intrusive and non-intrusive sensors with their usage have been shown in Table 9.2. These sensors have been used to detect traffic congestion, vehicle speed, parking place, accident, and location, etc. The SIoV vehicle messages contain dynamic sensory information along with other useful static information, i.e. vehicle identity, driver identity, vehicle's physical attributes, etc.

Considering WAVE standard, a number of physical layer protocols (i.e. Physical Medium Dependent [PMD], Physical Layer Convergence Procedure [PLCP], Physical Layer Management Entity [PLME]) have been implemented to support OBU communication with sensors over the fixed radio control channel (CCH). Intra-Vehicle Communication Layer offers communication technologies (Bluetooth, ZigBee, IrDaWi-Fi, etc.) for intra-vehicle communication through OBU. Radio Frequency Communication (RFCOMM), Link Manager Protocol (LMP), Service discovery protocol (SDP) TCP/IP are standard protocols at this layer.

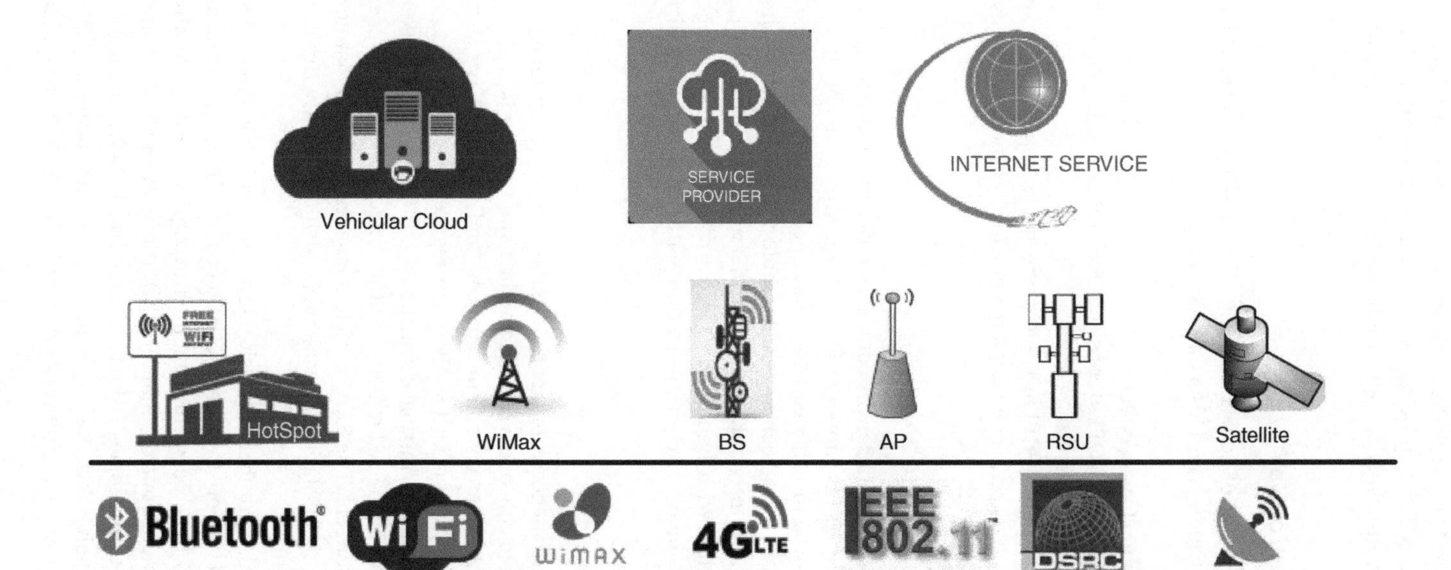


**Figure 9.6** Communication standards for IoV.

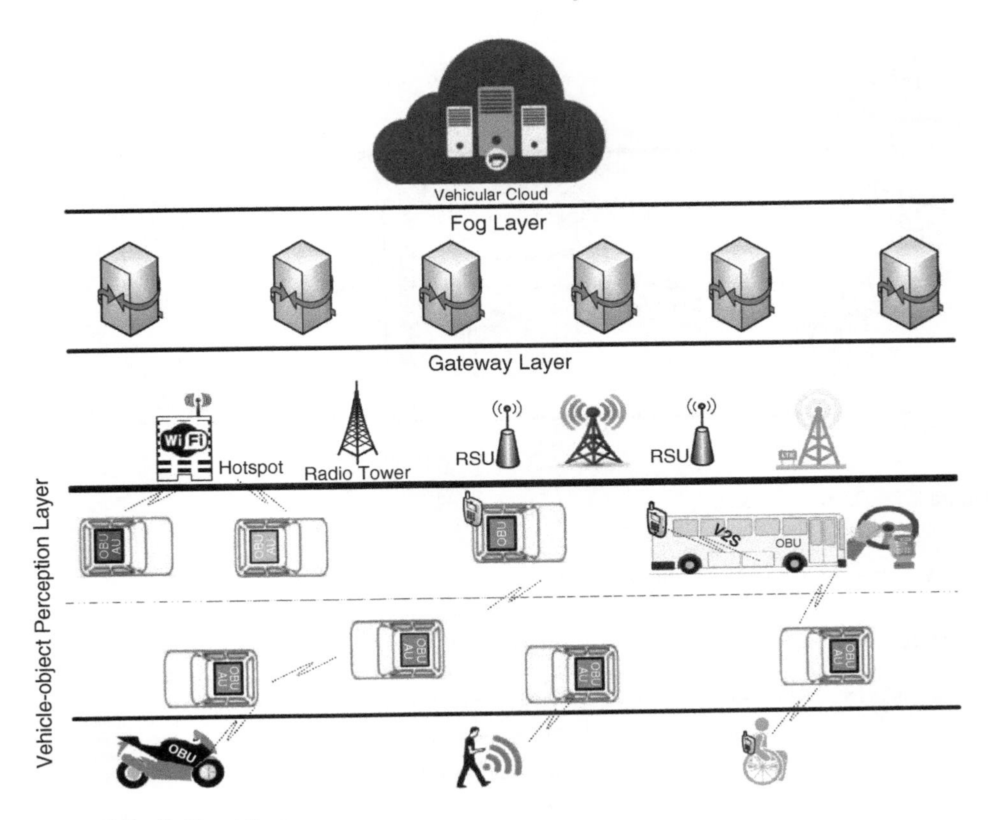


**Figure 9.7** SIoV architecture.

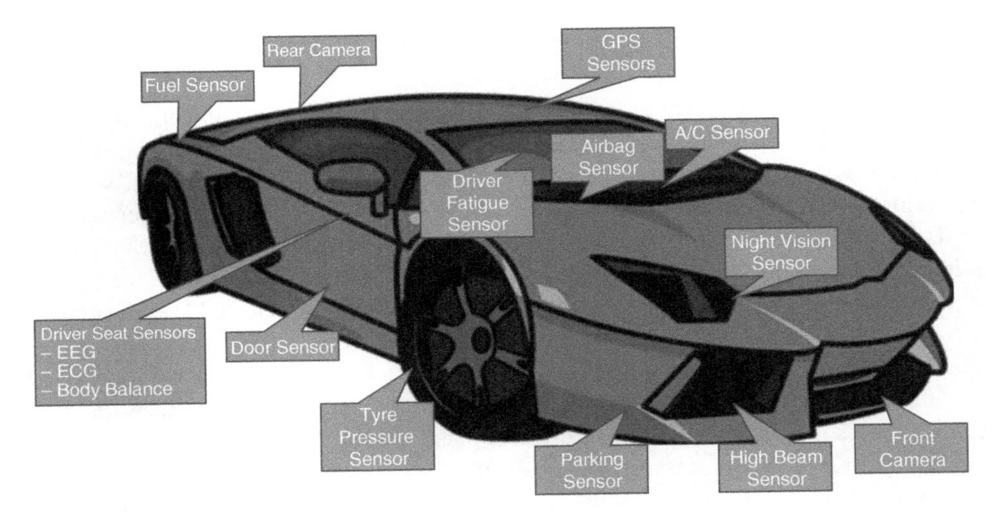


**Figure 9.8** In-vehicle sensors.

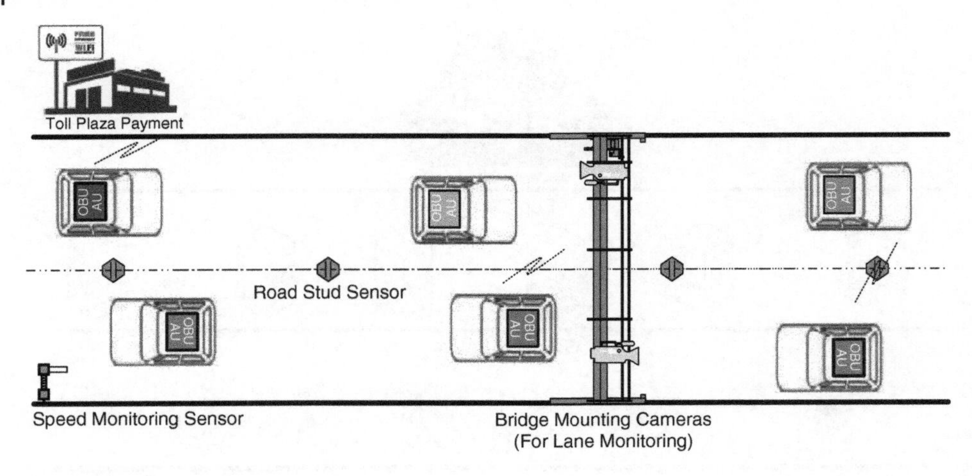


**Figure 9.9** On-road sensors.

**Table 9.2** Intrusive and non-intrusive sensors.

| Category | Sensor type | Usage |
|---|---|---|
| Non-Intrusive | RFID Tags | Vehicle identification for toll payment |
| | Infrared | For speed measurement and lane occupancy |
| | Ultrasonic | To track the number of vehicles |
| | Camera | To detect vehicle type and speed for this category |
| | Radar | To detect the direction of vehicles' movement |
| Intrusive | Magnetic | To sense the static and dynamic behavior of vehicles |
| | Piezoelectric | To measure vehicle speed and weight |
| | Pneumatic Road Tube | To count and tracking of vehicles |

OBU is the processing unit at Processing Layer, which has various components (i.e. CPU, a communication module, wired/wireless interfaces for in-vehicle sensors, Global Positioning System (GPS) module, etc.) and is responsible for the collection, storage, processing, and transmission of received data from social applications. OBU senses and builds vehicle messages using vehicles' static (UPC, Electronic Product Code [EPC], IPv6 address) and dynamic (sensory) information. Upon interactions with each other, SIoV vehicles exchange social messages and store social information in OBU–OBU social graph.

#### 9.6.3.2 The IoV Gateway Layer

The IoV Gateway layer consisting of three sublayers (i.e. Integration, Middleware, and Application) is responsible for the collection of data from SIoV vehicles and transfer it to the Fog layer for further processing. The Integration layer enables SIoV vehicles and RSUs to directly communicate with VANET vehicles and in-road sensors through different standards, i.e. DSRC, 6LoWPAN, Low Power Wide-Area Network (LPWAN), etc. The SIoV gateway middleware performs data pre-processing, management of web services, and creation/management of social relationships. The Application layer offers V2V and V2I safety and infotainment applications. OBU–RSU social graph information is shared and stored on RSU upon vehicles' interaction with RSU. RSU assigns social tags to received data before forwarding toward the Fog layer. The Fog agent at the application layer is responsible for social tagging and sending of social graph updates to the Fog layer.

#### 9.6.3.3 The Fog Layer

In SIoV architecture, the Fog layer offers social services of vehicular Cloud to the network edge, which is closer to end devices and consists of four sublayers, i.e. communication layer, data layer, service layer, and application layer. Communication sublayers provide several protocols to receive pre-processed data from the Gateway layer using heterogeneous technologies, i.e. DSRC, Bluetooth, ZigBee, Wi-Fi, 4G/LTE, etc. The data sublayer offers services for temporary storage, management, and analysis of social relationships between different vehicles at Fog nodes. The service layer is responsible for the provisioning of service APIs to deal with heterogeneous data. The Application sublayer provides various SIoV applications (i.e. local traffic information, parking, and local content delivery), which demand fog computing support.

#### 9.6.3.4 The Vehicular Cloud Layer

This layer deals with the dynamic management and provisioning of massive storage to store social vehicular data (received from in-vehicle/on-road sensors) in terms of OBU–OBU and OBU–RSU social graphs. Social graphs contain:

- vehicles' public information.
- vehicles' private information.
- friend structure of vehicles with respect to time.
- usual traveled routes information.
- Group joining information.

Moreover, the dynamic rendering of computing resources for BigData analysis is also part of this layer. The social data of SIoV vehicles contains vehicles' interaction information along with the necessary timestamp and helps to find certain patterns that affect traffic planning on roads through BigData analytics.

## 9.7 SIoV Application Services

The users of vehicular Clouds are intelligent software agents or humans that analyze information of vehicle profiles, vehicle user friendships, and usual and common routes, etc. to solve transport-related issues. Therefore, at application level, SIoV architecture deals with the provisioning of SIoV application-specific services to the user. The SIoV applications exploit the social graph of relationships among road vehicles for safety and infotainment services to support ITS.

## Review Questions

**9.1** What are the advantages of smart thing communication in SIoT?

**9.2** Explain the types of User–Thing relationships and Thing–Thing relationships.

**9.3** What are the applications of SIoT smart thing relationships?

**9.4** Explain the Architecture of SIoT.

**9.5** How reference architecture of IoV is different from SIoV?

**9.6** Predicting health status of chronic disease patients (through the monitoring of vital signs status and patient activities) is one of the basic objectives in IoT-based healthcare system. For efficient prediction, several sensors are required to monitor patients' activities. However, this method is costly, inconvenient, and even cannot correctly predict the health status of a patient. Propose an SIoT-based solution that offers efficient services to chronic disease people who need constant care and helps in predicting patient health status efficiently and correctly.

**9.7** The growing number of elderly people demands an efficient healthcare system. Advanced Information and Communication Technologies (ICT) techniques support elderly monitoring services at a high cost. Propose an SIoT-based solution for the monitoring of elderly people. In your proposed solution, heterogeneous devices (belonging to elderly people under observation) must be able to:
**A** collect environmental and physical user data.
**B** share data with each other.
**C** send social interaction data to respective physicians and caregivers.

## References


1 Ranta, E., Rita, H., and Lindstrom, K. (1993). Competition versus cooperation: success of individuals foraging alone and in groups. *The American Naturalist* 142 (1): 42–58.
2 Atzori, L., Iera, A., and Morabito, G. (2014). From" smart objects" to" social objects": the next evolutionary step of the internet of things. *IEEE Communications Magazine* 52 (1): 97–105.
3 Soro, A., Brereton, M., and Roe, P. (2018). *Social Internet of Things*. Springer.
4 Atzori, L., Iera, A., and Morabito, G. (2011). Siot: giving a social structure to the internet of things. *IEEE Communications Letters* 15 (11): 1193–1195.
5 Atzori, L., Iera, A., and Morabito, G. (2012). The social internet of things (SIoT) – when social networks meet the internet of things: concept, architecture and network characterization. *Computer Networks* 56 (16): 3594–3608.
6 Dutta, D., Tazivazvino, C., Das, S. et al. (2015). Social Internet of Things (SIoT): transforming smart object to social object. In: *National conference on Mathematical Analysis and Computation (NCMAC)*. Jaipur: Malaviya National Institute of Technology.

7 Roopa, M., Pattar, S., Buyya, R. et al. (2019). Social internet of things (SIoT): foundations, thrust areas, systematic review and future directions. *Computer Communications* 139: 32–57.

8 Evangelos, A.K., Tselikas, N.D., and Boucouvalas, A.C. (2011). *Integrating RFIDs and smart objects into a Unified Internet of Things architecture. Advances in Internet of Things* 1: 5–12.

9 Ortiz, A.M., Hussein, D., Park, S. et al. (2014). *The cluster between internet of things and social networks: Review and research challenges. IEEE Internet of Things Journal* 1 (3): 206–215.

10 Voutyras, O., Bourelos, P., Gogouvitis, S., et al. (2015). Social monitoring and social analysis in internet of things virtual networks. *IEEE 18th International Conference on Intelligence in Next Generation Networks.*

11 Voutyras O., Bourelos P., Kyriazis D., et al. (2014) An architecture supporting knowledge flow in social internet of things systems. *IEEE 10th International Conference on Wireless and Mobile Computing, Networking and Communications (WiMob).*

12 Alam, K.M., Saini, M., and El Saddik, A. (2015). Toward social internet of vehicles: concept, architecture, and applications. *IEEE Access* 3: 343–357.

13 Yang, L., Cheng, X., Ghogho, M. et al. (2019). Guest editorial special issue on IoT on the move: enabling technologies and driving applications for internet of intelligent vehicles (IoIV). *IEEE Internet of Things Journal* 6 (1): 1–5.

14 Gasmi, R. and Aliouat, M. (2019). Vehicular ad hoc NETworks versus internet of vehicles – a comparative view. In: *2019 International Conference on Networking and Advanced Systems (ICNAS)*. IEEE.

15 Baldessari, R., Bödekker B., Deegener M. et al. (2007). Car-2-car communication consortium-manifesto. DLR Electronic Library [http://elib.dlr.de/perl/oai2] (Germany).

16 Taleb, T. and Benslimane, A. (2010). Design guidelines for a network architecture integrating VANET with 3G beyond networks. Proceedings of IEEE Globecom.

17 Abdelhamid, S., Hassanein, H.S., and Takahara, G. (2014). Vehicle as a mobile sensor. *Procedia Computer Science* 34: 286–295.

18 Guerrero-Ibáñez, J., Zeadally, S., and Contreras-Castillo, J. (2018). Sensor technologies for intelligent transportation systems. *Sensors* 18 (4): 1212.

# 10

# Packet Tracer and IoT

**LEARNING OBJECTIVES**

After studying this chapter, students will be able to:

- explain the basics of Packet Tracer and Blockly programming language.
- design simple IoT projects in Packet Tracer.

## 10.1 IoT and Packet Tracer

This chapter guides you on how to make simple IoT projects in Packet Tracer. Packet Tracer is a visual cross-platform tool (can be run on Linux, Windows, Android, and also macOS) developed by Cisco Systems, which assists in understanding the concepts of computer networking through simulations [1, 2]. It follows the procedure of simple drag and drop method to add/remove network (including IoT) devices of all kinds. In addition, it allows programming in a variety of languages, i.e. JavaScript, Python, and Blockly, to allow a student to perform simulations in simple ways. The IoT project examples developed in Packet Tracer using Blockly language will be helpful for students to understand the basics of Things connectivity over the Internet. Real-life things embedded with electronic devices are able to communicate each other. IoT is an extension of the Internet, that with an aim of providing interconnectivity among our daily use objects and things over the Internet. Blockly is a programming language, which lets you create your own program in a much better and easier way using Packet Tracer.

Major components of Packet Tracer Integrated Development Environment (IDE) are highlighted in Figure 10.1.

You can select and drag any IoT device (as shown in Figure 10.2) into the workspace using the *Device-Type Selection Box* and *Device-Specific Selection Box*.

*Enabling the Internet of Things: Fundamentals, Design, and Applications*, First Edition.
Muhammad Azhar Iqbal, Sajjad Hussain, Huanlai Xing, and Muhammad Ali Imran.

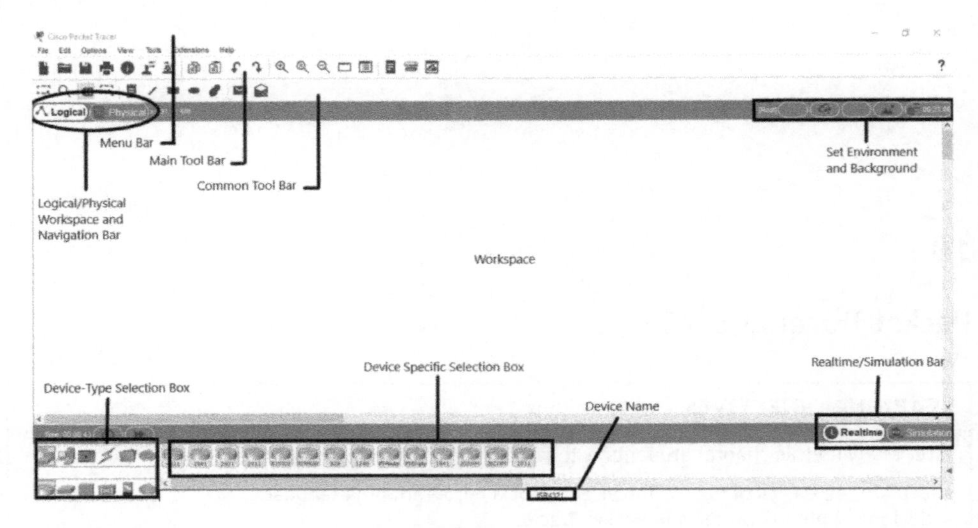


**Figure 10.1** The Cisco Packet Tracer user interface.

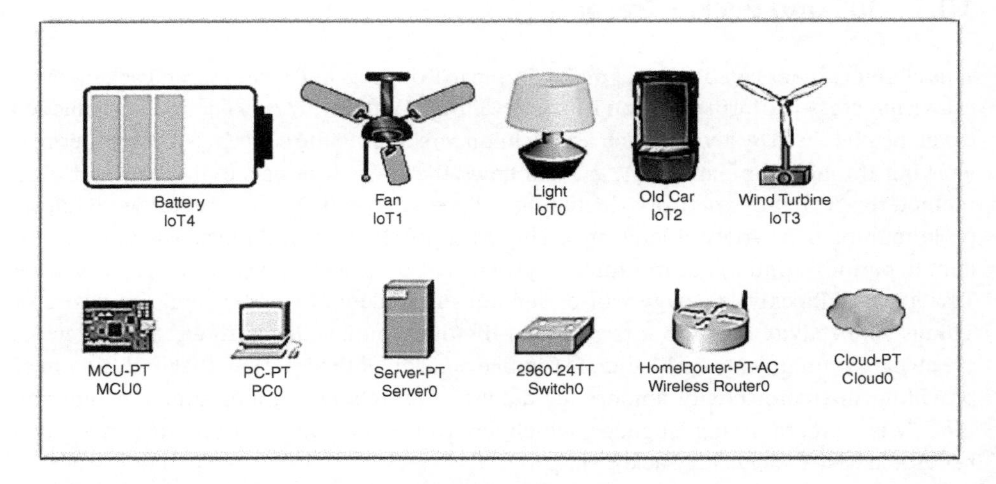


**Figure 10.2** Few IoT devices available in Cisco Packet Tracer.

## 10.2 Packet Tracer Programming Environment

By *double-clicking* on *any of the devices* (already dragged in workspace), a window appears as shown in Figure 10.3. For performing any programming, click on **Programming** tab. Packet Tracer supports three programming languages such as Python, JavaScript, and Visual (Blockly). In order to start new project or to select programming language, click on **New** button.

By clicking on the **New** button, **Create Project** window will appear (see Figure 10.4), which shows text field to enter the project **Name** and to select **template**. From the drop-down menu, here you can see three basic programming languages JavaScript, Python, and Visual as shown in Figure 10.5. In this chapter, we have used Visual programming (another name for Blockly Programming) because it is easy to understand at novice level.

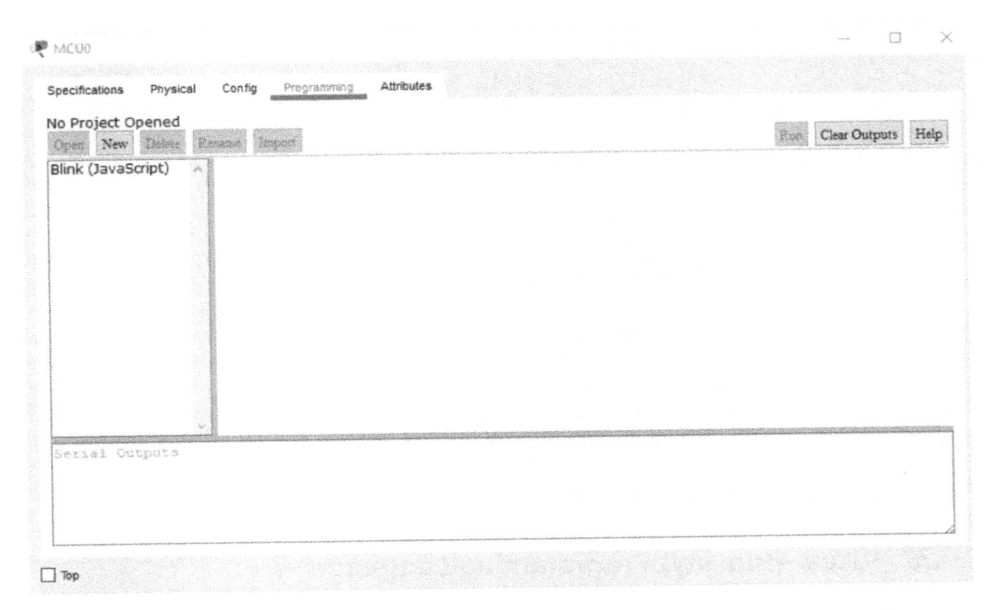


**Figure 10.3** Device selection window in Cisco Packet Tracer.

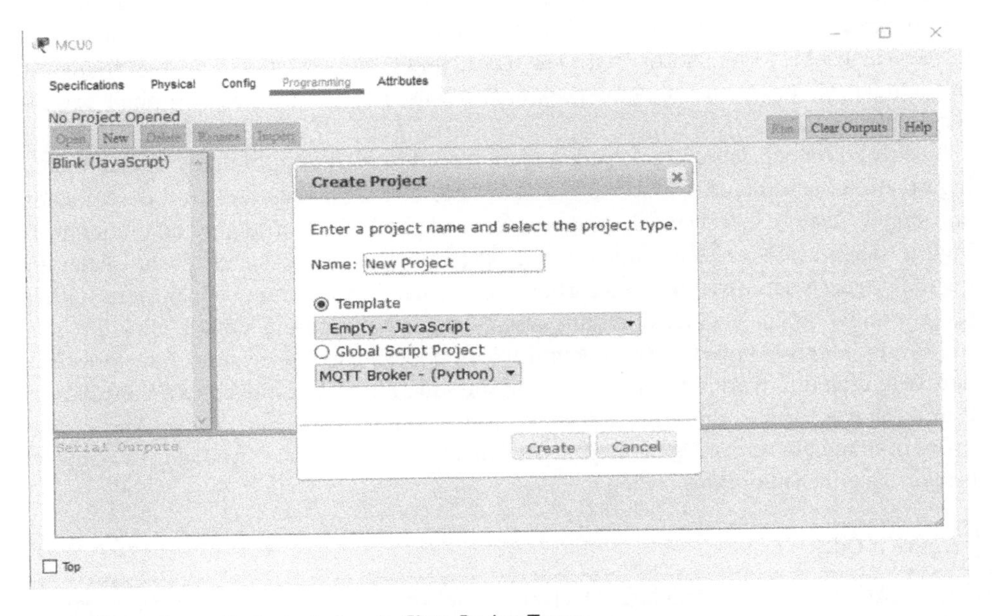


**Figure 10.4** Create Project window in Cisco Packet Tracer.

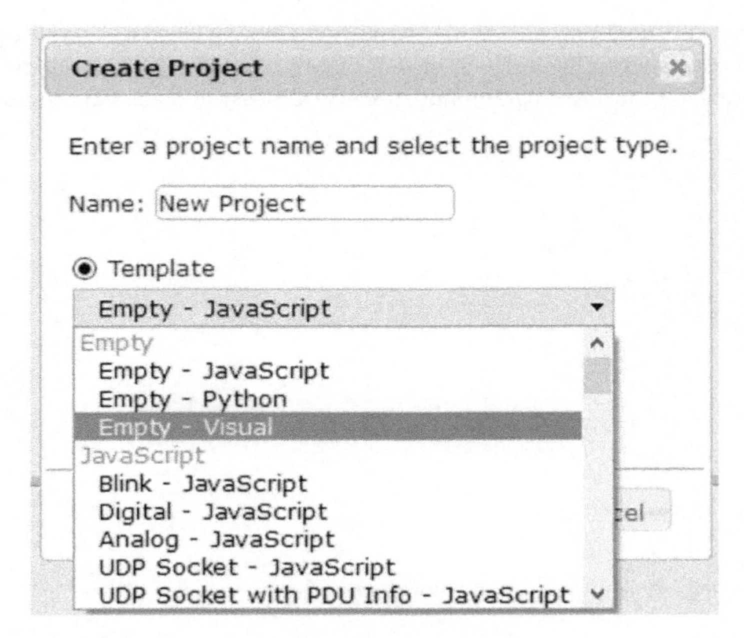


**Figure 10.5** Selection of programming languages in Cisco Packet Tracer.

## 10.3 Visual (Blockly) Programming Language

Visual or Blockly programming language provides visual coding blocks to make the program easier by connecting them together like puzzles. In this section, you will learn about the different blocks available in Packet Tracer, i.e. the names, reason for coloring, and usage. The goal is to get familiar with few blocks and terms, which will be used frequently in the next sections. You can come back to this section anytime and revise their use and recall their purpose.

Blockly editor consists of workspace and toolbox, which consists of many blocks (including functions, arithmetic operations, variables, networking blocks, etc.) in different colors and shapes. Users can drag blocks and generate code easily by rearranging different blocks. Figure 10.6 indicates programming interface step by step: start by selecting ① tab. After giving the project a name as discussed earlier, you can see the project name in ② along with its programming language (Visual in our case). To start programming, double-click the main file ③; another tab ④ appears showing multiple options. We can select any of the tools from here ④ to drag into workspace and later Run the Program ⑥. The output will then show in ⑤ and the execution result will be displayed on the main workspace. Packet Tracer comprises of many blocks for implementing code in a much easier way. To quickly overview all blocks, try different options from panel ④.

**Figure It Out!**

In *Program*, explore block palette, which comprises of different blocks of functions, variables, maths, logic, loops, etc. and other options such as pin access, networking, TCP, and many more. Blocks are of different colors to differentiate them from each other.

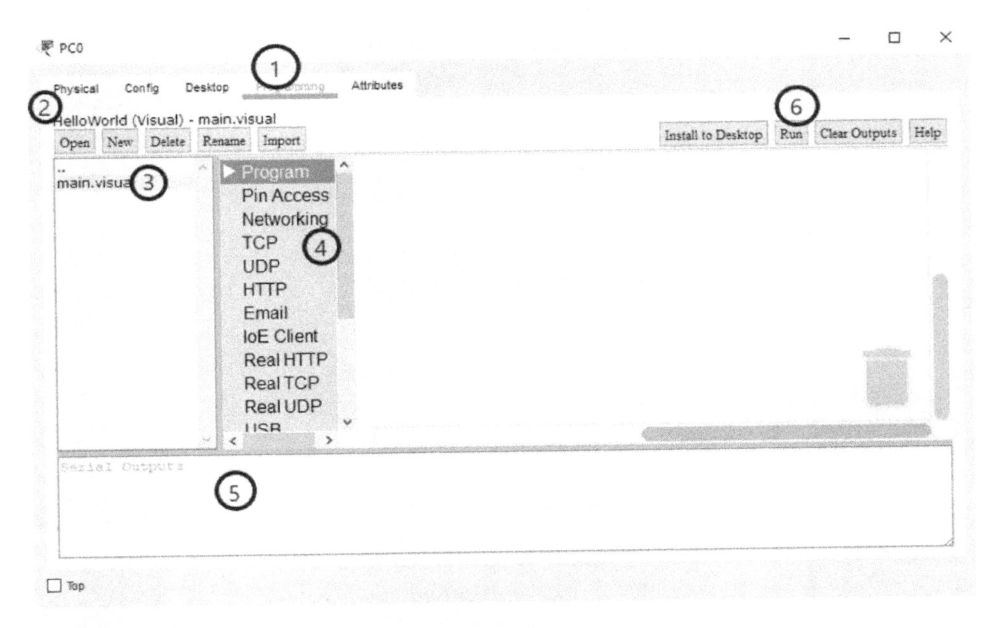


**Figure 10.6** Cisco Packet Tracer programming interface.

*The Program* block contains subsections of blocks: Functions, Variables, Logic, Loops, Math, Text, Lists, and Dictionaries as shown in Figure 10.7. As illustrated in Figure 10.8, some blocks have inside cuts, some with outside edges, whereas some blocks have both inside and outside cuts. Few blocks have inside puzzle-shape marks, indicating that these blocks require another block to connect with. Block alone has no meaning; every block must connect with any of the blocks in order to execute. The color of the blocks indicates the type of data they return. For example, function blocks are in purple color, all type of variables are in pink-reddish color, and so on. Thereafter, complex and lengthy code can be easily debugged with the type of color they have. Other than Program block type, there are 17 kinds of blocks (i.e. Pin Access, Network, Transmission Control Protocol [TCP], UDP, HTTP, Email, Internet of Everything [IoE] Client, Real HTTP, Real TCP, Real UDP, USB, Bluetooth, File System, Physical, Environment, Workspace, and GUI) are available in Packet Tracer (version 7.2.1.0.218) as shown in Figure 10.9.

### 10.3.1 Hello World Program

To start the first program *"Hello World,"* click on the **Program** tab; select the **Textfield**. A tab appears that represents different blocks from text field. There are blocks available in square and in oval shape. Click on the **print** block as shown in Figure 10.10. It will automatically appear in your workspace. You can observe that this print block have inner blended edge at the top, puzzle mark on right, outer edge at the bottom, and is in light green color. This is because print block requires something to print and that block requires

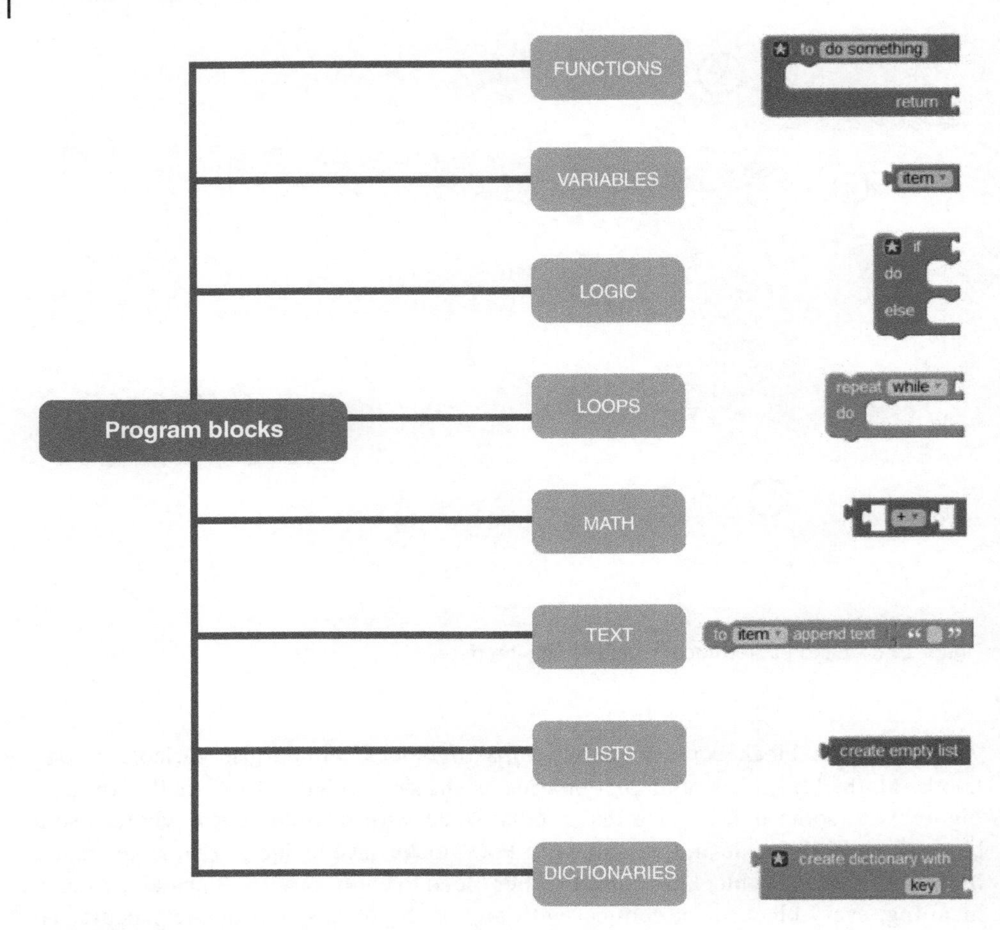


**Figure 10.7** Types of program blocks.

another block to connect with it. Just like puzzles require another puzzle to link with to display a picture, similarly blocks in Blockly programming required other blocks to connect with to execute something.

After that click on the **Text** field again and select the very first block having **double quotes** on it, which indicates a string type data can be placed in it. Click the block and write text Hello World in it. Using a mouse to hover any block helps you getting tooltip, which can provide guidance related to all blocks as shown in Figure 10.11. Drag and join the two blocks together as shown in Figure 10.12. Here, yellow borderline indicates that the block can be fixed with another block.

Now you need to execute the program. For this, click on the **Run** button as shown in Figure 10.6 at ⑥. The program will start executing without errors and can be seen under output window ⑤ see Figure 10.13.

To stop executing the program, click the Stop button ⑥ shown in Figure 10.6.

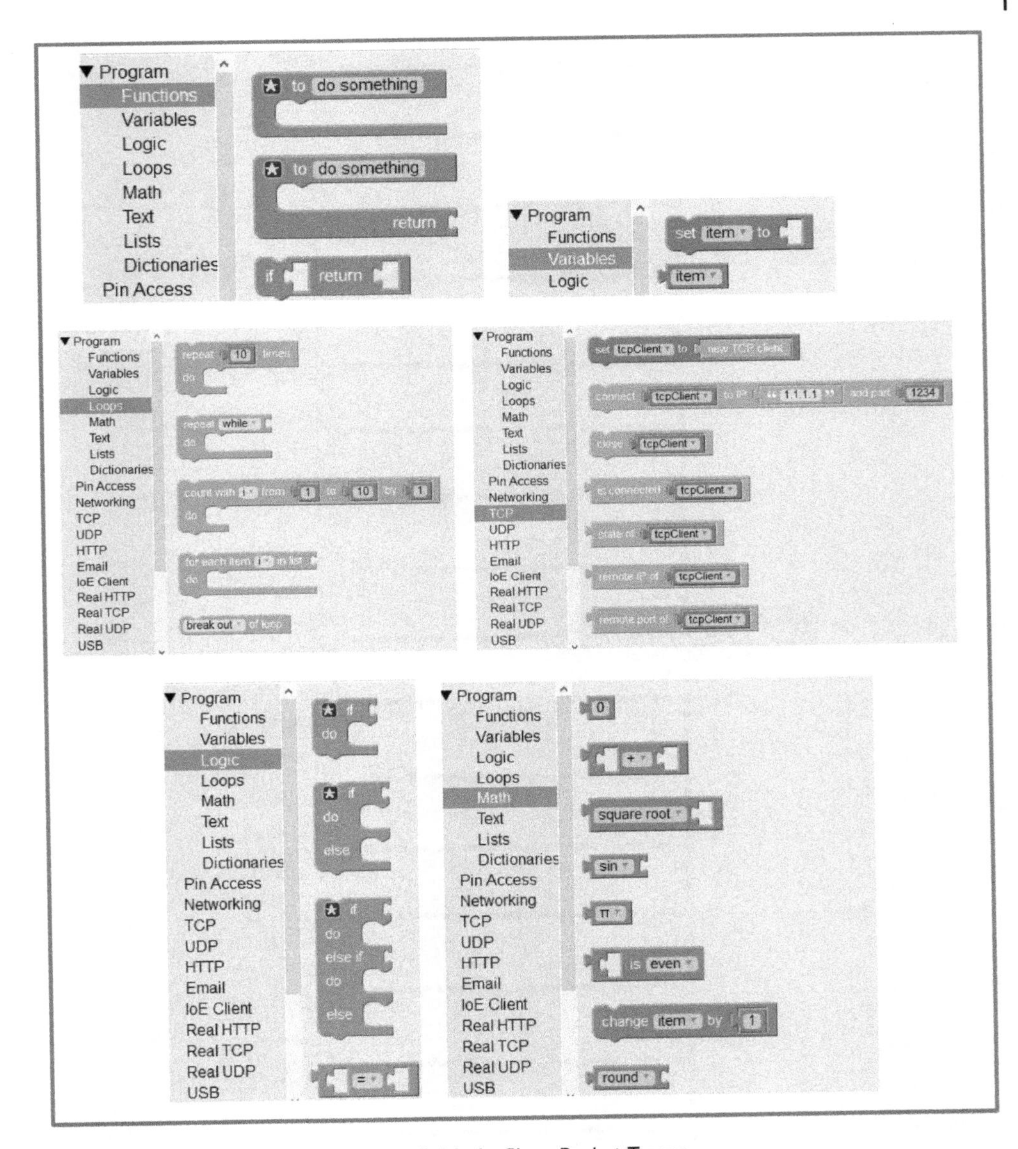


Figure 10.8 Programming blocks available in Cisco Packet Tracer.

## 10.4 Simple Smart Light Project

As you already know the programming interface and its environmental tools at this stage, you are now ready to use these tools in a new way by developing a new program. You have also learned different blocks and from now onward you can easily use them. In this section, you will learn:

- How to use workspace in Packet Tracer
- How to build blocks together to create a Blockly program

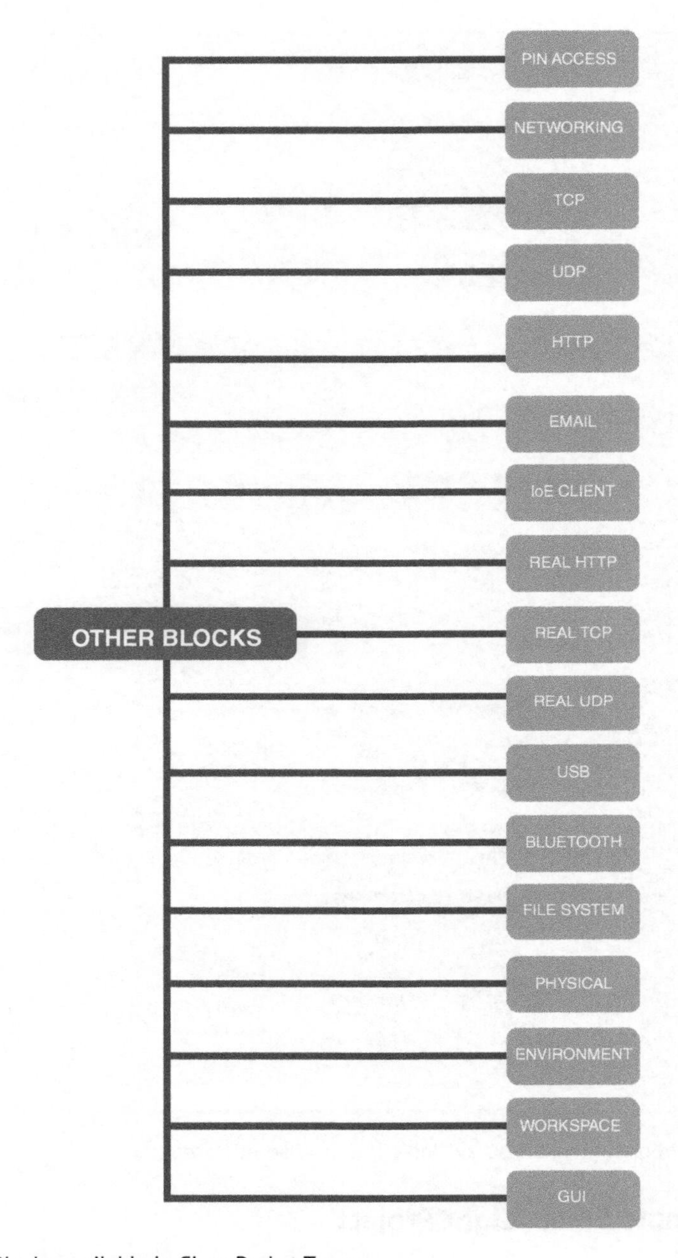


**Figure 10.9** Blocks available in Cisco Packet Tracer.

- How to execute a smart light program in a realistic way
- How Blockly makes programming easy
- How to use loops

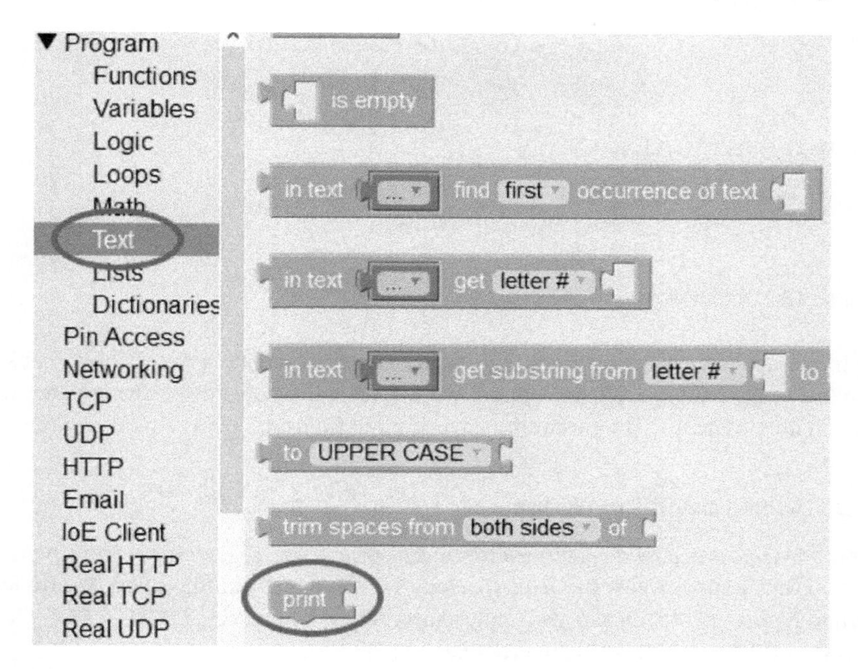


**Figure 10.10** Blocks to print and format text.

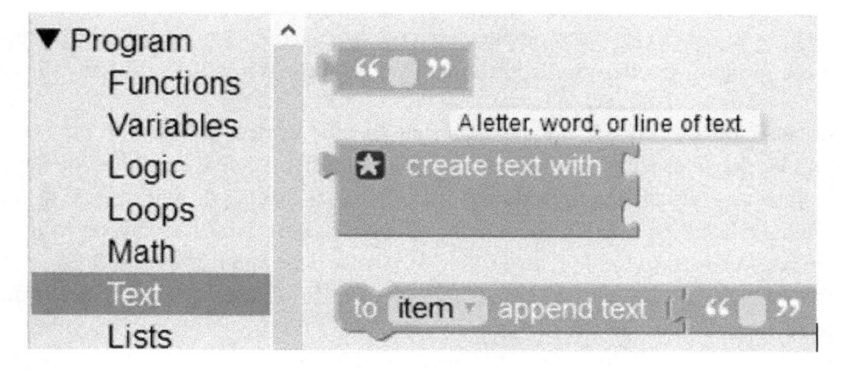


**Figure 10.11** Text field block with tool tip.

The word *smart* means to convert ordinary things into being smart, which can take decisions based on certain circumstances. Being a *smart device* means it must have underlying technology, power computation, some backend programming, or communication features. This section will help you create a smart light, which will be controlled with the help of a power button that covers these two basic features:

- When the power button turns ON, the lamp is turned on and the light starts blinking.
- When the power button turns OFF, the lamp is turned off.

**Figure 10.12** Hello World block.

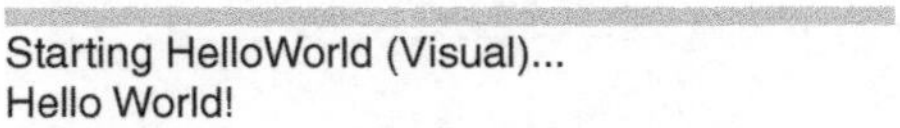

Starting HelloWorld (Visual)...
Hello World!

**Figure 10.13** Output window.

To implement this, start by executing Packet Tracer. To create a new project, click from the menu toolbar **File → New**. A blank workspace will appear. (See Section 10.1 for the details of all toolbar and programming environment interface.)

### 10.4.1 Adding Devices to Workspace

A list of all devices can be seen at the left-bottom corner. There you will see multiple devices of various categories. Select the **End Devices** section and from this select **Home** icon as shown in Figure 10.14. You can also apply shortcut keys by pressing **CTRL + ALT + V** (End Devices) or **CTRL + ALT + H** (directly selecting Home section).

From there, you will see a variety of home icons used in our daily lives, such as air conditioners and home appliances such as coffee makers, batteries, fans, lights, doors, and so on. Select the **Light** device  and click on workspace; you will note a plus sign appear on workspace indicating where to drop light in workspace. While selecting any device, you will notice a cancel mark  appear on that specific device, in case you do not want to drag it on workspace. Simply cancel or delete it by pressing delete key from the keyboard.

Now, you have dragged a light device into the workspace, you will notice that a light device has a label with it IoT0. By default, all devices have labels according to their category. This device is from the IoT, also called Smart Devices and number as 0. The more you drag, the label would be like IoT1, IoT2, and so on. To rename a label click on the label (IoT0) and rename it as shown in Figure 10.15.

Now, delete the extra lights as we only need a single light to deal with. The next device we need is the power button, which has two options ON and OFF. From now onwards we name power button to be as *Rocker Switch* button. To drag a rocker switch into workspace, click the **Components** section **(CTRL + ALT + B)** and then select the **Sensors (CTRL + ALT + X)** option as shown in Figure 10.16, and select the **Rocker Switch**  from the list having label 1 and 0.

If you click on the device present on a workspace you will see its features and all the related information in detail. For example, in case of light, click on a light device; a window will appear that will define light, its features, and the role it played. Similarly, if the rocker switch is pressed, you will see its relevant details in the **Specifications** tab as shown in Figure 10.17.

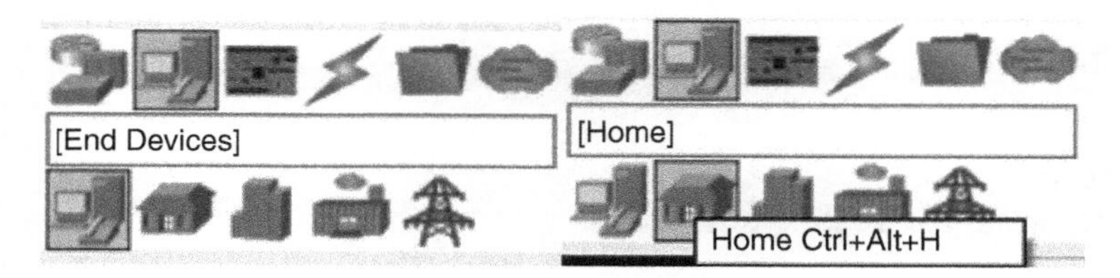


**Figure 10.14** Selection of devices.

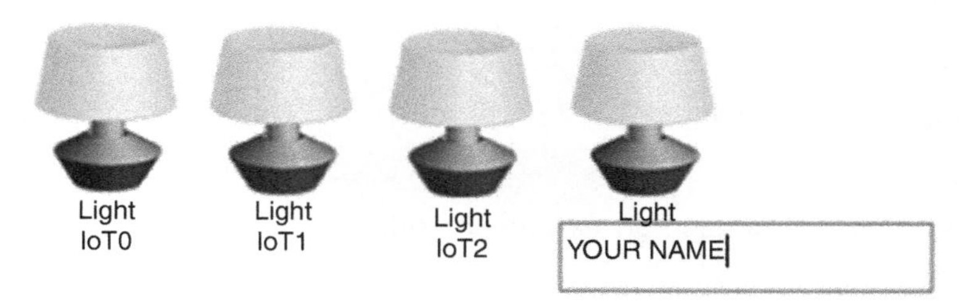


**Figure 10.15** Rename a device.

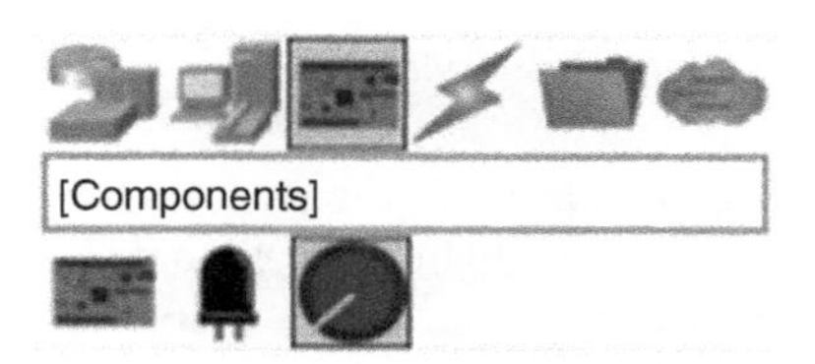


**Figure 10.16** Selecting rocker switch from the components.

To see every device's role and animation, press **ALT + CLICK** together to toggle between device options. A light can be turned on and off. In our case, a light can be dim and high. Similarly, fan speed can be slow or high. To observe such behavior, press **ALT + CLICK** on lamp; you will see that the first click makes lamp light *DIM ON*, second click makes lamp light *HIGH ON*, and third click turns the lamp *OFF* as shown in Figure 10.18.

| **Figure It Out!** |
|---|
| Try these toggle options against multiple devices such as fan, switches, coffee maker, portable music player, and so on and find out those devices who don't have such options. |

Now, we have two devices on the workspace, light and rocker 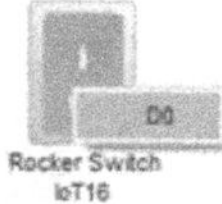

switch working individually. The problem here is we want to control light through switch. If the switch is ON, the light turns on automatically, whereas, if the switch is

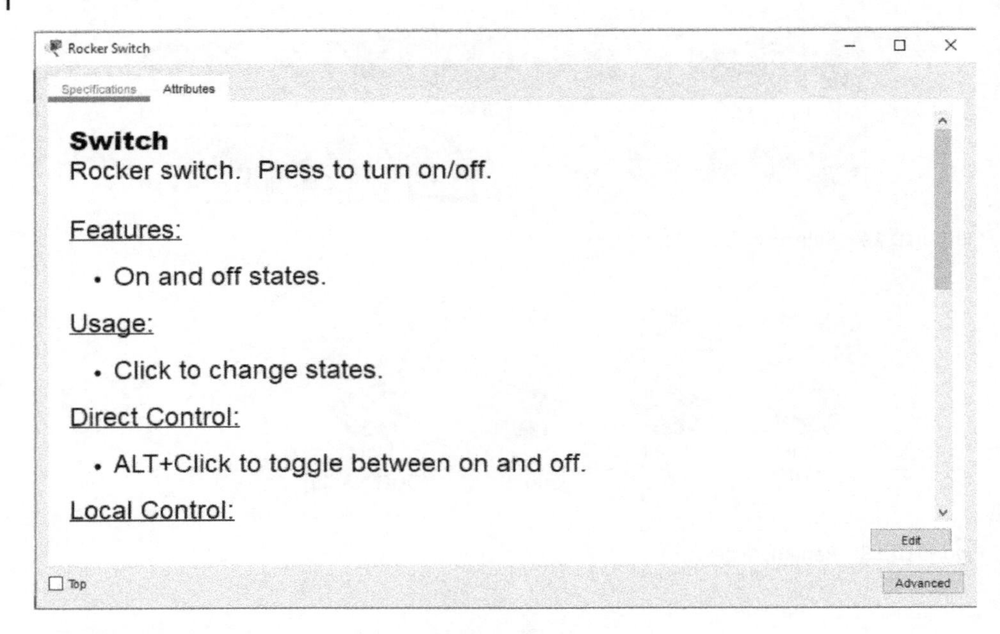


**Figure 10.17** Description for rocket switch device.

turned off, the light goes off smartly. For this, there comes the programming part on a third device, which can communicate with light as well as switch. From the devices, select **Board (CTRL + ALT + B)** devices from the **Components** section. There you will see two types of boards – Microcontroller (MCU) board and Single Boarded Computers (SBC) Board. We will deal with **MCU Board** . Drag the board in the workspace (Figure 10.19).

MCU is a microcontroller board that supports multiple programming languages used mainly for controlling other devices through programming.

### 10.4.2 Connecting Devices

The three devices on the workspace are still not connected with each other. For this purpose, we need a cable that connects all the three devices together. Select the **Connections** tab **(CTRL + ALT + O)** as shown in Figure 10.20 and select **IoT Custom Cable** from the list.

**Figure 10.18** States of light (OFF, DIM, HIGH).

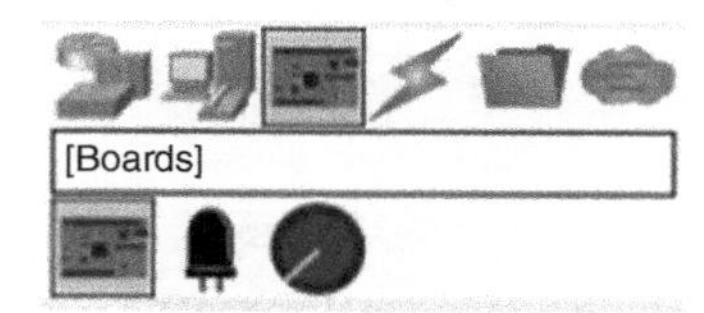


**Figure 10.19** Selecting the MCU board from the components.

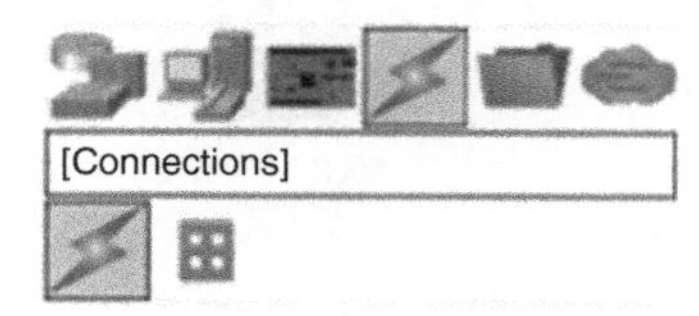


**Figure 10.20** Selecting cables from the connections.

The mouse icon changes, which allow you to click on the rocker switch device and select the **D0** port. Now the cable requires the other device to connect with. Click on MCU board and select **D0** port as shown in Figure 10.21. The two devices are now connected.

Now, repeat the procedure for the light device. Select the IoT custom cable again and click on the light **D0** port with MCU board **D1** port. As D0 port is already busy with the rocker switch button, the final look of the connections will be like the one shown in Figure 10.22.

That's not enough. The functionality is not yet complete. To verify ALT+CLICK on switch button which toggles switch states but light is not affected. Here starts the programming part for Blockly. Now, double-click the MCU board; a window will appear, which allows you to switch between Specifications, Physical, Config, Programming, and Attributes tab. Select the programming tab as shown previously in Figure 10.3. Click on the New

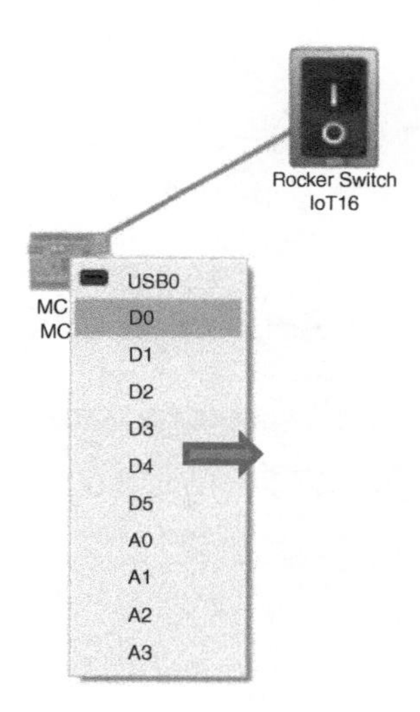


**Figure 10.21** Connecting devices through connection cable in Cisco Packet Tracer.

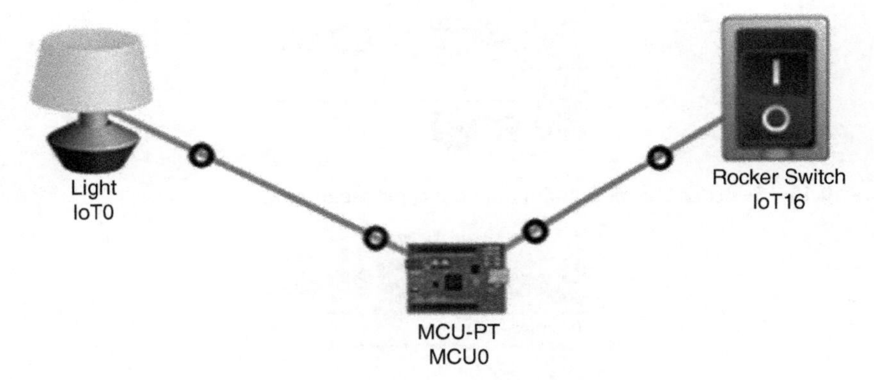


**Figure 10.22** Connecting all the three devices together.

button to start creating new project. Give your project a name. The Template option is selected by default. From the drop-down menu of the Template section, select the Empty - Visual option (shown in Figure 10.23) and click on Create button. From the drop-down menu, here you can see three basic programming languages JavaScript, Python, and Visual as shown in Figure 10.5. Our focus will be on Visual, another name for Blockly Programming.

On the left side of the panel, you will see your project name and main.visual. Double-click main.visual and the block panel will appear as shown in Figure 10.6, where the interface is explained in detail.

**Figure 10.23** Creating new project.

### 10.4.3 Using Program Blocks and Pin Access

In order to create a program that requires mathematical computation or requires some sort of logical implementation or repetition of any task or in any case just to create variables to handle the programming environment, we use **program** blocks from the block palette. You will find out how in this section.

Every program must contain the main function, which allows program to start. For this purpose, from the program tab under function option, select block. This will appear on the workspace of programming interface window. Always assign functions name that reflects its nature; therefore, rename and do something with main like this to make function name as user friendly. You can drag the blocks anywhere on the workspace and adjust accordingly. This main function is defined now and is ready to call. To call a function, the new block of main is added automatically in the Function section. Add this block anywhere into workspace. So, whatever blocks we put into this function main will be called by this .

From the Pin Access section, select . As you have observed this block has inside cut at the top and outside edge from the bottom, which indicates that this block requires to connect with two more blocks to show its functionality, which matches its edges accordingly. Also, pinMode has two parameters by default, slot 0 and mode INPUT. You can change these two settings according to the cable connections you made earlier in this chapter. Now repeat this process again by selecting again from Pin Access section into workspace. However, this time change slot 0 to 1 and from the drop-down menu change mode to OUTPUT. Your workspace would look like the one shown in Figure 10.24.

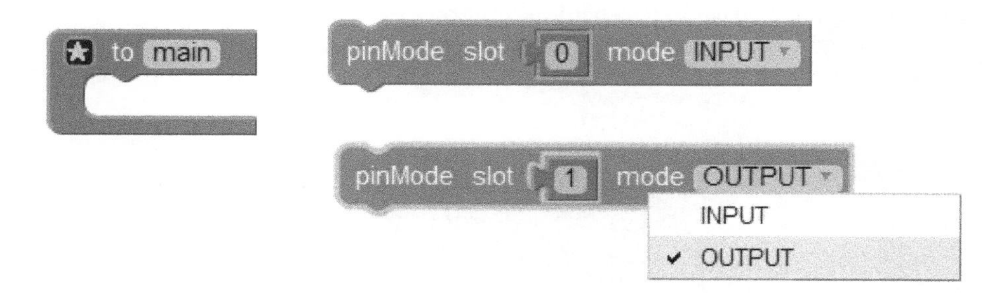


**Figure 10.24** Settings pinMode parameters.

Now, what is this all about? As shown in Figure 10.22, the MCU board has two cables attached, one with rocker switch connected with port D0 and the other with light port D1. So, we have to configure two pins against two cables attached. We want signals to come from rocker switch as INPUT to MCU board and after some operations pass those signals to light as OUTPUT from MCU board. Remember, we are in MCU board programming section; therefore, input and output modes will set according to MCU board. That's why we set pinMode D0 as Input (Don't write D with 0) for rocker switch device cable and pinMode 1 as Output for light device. Drag the two blocks one by one into main function block. The yellow line indicates that the two blocks can join together as their cut marks matched as shown in Figure 10.25. Repeat the same with the other block as shown in Figure 10.26.

We need now to get signals from rocker switch in a way that if the signal indicates 1 light should turn on and if the signal indicates 0, the light should turn off. Therefore, it indicates that rocker switch deals with digital signals also explained in the Specification tab of rocker switch. In order to read signals digitally from rocker switch, select  digitalRead slot 0 from Pin Access section. Set slot to 0 (already set by default). This will return the signal from port D0 and will give its value, which should be stored somewhere. In programming, we used to store values in variables. Thus, a new variable must be created first, which can store the resultant signal value. To create a new variable, select 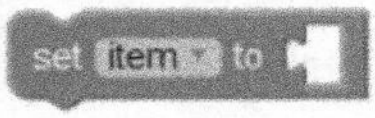 set item to block from Variable section and rename it or simply create New variable as shown in Figure 10.27 as follows:

Rename a variable to be as *switchValue* or a name of your choice and press **OK**. You will notice the change in the variable palette with the addition of two new blocks as shown in Figure 10.28.

Set *switchValue* to what? Whenever we create a variable at the start, we used to assign its value to 0 to avoid garbage results. Therefore, from **Math** section, add  0 and attach it

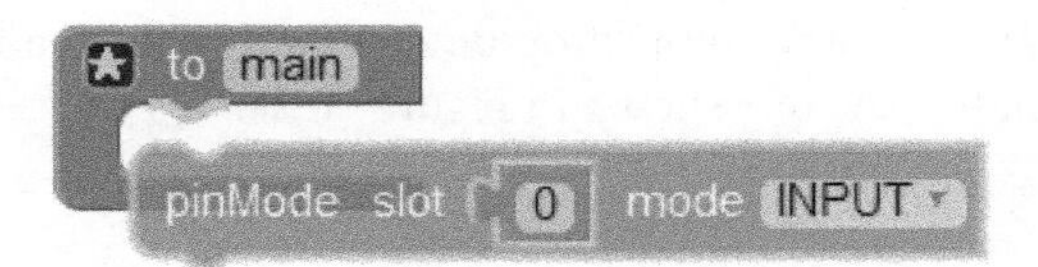


**Figure 10.25** Connecting blocks together. Source: Cisco.

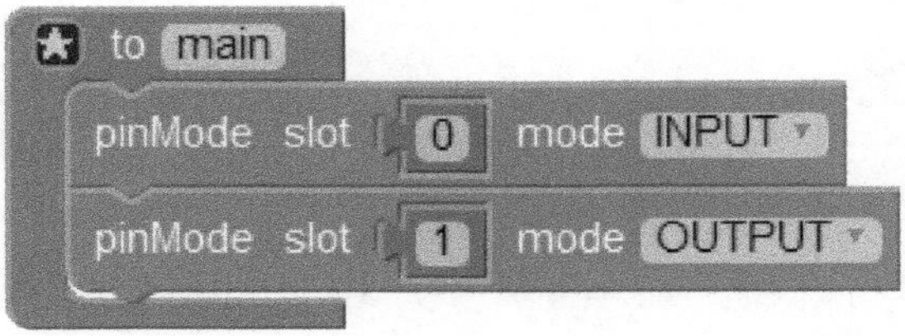

to main
pinMode slot 0 mode INPUT
pinMode slot 1 mode OUTPUT

**Figure 10.26** Inserting blocks into function.

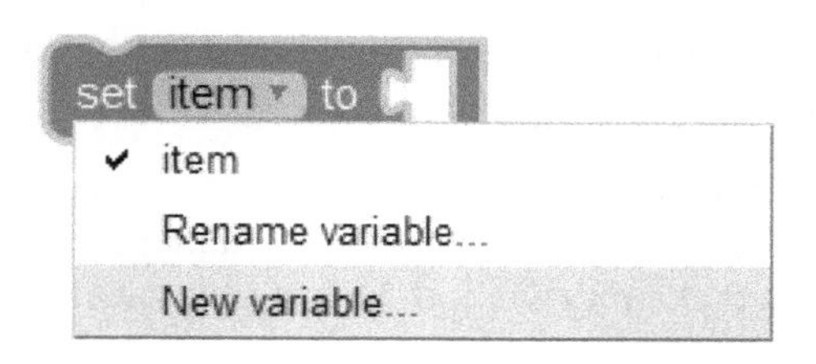


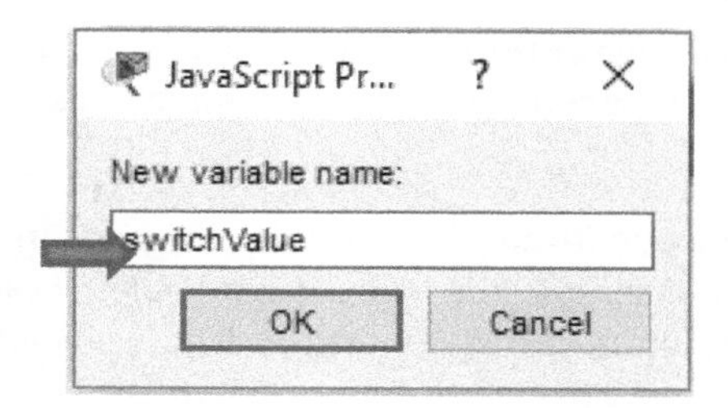


**Figure 10.27** Creating a new variable.

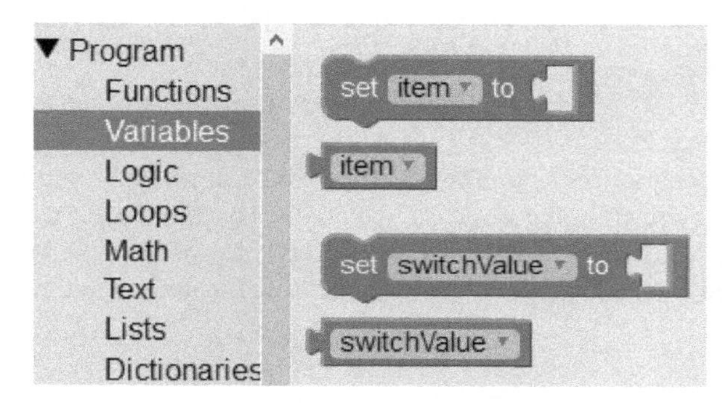


**Figure 10.28** Addition of new variables.

to the variable added before like set switchValue to 0. Initially, the *switchValue* is set to 0. As soon as we get value from digital signal, the value needs to be updated; therefore, from the **variable** section, add set item to again and attach it with the brown block (we added previously) set switchValue to digitalRead slot 0. This means whenever Pin 0 (cable D0) sends signal, it will be read and the value is assigned to a defined variable.

Now, whenever we press the rocker switch sensor button to be ON, the light should turn on and off accordingly. This process must be repeated again and again. Therefore, here we need a loop, which constantly checks the values and repeats this process until the condition is satisfied. For this, select while loop repeat while do from Loops section. The loop required to execute until the condition is true; therefore, from the Logic section add block true into workspace and attach in front of the while loop repeat while true do. Drag the digitalRead blocks into loop repeat while true do set switchValue to digitalRead slot 0. Now the values from the sensors (rocker switch) will be checked constantly without any delay, which creates a program to be stuck sometimes. Therefore, adding delay to the program resolves this issue. To add delay of 1000 ms, go to Pin Access and drag it delay ms 1000 . Add this block into the loop after variable blocks. To test whether the program is reading digital values or not, we need to display something on the output window. For this purpose, add print from the Text section. This block demands something to print. You can add any string with this or add a variable to be print. Let's check the signal value to be displayed. Thus, from the Variable section, add variable and attach the two blocks together print switchValue . Your programming environment workspace will look like the one shown in Figure 10.29.

To execute the program, click on the Run button. You will start getting 0 output in the *Output* window as shown in Figure 10.30, which displays the project name and its output. Now, go back to your main workspace of Packet Tracer where the sensors are placed. **ALT + CLICK** on the rocker switch sensor once. It will change its state from 0 to 1 and a

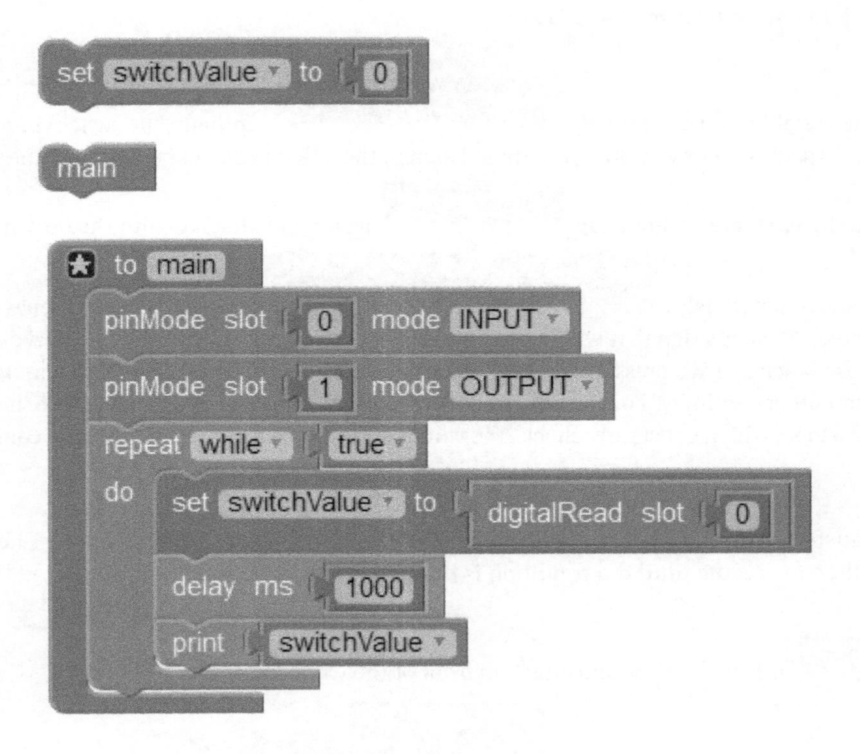


**Figure 10.29** Light program test mode.

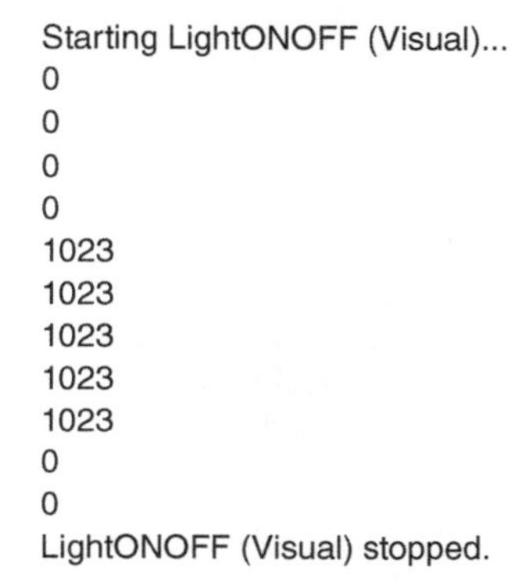

```
Starting LightONOFF (Visual)...
0
0
0
0
1023
1023
1023
1023
1023
0
0
LightONOFF (Visual) stopped.
```

**Figure 10.30** Light program output window.

green light turns on at the sensor . Now go back to output window; you will start getting signal values constantly with the delay of 1 s which is 1023 as shown in Figure 10.30. Toggle between the ON and OFF button of the sensor to get values 1023 and 0. To stop the execution click on Stop and to clear output window click Clear Outputs.

You can also print a text that defines the current situation instead of values 0 and 1023. Like when you press OFF, the output would be like "button off," and when you press ON, output would be like "button on." For this add " " from the Text section and write the text "button off." Replace the block print switchValue with the text block print " button off ". In this case, you will need to handle if-else statements too (we will discuss in the coming sections).

The functionality of handling the light is not yet complete. However, the reading commands from the sensor is completed as we are getting proper signals from the sensor, reading and storing it in a proper variable. Now, our aim is whenever we read 0 signal from the sensor, the light remains off; however, the light turns on as soon as we get signal 1023 from the sensor. To implement this functionality, we have to write some commands for light. For writing commands, we have *customWrite* blocks available in the palette. To add, go to **Pin Access** section, find the block customWrite slot 0 value and add it to the programming workspace. Change the slot from 0 to 1. As the connected port from MCU board to light is 1 as shown in Figure 10.31, we need to write the commands on port 1. Set its value to 2 by adding a number block 0 from **Math** section like this customWrite slot 1 value 2. You can get port number information by just hovering the black circles on the cable as shown in Figure 10.31.

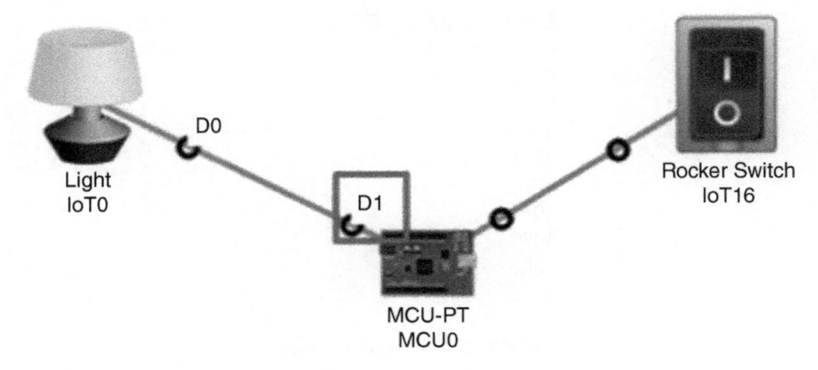


**Figure 10.31** Writing D1 port.

There arise two situations: (i) if the signal is 0 then the light remains off and (ii) else if the signal is 1023 the light turns on. This means we need if-else block, which will handle the situation. For this, go to **Logic** section, and add block . Now *if* statement follows what *condition*? Yes, we need here a variable to compare the signal reading with; that's why we store the signal values in a variable named *switchValue*; thus if the switchValue is equal to zero, light will be off else on. Therefore, add variable *switchValue* in if statement by adding block from the **Logic** section. Join the block to be first parameter and add numeric value 1023 to be second from the **Math** section. The block will look like . Join the block with the *if* statement of the block. Now, if this statement is satisfied, the if block executes the *do* statements. Join the block into if block. What *else* part will do? Surely, the else block makes the light off; therefore add another and set the value to be as 0. The if-else block will look like the one shown in Figure 10.32 and the complete program is shown in Figure 10.33.

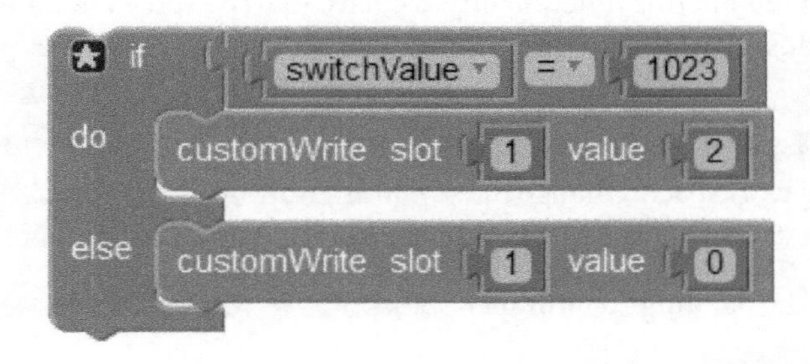


**Figure 10.32** If-else block statements.

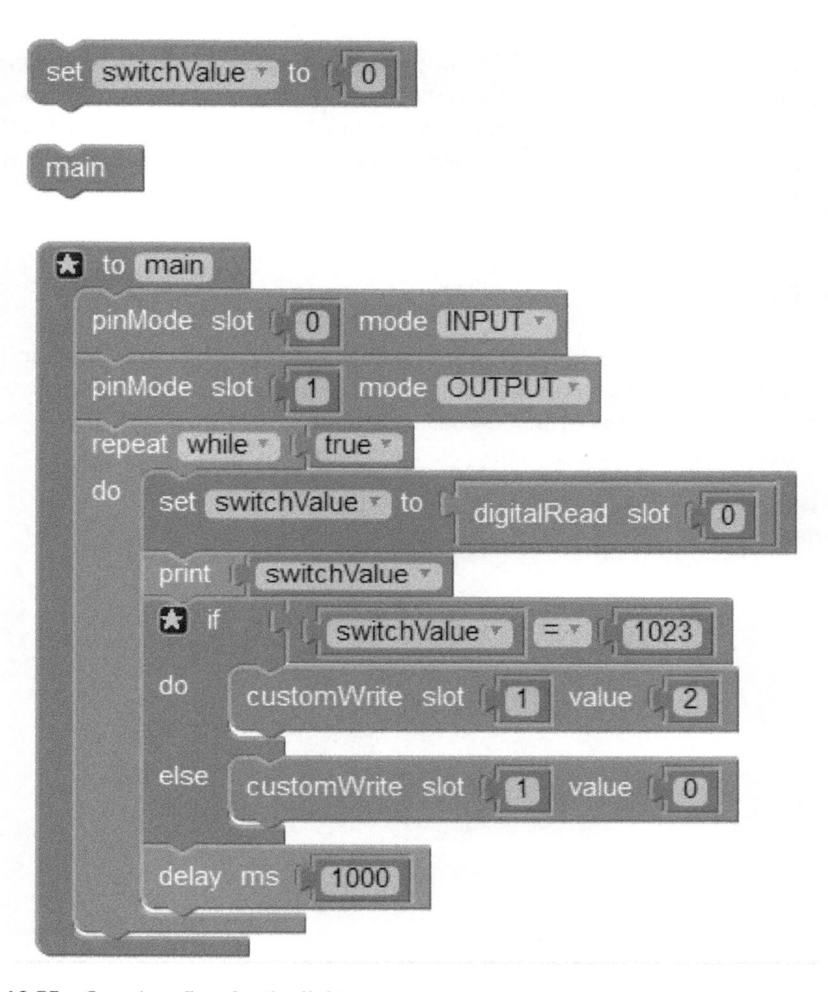


**Figure 10.33** Complete flow for the light program.

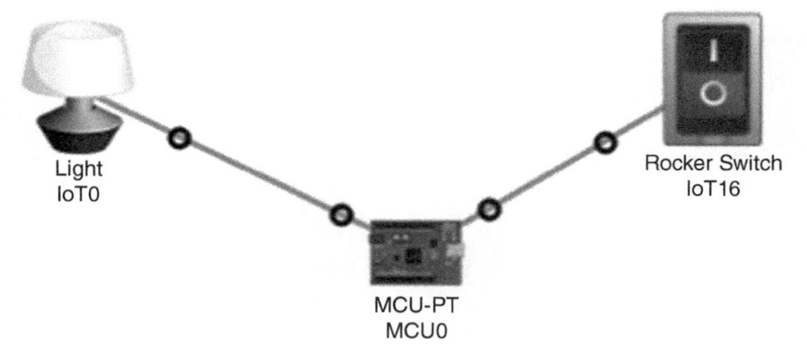


**Figure 10.34** Light program in execution mode.

Test the program by clicking on Run tab and toggle between the sensor switch as shown in Figure 10.34. Now, the light turns on as the sensor gives readings and off otherwise. The light in Packet Tracer has three states: light on but DIM = 1, light on but HIGH = 2, and light off = 0. To implement these three states, the slots' values are required to be set accordingly.

## References

**1** Janitor, J., Jakab, F., and Kniewald, K. (2010). Visual learning tools for teaching/learning computer networks: Cisco networking academy and packet tracer. In: *IEEE 2010 Sixth International Conference on Networking and Services*, 351–355. IEEE.
**2** Jesin, A. (2014). *Packet Tracer Network Simulator*. Packt Publishing Ltd.

# 11

# IoT Projects in Packet Tracer

| LEARNING OBJECTIVES |
|---|
| After studying this chapter, students will be able to:<br>• develop IoT projects in Packet Tracer. |

## 11.1 IoT Projects in Packet Tracer

In Chapter 2, it is mentioned that in Packet Tracer, depending on the scenario, typical IoT systems can be designed or configured in three ways:

- First, IoT smart things can be directly connected to the gateway.
- Second, IoT components and smart things can be connected to microcontroller (MCU) for automatic working without using gateway connectivity.
- Third, IoT smart things can be connected either directly to the gateway or indirectly through MCU. However, basic IoT components, i.e. sensors and actuators (due to lack of ethernet port), can only be connected to Gateway through MCU. Gateway is dependent on the programmed MCU to receive the status of the attached sensors and actuators.

In this chapter, we have discussed the implementation of first two scenarios.

## 11.2 Smart Things Directly Connected with Gateways

This section will help you to learn the controlling of two appliances, i.e. fan and light in smart home with mobile phone through gateway connectivity as shown in Figure 11.1.

Two smart home appliances (Ceiling Fan and Light), one home gateway, and one smart phone are required for this Packet Tracer project:

- Find Ceiling Fan and Light appliances by pressing combination of **ALT + CTRL + H.**
- Find Smart Phone by pressing key combination of **ALT + CTRL + V** and then add it from list.

*Enabling the Internet of Things: Fundamentals, Design, and Applications*, First Edition.
Muhammad Azhar Iqbal, Sajjad Hussain, Huanlai Xing, and Muhammad Ali Imran.

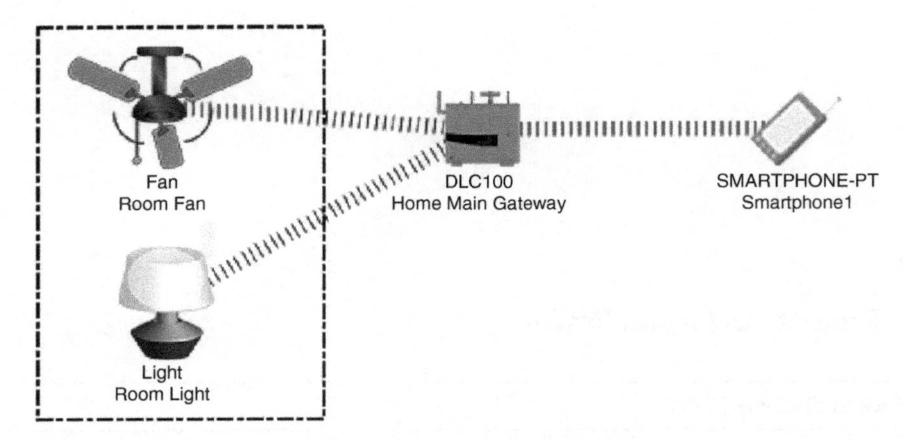


**Figure 11.1** Controlling smart things with mobile phone through gateway.

- Gateway can be added by pressing **ALT + CTRL + W** and then add it from list.
- Connect home appliances to home gateway by using wireless connection. Click on Fan to view properties, and go to advance settings as shown in Figure 11.2.
- Go to I/O Config tab to change network adapter to choose wireless card, which is PT-IOT-NM-1W (Figure 11.3).
- Press Config tab and select Home Gateway radio button to control it with smart phone via gateway (Figure 11.4).
- Following same steps, connect Light and Smart Phone objects to gateway.
- Copy Service Set Identifier (SSID) of gateway and click on Smart Phone to configure mobile to gateway. Change SSID in wireless as shown in Figure 11.5 and Figure 11.6.
- Configure IP Address of Smart Phone with Gateway (Figure 11.7).

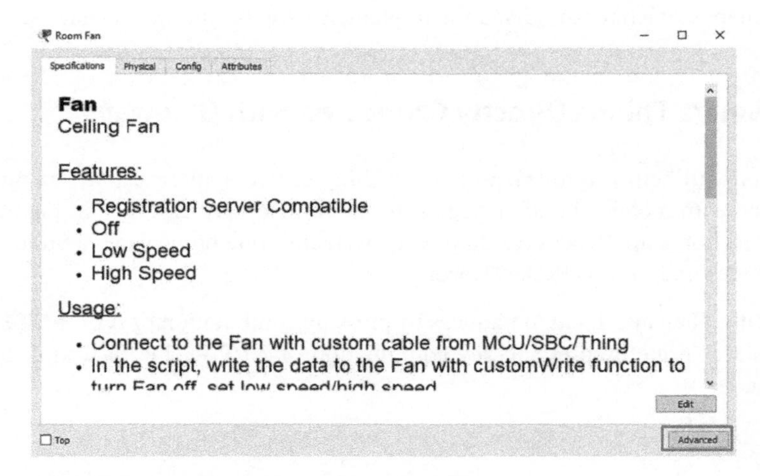


**Figure 11.2** Ceiling fan properties in Cisco Packet Tracer.

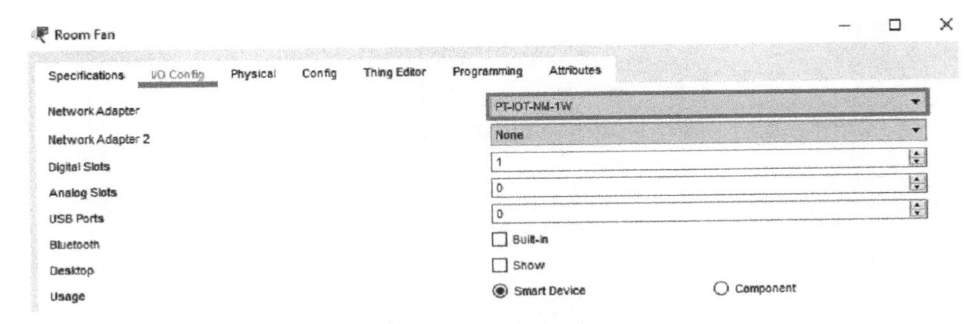


**Figure 11.3** Network adapter card setting in Cisco Packet Tracer.

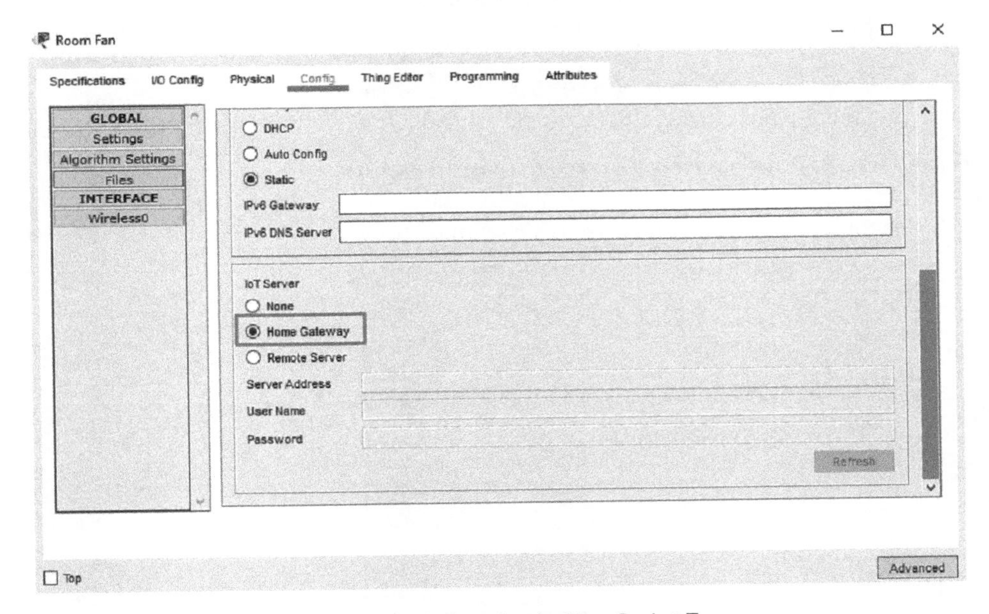


**Figure 11.4** Selection network adapter card setting in Cisco Packet Tracer.

- Next, open Smart phone Desktop and click on IoT monitor (Figure 11.8) and Login screen will appear (Figure 11.9). Provide IoT server address, which should be the IP address of gateway.
- Devices interface will be shown (Figure 11.10) from where you can control IoT devices (Fan and Light in this example) with smart phone.

## 11.3 Smart Things and Sensors Directly Connected with MCUs (Without Gateways)

This section will help you create a smart room. A room consists of multiple smart objects surrounding; some are independent whereas some objects depend on other objects'

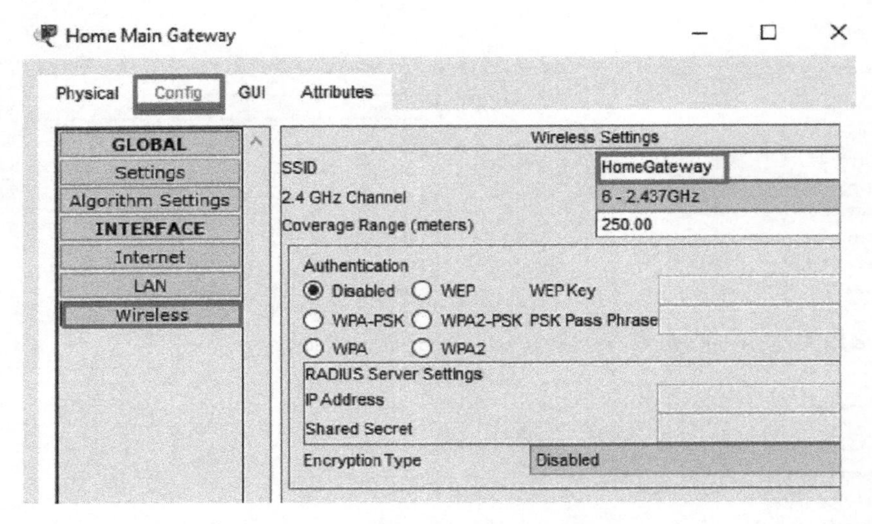


**Figure 11.5** SSID settings for home gateway in Cisco Packet Tracer.

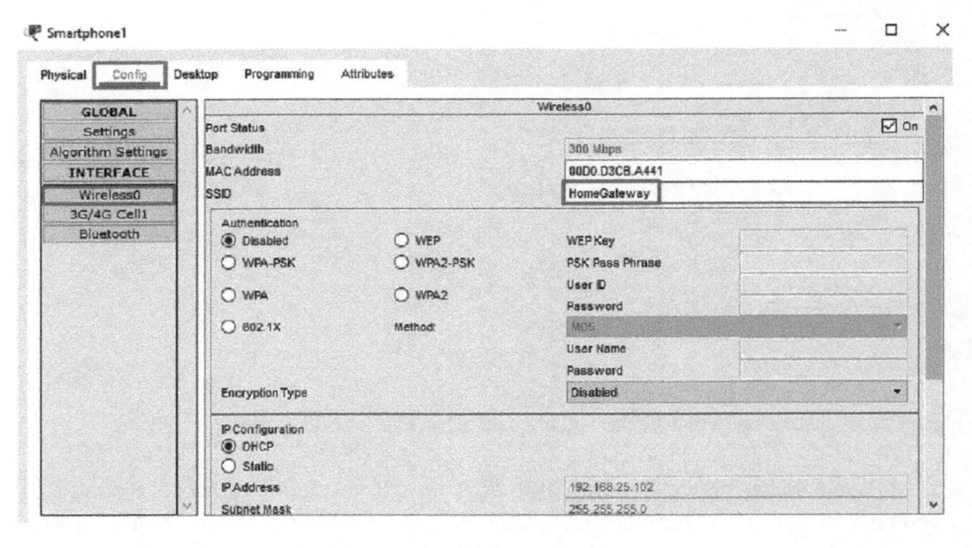


**Figure 11.6** SSID settings for smart phone in Cisco Packet Tracer.

behavior. Therefore, we are about to create a smart room, which consists of the following smart objects:

- Smart Light
- Smart Fan
- Smart Coffee maker
- Smart Air Cooler
- Smart Heating Element

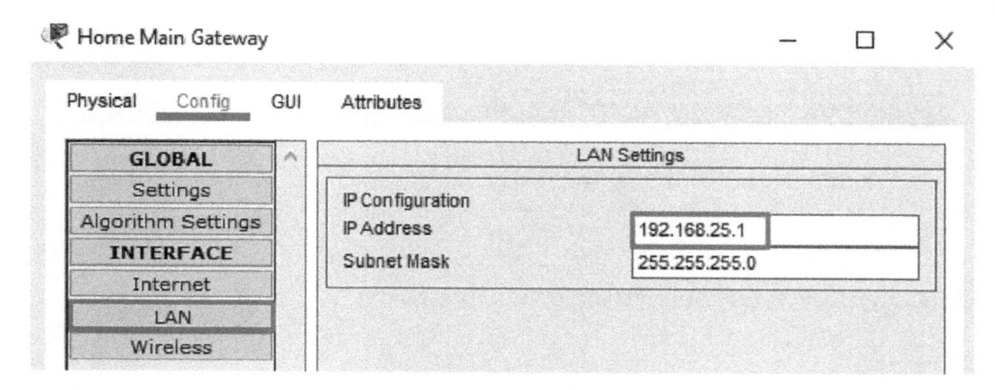


**Figure 11.7** IP address configuration of smart phone with gateway in Cisco Packet Tracer.

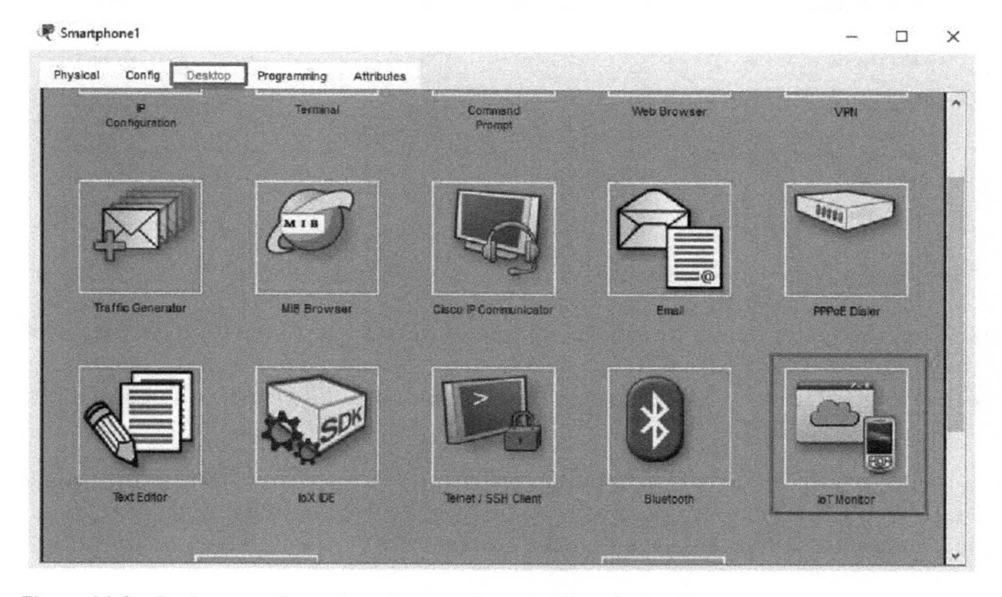


**Figure 11.8** Desktop configuration of smart phone in Cisco Packet Tracer.

The functionality of these IoT things include:

- When the power button turns on:
  - The lamp turns on
  - The coffee maker starts making coffee
  - Temperature sensor starts sensing temperature in the room:
- If the room temperature is normal (between 15 and 18), the fan turns on.
- If the room temperature is cool (less than 15), the fan turns off, air cooler turns off (if on), and heater turns on.
- If the room temperature is hot (greater than 25), fan remains off, air cooler turns on, and the heater turns off (if on).

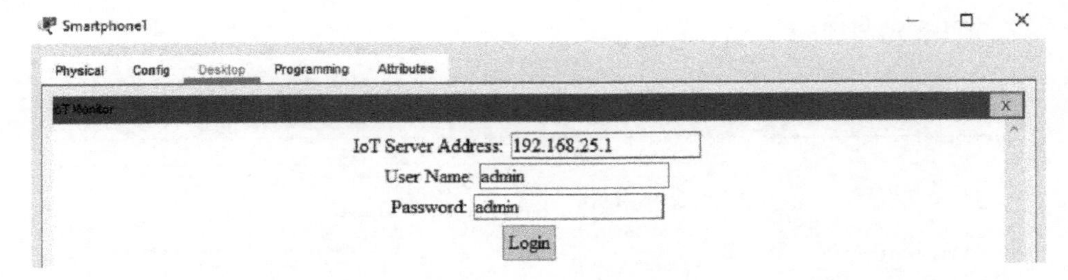


**Figure 11.9** Login screen of smart phone in Cisco Packet Tracer.

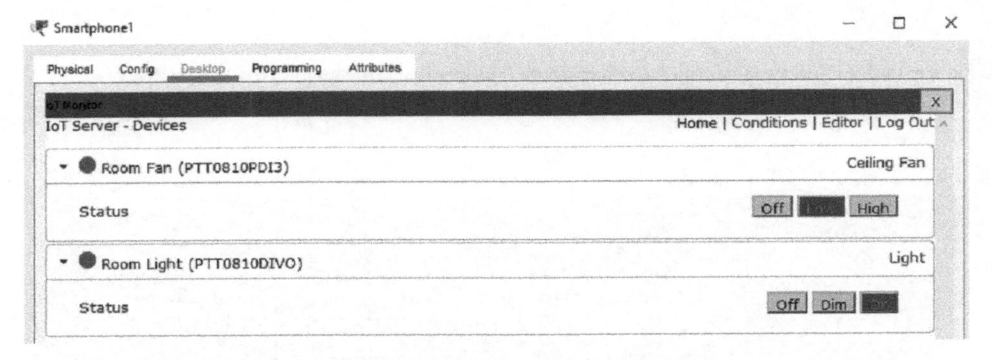


**Figure 11.10** Interface to control IoT devices in Cisco Packet Tracer.

- When the power button turns OFF, light and coffee maker turns off. The rest of the devices stop working.
- Temperature Sensor senses temperature from environment and temperature is shown on Temperature Monitor device.

To implement this, start by creating a new project; click from the menu toolbar **File → New**. A blank workspace will appear. (See Section *1 of Chapter 10 for the details about all toolbar and programming environment interface.*)

### 11.3.1 Adding Devices to Workspace

A list of all devices can be seen at the left-bottom corner (as discussed in detail in Section 1 of Chapter 10). Select the **End Devices** section and then select **Home** icon. You can also apply shortcut keys by pressing **CTRL+ALT+V** (End Devices) or **CTRL+ALT+H** (directly selecting Home section). There you will see a variety of home smart devices as shown in Figure 11.11.

Select the **Light** device and click on the workspace. Add **Appliance** (coffee maker) into the workspace. Now add **Ceiling Fan** by clicking on the icon . We need a Temperature Monitor on the workspace, which senses the room temperature; therefore

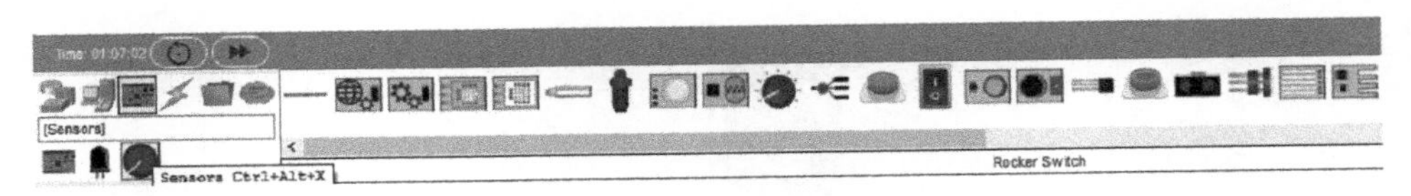


**Figure 11.11** Selection of devices in Cisco Packet Tracer.

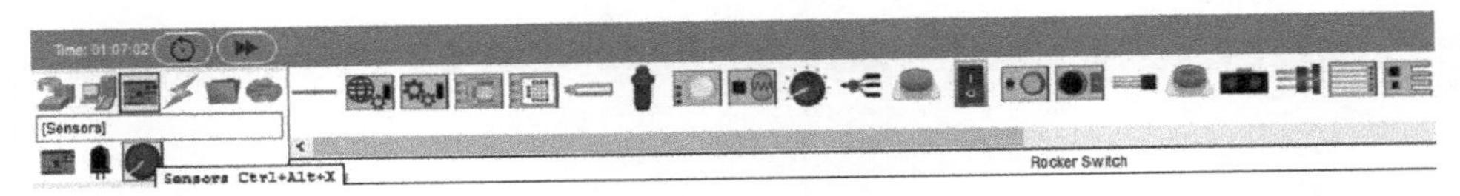


**Figure 11.12** Adding sensors in Cisco Packet Tracer.

add it by clicking Temperature Monitor device. Now Add **Rocker Switch** device from the *Component* devices or by pressing shortcut keys **CTRL + ALT + X** as shown in Figure 11.12. From here, add **Temperature Sensor** device too.

Now, we have five devices on the workspace. To add air cooler and heating sensor devices, go to *Components* and select *Actuators*. Select the **Air Cooler** and **Heating Element** as shown in Figure 11.13.

Now add the final device, which actually controls all the devices, by clicking on **Component** section, add **MCU board** or press **CTRL + ALT + B,** and add MCU Board.

## 11.3.2 Connecting Devices Together

Your workspace seems messy because of so many devices and none of the devices are linked together. Therefore, here we need an **IoT custom cable**, which can attach all the components, devices, and sensors together to make some sense. Your workspace should look like the one shown in Figure 11.14. You can add any cable port with any device. However, in our case the description of the ports is given as follows:

- MCU Board has six digital ports representing D0, D1, D2, and so on till D5. However, it has four Analog ports A0, A1, A2, and A3.

**Figure 11.13** Adding cooling/heating elements on the workspace in Cisco Packet Tracer.

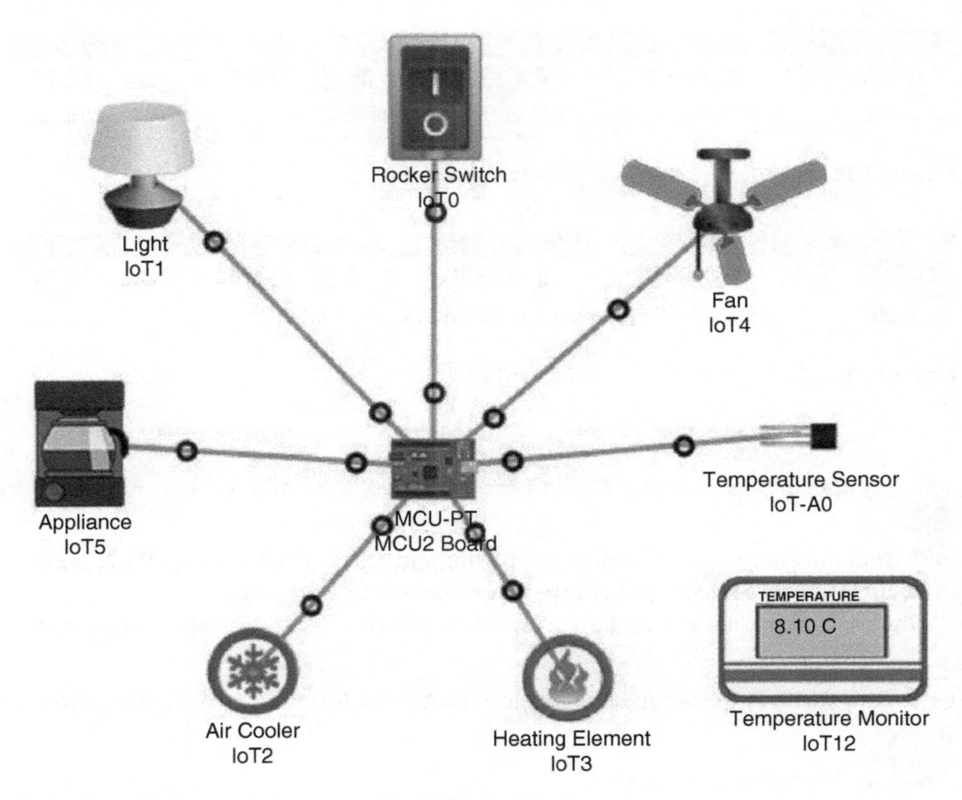


**Figure 11.14** Smart room devices connecting with MCU board.

- Connect the D0 port of the MCU board with D0 port of Rocker Switch.
- Connect the D1 port of the MCU board with D0 port of Light.
- Connect the D2 port of the MCU board with D0 port of Air Cooler.
- Connect the D3 port of the MCU board with D0 port of Heating Element.
- Connect the D4 port of the MCU board with D0 port of Fan.
- Connect the D5 port of the MCU board with D0 port of Appliance.
- Connect the A0 port of the MCU board with A0 port of Temperature Sensor.

To observe room temperature, the *temperature monitor* will constantly be updating and displaying the values as the temperature changes as shown in Figure 11.14.

### 11.3.3 Blockly Programming for Smart Room

To start programming at MCU board now, double-click on the MCU board. A window will appear; click on *programming* tab (details available in Section 10.3 of Chapter 10). Here we will start blockly programming.

First of all, make the **main** function and call this function by adding into the workspace of programming environment. We want to *read* values from *rocker switch* button and *write* it on MCU board, which can control other devices accordingly. For this make a new variable and set it to value 0 set switchValue to 0 as shown. Now, in the main function, we need to set all the ports' value according to the IoT custom cable.

- As the D0 port of the MCU board is connected with D0 port of Rocker Switch, we need to set pin 0 to mode Input pinMode slot 0 mode INPUT. This is the reason for setting mode INPUT as we are reading signals, and input is coming from rocker switch.
- The D1 port of MCU board is linked with D0 port of Light; so, set pin 1 as pinMode slot 1 mode OUTPUT.
- The D2 port of MCU board is linked with D0 port of Air Cooler; so, set pin 2 as pinMode slot 2 mode OUTPUT.
- The D3 port of MCU board is linked with D0 port of Heating Element; so, set pin 3 as pinMode slot 3 mode OUTPUT.
- The D4 port of MCU board is linked with D0 port of Fan; so, set pin 4 as pinMode slot 4 mode OUTPUT.
- The D5 port of MCU board is linked with D0 port of Appliance; so, set pin mode 5 as pinMode slot 5 mode OUTPUT.
- The A0 port of the MCU board is linked with A0 port of Temperature Sensor; we cannot read analog signals like the aforementioned reading digital signals.

Connect these pins together and to the main function as shown in Figure 11.15.

Now, in the main function, we need to read value from the sensor button in such a way that if the button is OFF, all the devices stop working, whereas if the button is ON, devices will start working accordingly. For this, set *switchValue* variable to the value read from sensor like this

. Now it will read

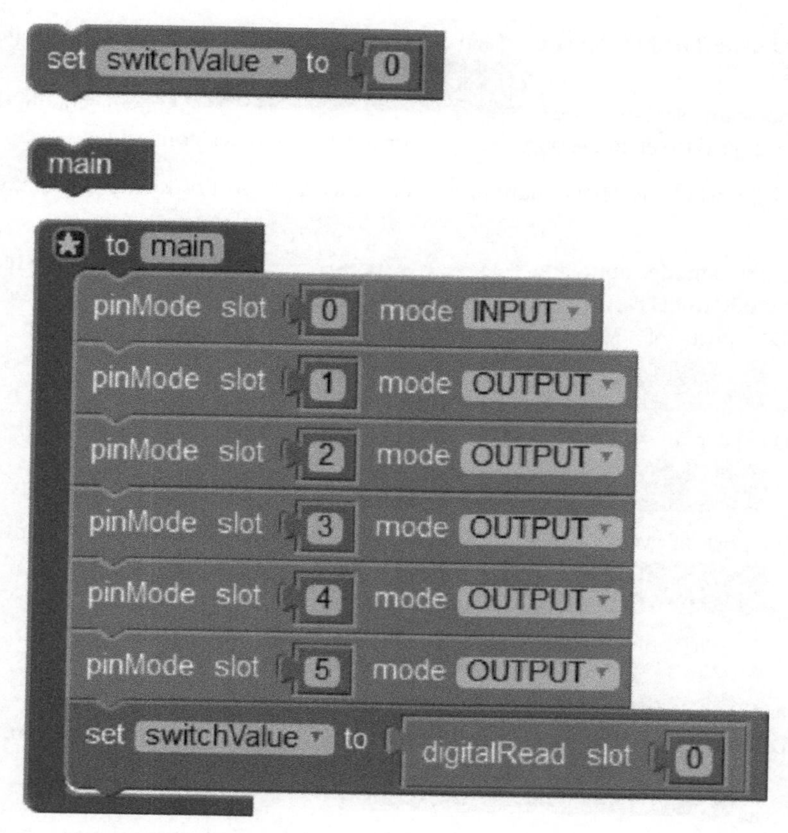


**Figure 11.15** Setting all the pin modes.

digital signals from the rocker switch port 0 and set its value to *switchValue* variable. Now, if the switchValue is equal to 0, it means button is OFF and ON in other case as shown in Figure 11.15. This scenario will work only once; to keep it working we need to constantly read the signals from the sensor as you never know when the button is turned off by the user. For this we need loop.

To insert loop, add block repeat while do from the **loop** section and set while condition to true repeat while true do. Insert the *switchValue* block into the while loop repeat while true do set switchValue to digitalRead slot 0. Now the program will constantly be checking the value of the rocker button if either the button is ON or OFF. Now, if you execute the program, it will start reading values from the button; however, you cannot see any output nor observe any differences in the workspace. It is because we have not included anything to display values. We are just reading values and setting the value to a variable. In order to display values, you can print the variable *switchValues* value print switchValue and then execute the

**Figure 11.16** Reading values from the rocker switch sensor button with the delay of 1 s.

program by clicking on Run button. You will start reading values such as 1023 continuously. The program might hang or not stop because of not adding delay. The values are reading so quickly that might stuck the program, so here we need a delay of 1 s, i.e. 1000 ms as shown in Figure 11.16. If you execute the program, the values will display with the delay of 1 s.

We need now to write on the IoT devices depending upon the rocker switch state readings we get earlier. So, if it is ON, turn on the light and the coffee maker and set room temperature accordingly. If the rocker switch is OFF, turn off everything. For this, we need logical if-else statements. If *switchValue* is equal to 1023 value (i.e. it is ON), then set pins light and coffee maker to states ON by adding block *customWrite* from the **Pin Access** tab; otherwise, set them to off. You can copy the blocks that need to be placed again and again by simply using shortcut keys **CTRL + C** and **CTRL + V**. Set light slot 1 to state ON (1). Set coffee maker slot 5 (in this case) to state ON (1) as shown in Figure 11.17. You can set light state to 2 as well if it requires full light. Place the **if** block into **while** block between the **print** and **delay** blocks.

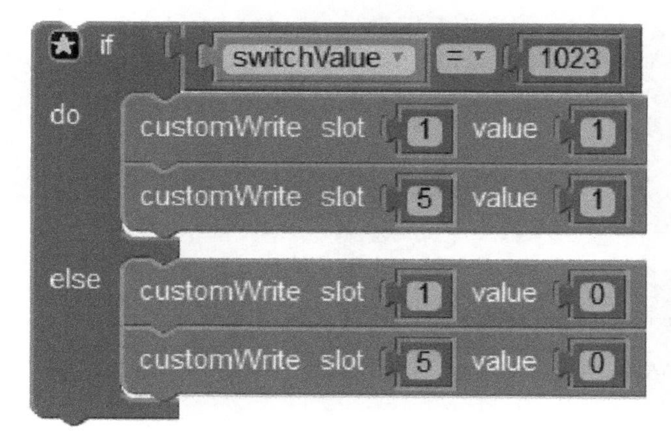


**Figure 11.17** If-else statements for switching the light and coffee maker to ON/OFF.

Execute the program and try toggle between the two states of the rocker switch button. You can see that when the rocker button turns ON, green light turns ON as well. As the dim light turns on, red light turns on of the coffee maker, indicating its on state . The while block will look like the one shown in Figure 11.18.

As you can see, the length of the program keeps increasing, and it will surely turn into messy programming if we do not arrange the blocks at this stage. Till now we are working on structured programming which is not an efficient way to program. Therefore, here comes the use of functions. Now, we will practice first to make functions and arrange our program. Then, we will complete the rest of the functionality of switching on the fan, air cooler, and heater sensor, which depends on the temperature of the room.

We have already made the *main* function many times. So, this time we make a new function named as settingPins and copy all the pins mode setting to this function. Call

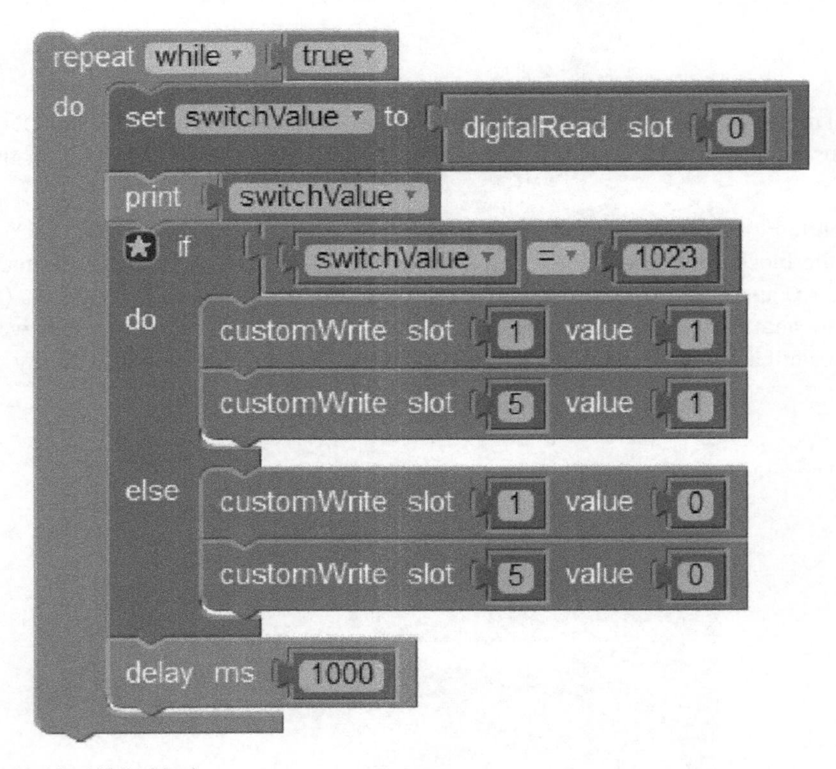


**Figure 11.18** While block.

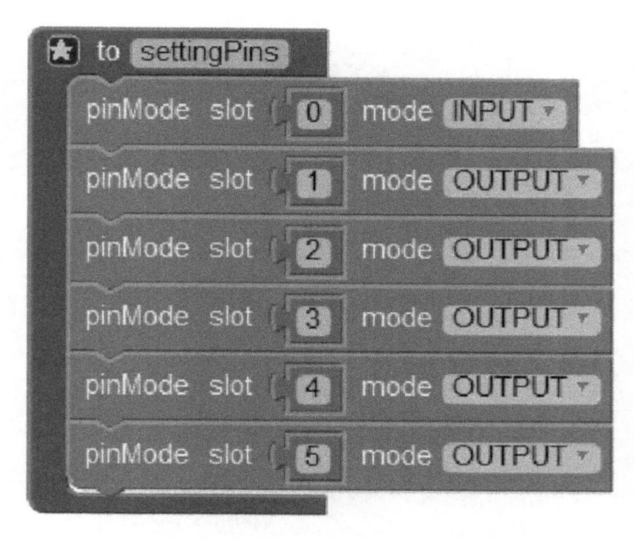


**Figure 11.19** Setting pin function flow of blocks.

the function from the main program. Now make another function named as *readingFromSensor,* copy the blocks other than while loop, and drag into the function definition. Call this function readingFromSensor in main program within while loop. The complete function *settingPins* is shown in Figure 11.19, whereas the complete definition of the *readingFromSensor* function is shown in Figure 11.20. Now you can see how main function becomes so neat, small, easy to read, easy to debug, and easy to control as shown in Figure 11.21.

To make it simpler, split the reading and writing functions by making another function writingSwitchSensorStates, which contains if-else block statements and call it to main program as shown in Figure 11.22. Set your blocks accordingly if not.

As you can see now, we only write on ports 1 and 5 but not for the rest of the ports. This is because these ports depend on room temperature. For this we need to monitor room temperature constantly and the conditions are based on the following:

- If the room temperature is less than 15 °C, then switch on the heating element whereas switch off the fan and air cooler (if on).
- If the room temperature is greater than 25 °C, then switch on the air cooler and switch off the fan (if on).
- If the room temperature is between 18 and 25 °C, then switch on the fan while switching off both air cooler and heating element (if already on).

For this, first, we need a separate function to monitoringTemperature, which will perform all these if-else blocks statements. Do not forget to call the function

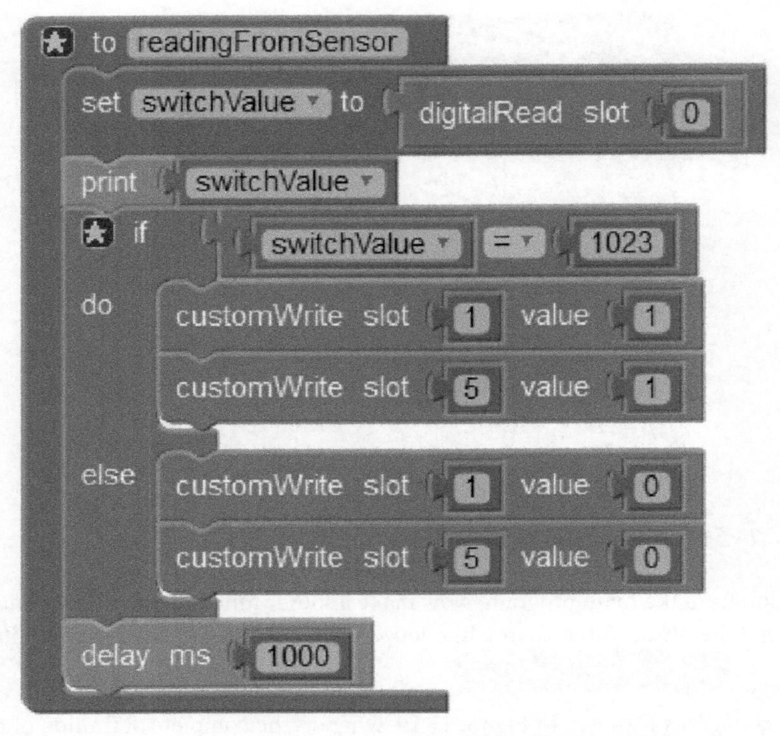


**Figure 11.20** Flow for the function definition of readingFromSensor (Rocker Switch).

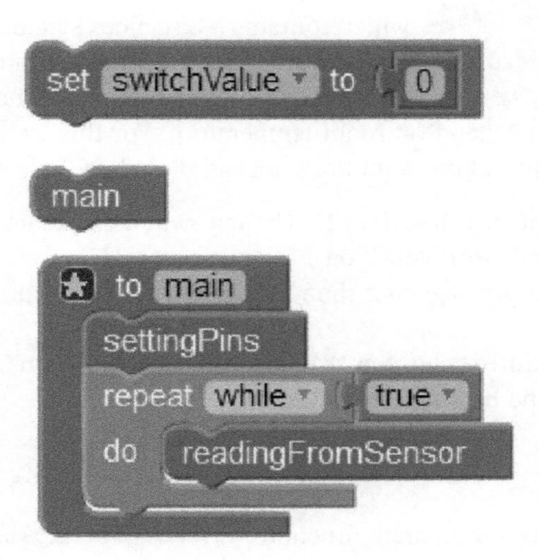


**Figure 11.21** Main program.

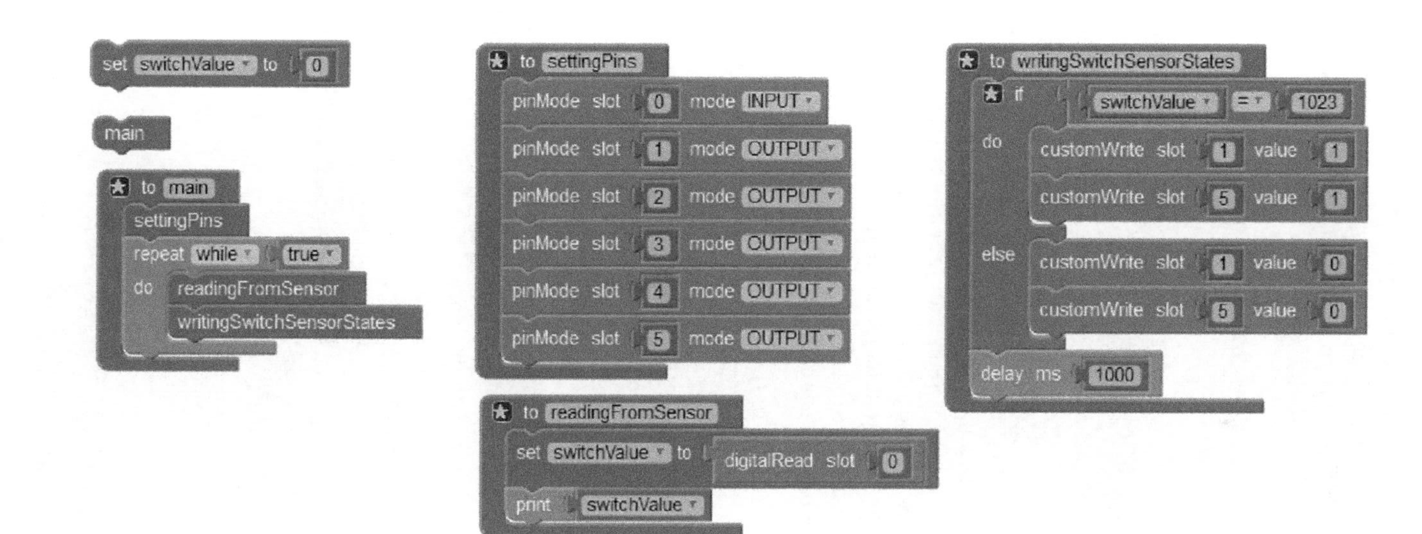


**Figure 11.22** Overview of the workspace including functions and the main program.

*monitoringTemperature* in the main program ; otherwise it will not execute. From now onward, we will perform every task into separate functions to make programming simpler. Therefore, make a new function that will sense the value of room temperature

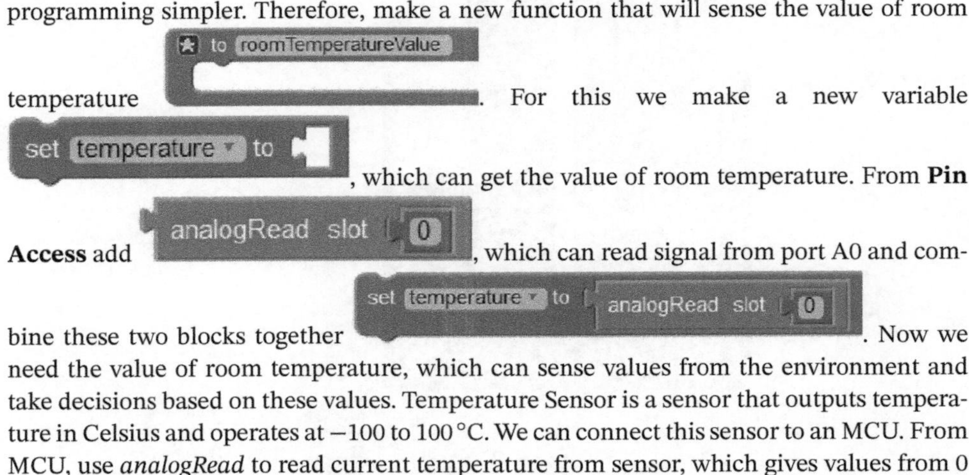

. For this we make a new variable , which can get the value of room temperature. From **Pin Access** add , which can read signal from port A0 and combine these two blocks together . Now we need the value of room temperature, which can sense values from the environment and take decisions based on these values. Temperature Sensor is a sensor that outputs temperature in Celsius and operates at −100 to 100 °C. We can connect this sensor to an MCU. From MCU, use *analogRead* to read current temperature from sensor, which gives values from 0 to 1023, which will be mapped to −100 to 100 °C using the *Interpolation* formula shown in Eq. (11.1) (*where X means original temperature*) and Figure 11.23.

$$X = \frac{\left[100 \times (-100) \times (\text{temperature}(\text{A0}) - 0)\right]}{1023 - 0} - 100 \tag{11.1}$$

Now if the temperature is less than 15 °C, we need to switch on the heating element and switch off the air cooler and fan. Thus, make a separate function for this . In this function, set all the *pinModes* that should be on and that should remain off. Therefore, add the block and set the slot 3 to be as 1 (i.e. slot 3 stands for the heating element and 1 indicates its state ON). If the temperature is less than 15, heating element will turn on. Add remaining blocks that turn the air cooler state to be as OFF and set fan port as OFF . Your function will look like the one shown in Figure 11.24.

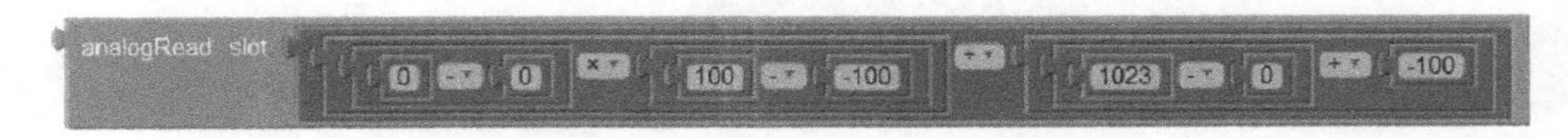

Figure 11.23 Reading temperature sensor and mapping to Celsius degree.

**Figure 11.24** Function for the heating element to turn on.

Now if the temperature is greater than 25 °C, we need to switch off the heating element analogWrite slot 3 value 0 and switch off the fan analogWrite slot 4 value 0 and power on the air cooler analogWrite slot 2 value 1. This function will now look like the one shown in Figure 11.25.

Now, the last condition, if temperature is between 18 and 25 °C, it will turn the fan on analogWrite slot 4 value 1 and turn off air cooler analogWrite slot 2 value 0 and heating element analogWrite slot 3 value 0. The function block for this function is shown in Figure 11.26.

**Figure 11.25** Temperature block for value greater than 25 °C.

Figure 11.26 Block for temperature between 18 and 25 °C.

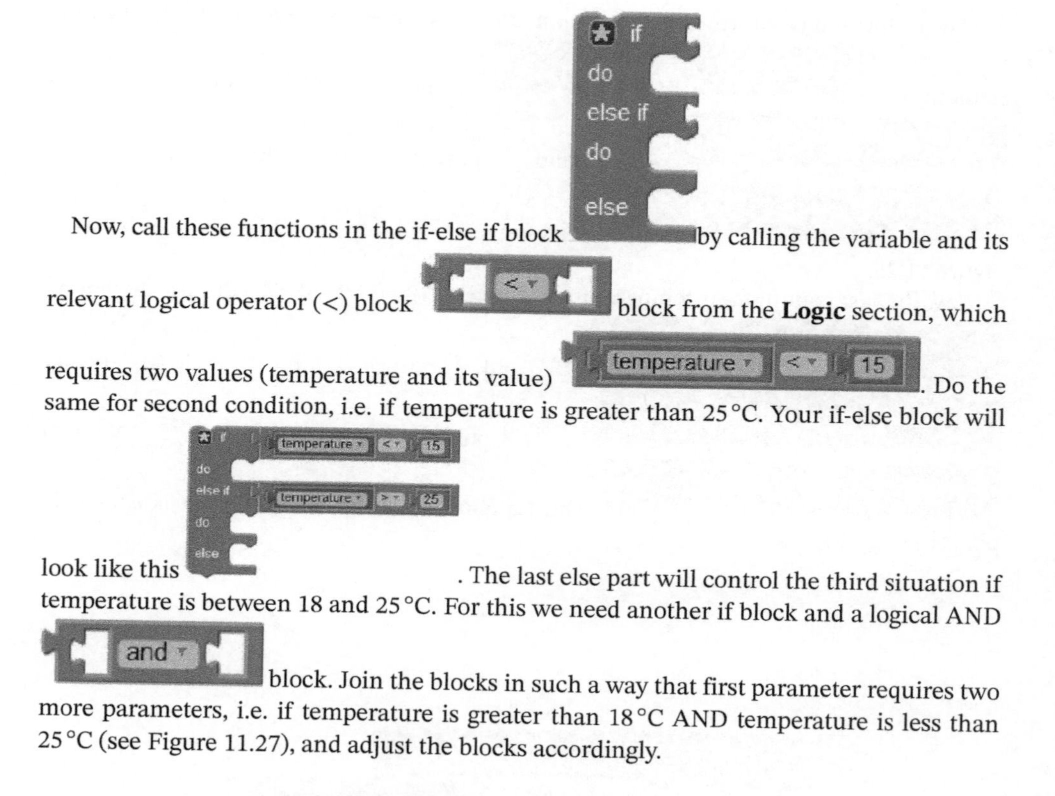

Now, call these functions in the if-else if block by calling the variable and its relevant logical operator (<) block block from the **Logic** section, which requires two values (temperature and its value). Do the same for second condition, i.e. if temperature is greater than 25 °C. Your if-else block will look like this. The last else part will control the third situation if temperature is between 18 and 25 °C. For this we need another if block and a logical AND block. Join the blocks in such a way that first parameter requires two more parameters, i.e. if temperature is greater than 18 °C AND temperature is less than 25 °C (see Figure 11.27), and adjust the blocks accordingly.

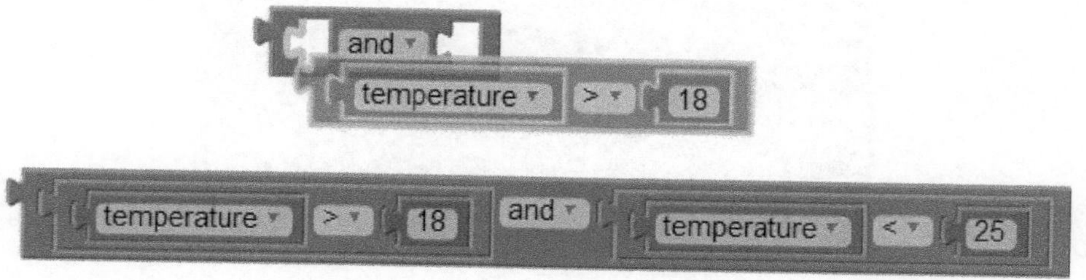


Figure 11.27 Handling logical AND operator for temperature between 18 and 25 °C.

Now, call the functions in the if block such that if condition 1 satisfies *do this* and else if condition 2 satisfies *do this*, else *do this*. To call a function, add functions tempLessThan15, tempGreaterThan25, and tempBetween18to25 from the **Function** block in the if statements as shown in Figure 11.28.

Now call the function *roomTemperatureValue* to monitoringTemperature roomTemperatureValue in the function *monitoringTemperature*, and call this function in the main program as shown in Figures 11.29 and 11.30. You can observe now that the main program looks so simple and easy to read just because of handling functions and using it in proper way.

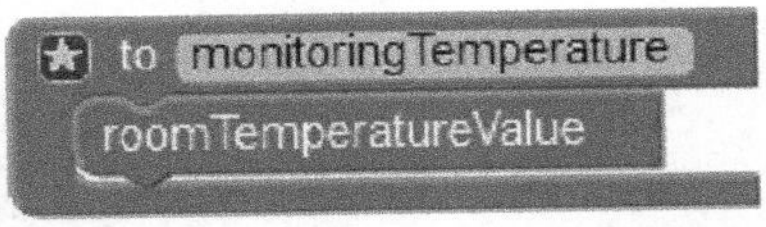

Figures 11.31, 11.32, and 11.33 show the working of different devices against different Temperature conditions.

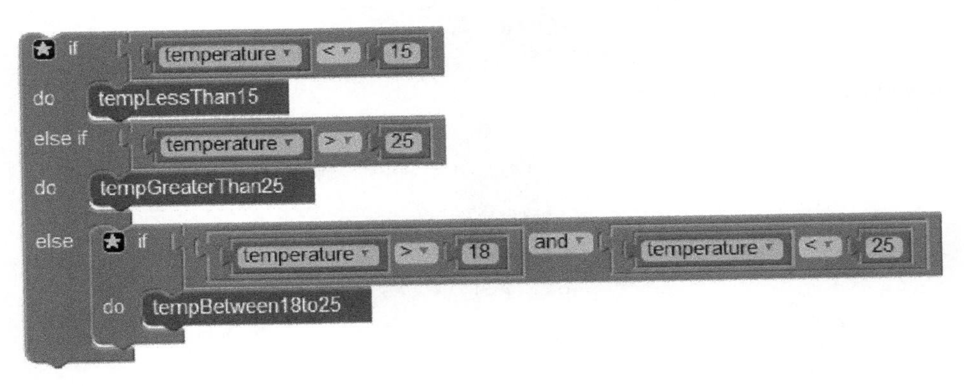


**Figure 11.28** Complete overview of if-else-if block.

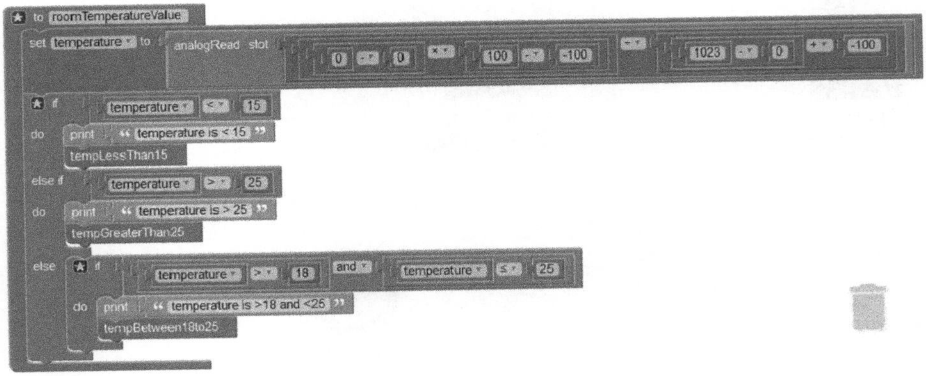

**Figure 11.29** Function block for getting the room temperature value.

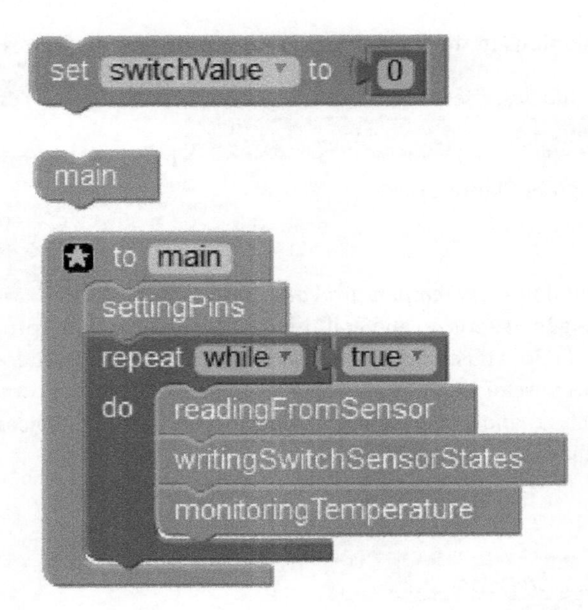


**Figure 11.30** Final main program.

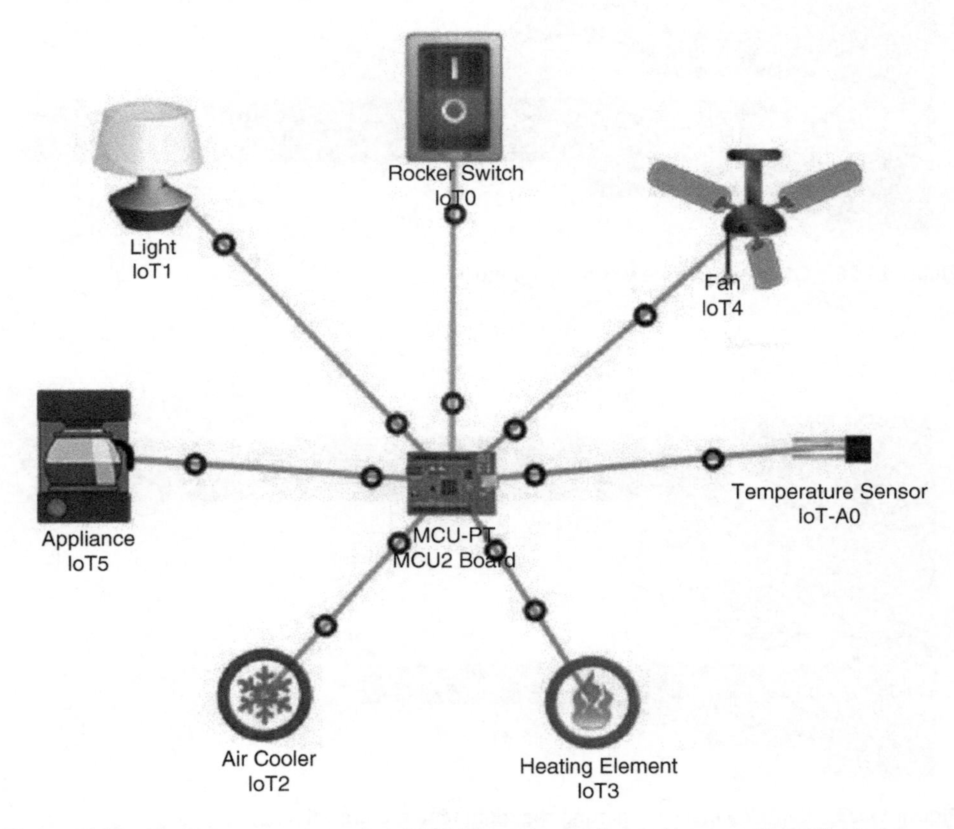


**Figure 11.31** Temperature greater than 25 °C (air cooler on, fan and heater off, and light and coffee maker on).

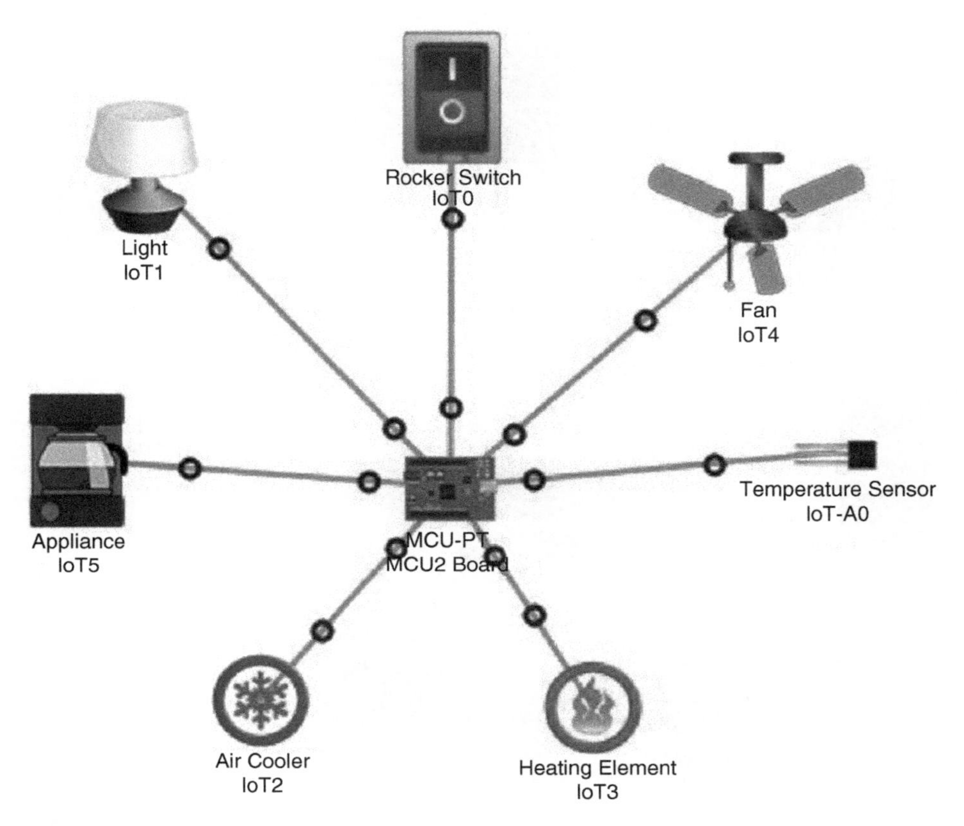


**Figure 11.32** Temperature less than 15 °C (heating element on, fan and air cooler off, and light and coffee maker on).

## Review Questions

**11.1** Implement smoke detection IoT project (using Block, Javascript, or Python language) in such a way that if cars started their engines, smoke detected leads to ringing alarm and fire sprinkler starts sprinkling water as shown in Figures 11.34 and 11.35.

**11.2** Implement the scenario shown in Figure 11.36 in Packet Tracer
- **A** Control fan speed through Rocker Switch via MCU.
- **B** Control light through Rocker Switch via MCU.

**11.3** Implement the scenario shown in Figure 11.37 in Packet Tracer (using JavaScript or Python language).
- **A** Control fan speed through laptop or mobile devices.
- **B** Sensed temperature should be shown on mobile phone or laptop.

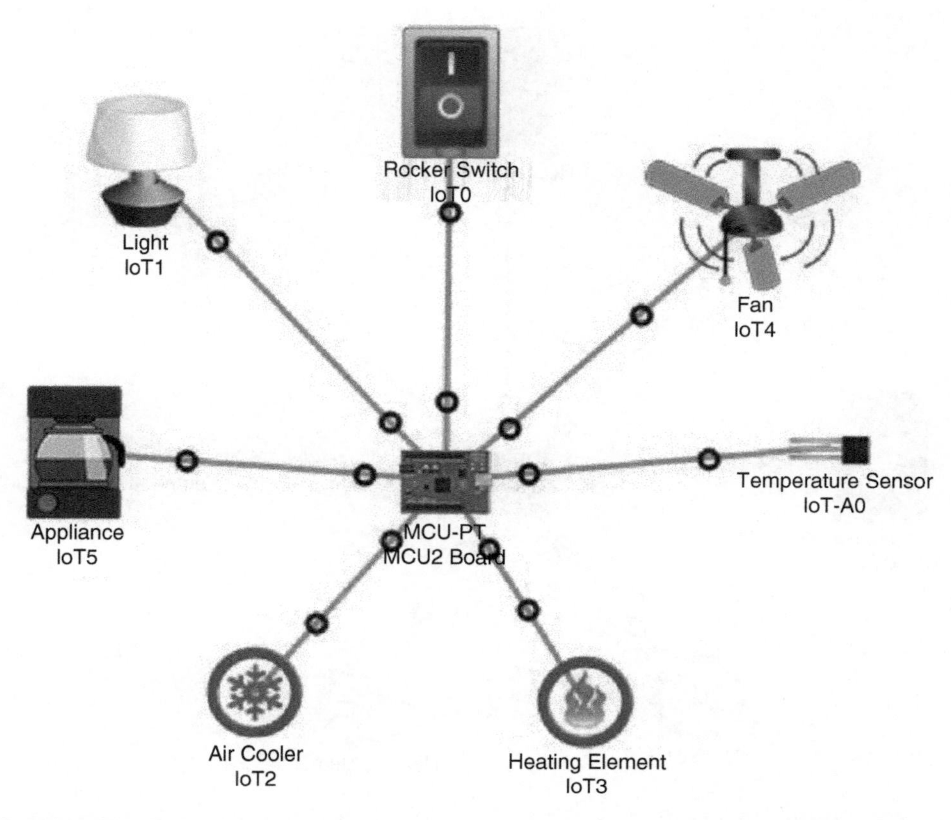


**Figure 11.33** Temperature between 18 and 25 °C (air cooler off, heating element off and fan on at full speed, and light and coffee maker on).

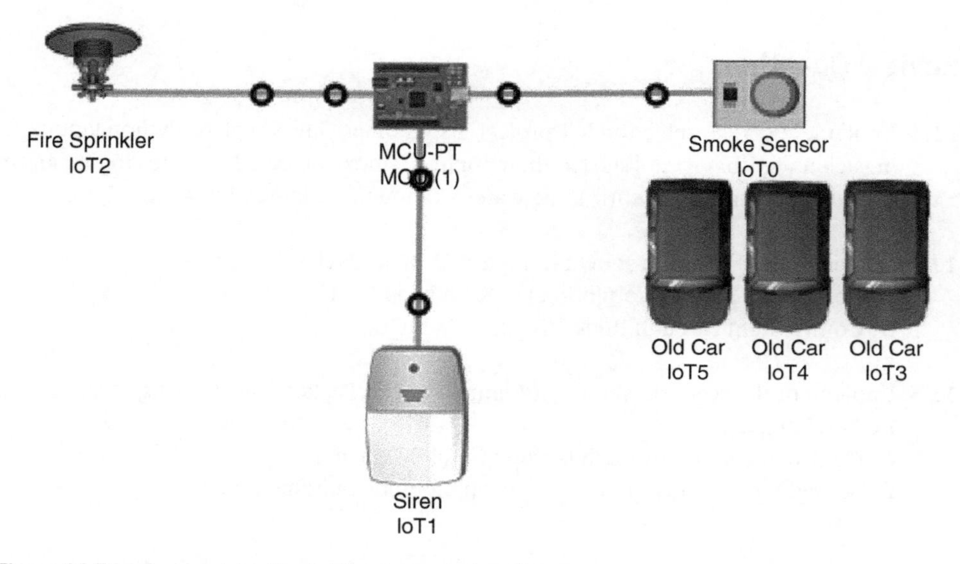


**Figure 11.34** Smoke detection IoT implementation diagram.

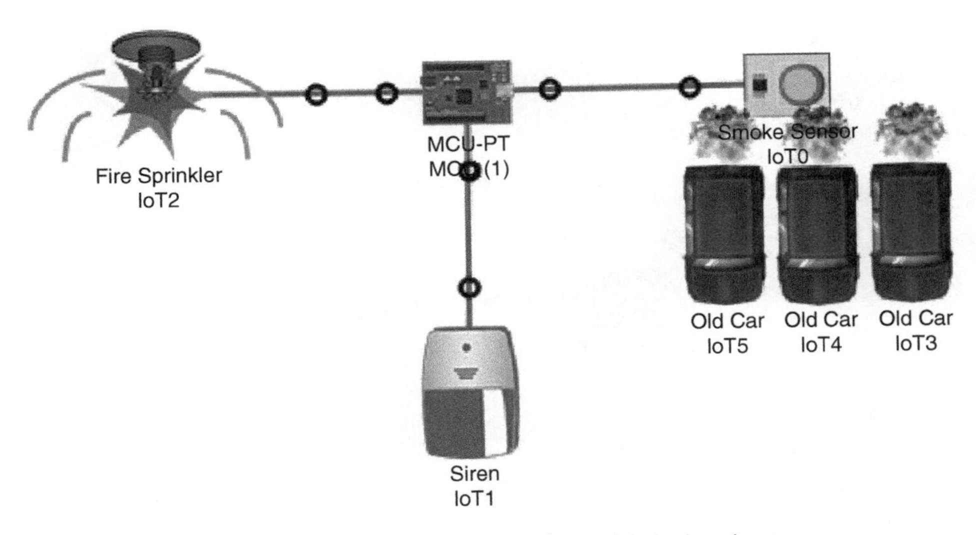


**Figure 11.35** Detecting smoke and sprinkling water along with ringing alarm.

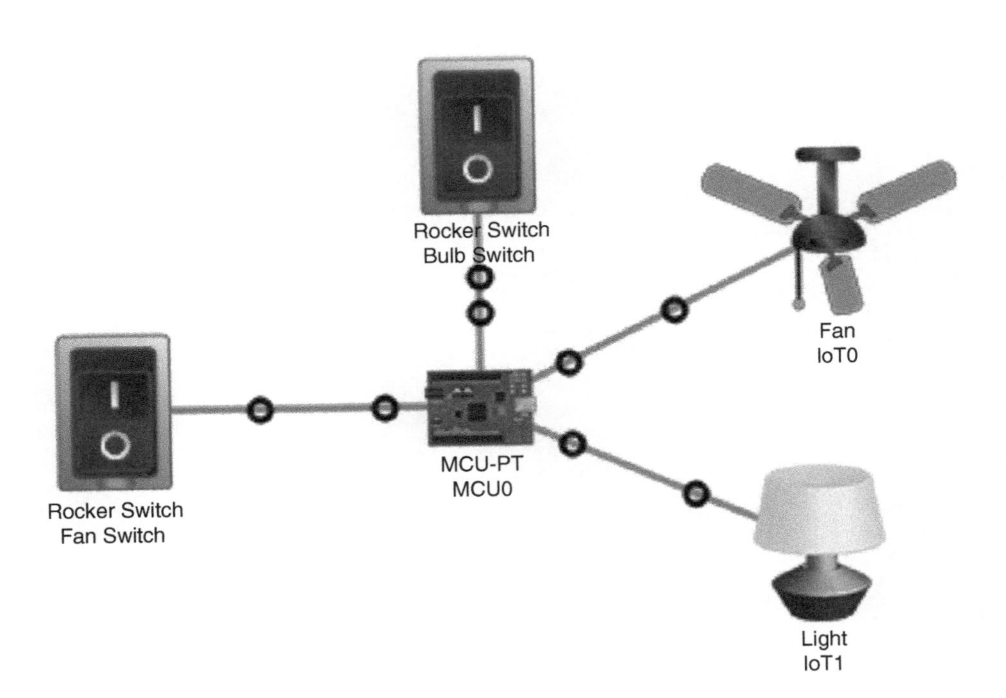


**Figure 11.36** IoT Scenario for Question 11.2.

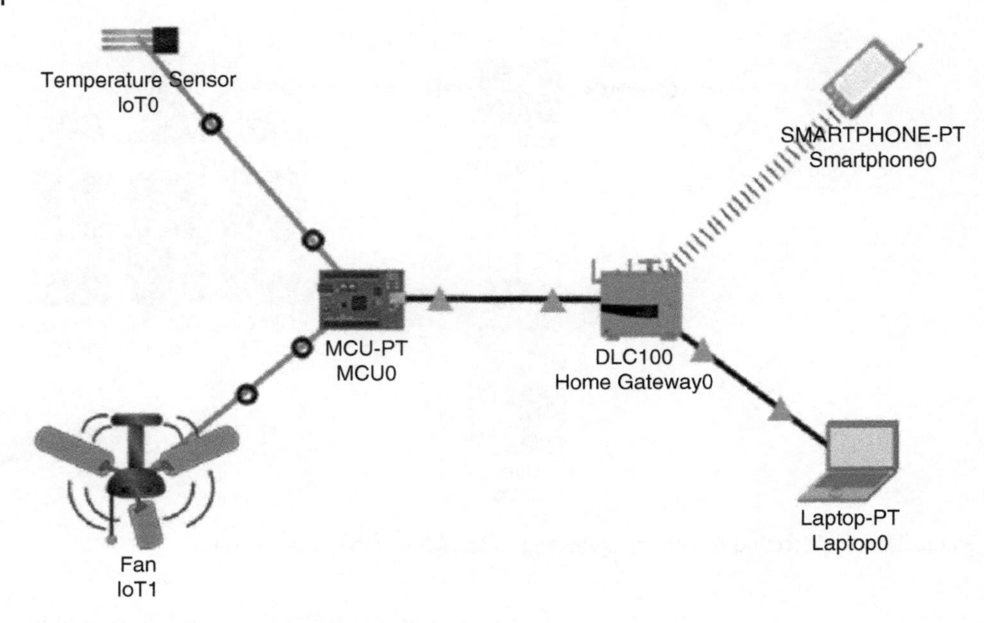


**Figure 11.37** IoT Scenario for Question 11.3.

# Index



*Enabling the Internet of Things: Fundamentals, Design, and Applications*, First Edition.
Muhammad Azhar Iqbal, Sajjad Hussain, Huanlai Xing, and Muhammad Ali Imran.